TEXTBOOK OF
Mineral Processing

D.V. SUBBA RAO

Formerly Head of the department of Mineral Beneficiation
S D S Autonomous College, Garividi
Vizianagaram district, Andhra Pradesh, India

SCIENTIFIC
PUBLISHERS

Published by
SCIENTIFIC PUBLISHERS (INDIA)

Jodhpur –

5 A, New Pali Road
P.O. Box 91
Jodhpur - 342 001 INDIA

Delhi –

4806/24, Ansari Road
Daryaganj
New Delhi - 110 002 INDIA

ISBN: 978-93-83692-56-9

Visit the Scientific Publishers (India) website at
http://www.scientificpub.com

Printed in India

Digitally Printed at Replika Press Pvt. Ltd.

To my Wife

Krishna Veni

Contents

Preface

This book "TEXTBOOK OF MINERAL PROCESSING" has been written to cater the needs of undergraduate students of Mining and Metallurgy Engineering and postgraduate students of Applied Geology as most of the universities have a compulsory paper of Mineral Processing in their courses of study. This is first of its kind and a concise book for Mineral Processing to cover their syllabus otherwise one need to refer several books to deal with the syllabus.

Various aspects of the Mineral Processing have been dealt and explained in an easy to understand manner in this book. Simple numerical calculations necessary for evaluation of the processes have also been incorporated. A chapter "Coal Characterization" has also been included in this book for the benefit of those students who have washability characteristics of coal in their syllabus. Further, the present practices of processing of ores and coal in India have been dealt.

It is believed that the subject is presented in this book will be appreciated by the students and realize the importance of mineral processing and motivate them to choose Mineral Processing as one of the stepping stones in their professional carrier. It is hoped that this book will serve as a first book for Mineral Processing students to understand the basic principles of mineral processing and their applications.

I sincerely appeal the readers to communicate their feedback on the contents and suggestions for improvement which will be useful to incorporate in future editions.

D.V. SUBBA RAO
Formerly Head of the Department of Mineral Beneficiation
S D S Autonomous College
GARIVIDI – 535 101
Vizianagaram district, Andhra Pradesh, India

Acknowledgements

First of all I would like to thank the management of GODAVARI INSTITUTE OF ENGINEERING AND TECHNOLOGY, Rajahmundry for giving me an opportunity to teach Mineral Processing for the students of Mining Engineering which prompted me to think of writing a separate book on Mineral processing for Mining and Metallurgy Engineering and Geology students.

I am grateful to Dr. T.C. Rao, Formerly Director, Regional Research Laboratory, Bhopal, and Professor and Head of the department of Fuel and Mineral Engineering, Indian School of Mines, Dhanbad, Sri A.L. Mohan, Formerly Principal, S D S Autonomous College, Garividi, and my colleagues Sri Y. Ramachandra Rao and Sri K. Satyanarayana for their encouragement and help in writing this book.

I thank my students who helped me in identifying the extent of coverage of each topic in this book suitable to those this book is primarily intended. I am thankful to Scientific Publishers India for accepting this book for publication and took pains and all the care in publishing in short period.

Support and help from my daughters Mrs. Radha Rani and Mrs. Lalitha Rani while preparing this book is invaluable.

I am indebted to Dr. C. Raghu Kumar, Head, Ore Beneficiation, Process Technology group, Tata Steel Ltd., Jamshedpur, for going through the script patiently and help to finalise with his critical remarks.

I thankfully acknowledge the following organizations for permitting to use respective photographs:

edart, South Africa	Poppet sampler
FRITSCH Milling and Sizing	Rotary Cone splitter
Lavallab.com	Rotary table splitter
Jayant Scientific Industries, Mumbai	Table model & Ro-tap sieve shakers
Mogensen	Divergator, Sizer, Grizzly
Deister Concentrator, LLC	DSM Sieve bend
JXSC Mining, China	Shaking screen, Log washer
Deister Machine Co. Inc	Vibrating screens

Pennsylvania Crusher Corporation	Rotary breaker, Hammer mills
aimehq.org	Single toothed roll crusher, Ball mill, Rod mill
Balling machinery.com	Spiral classifier
Dione oil Ltd	Rake classifier
MBE Coal & Mineral Technology India Pvt Ltd, Kolkata	Baum jig, ROM jig
Westech Process Equipment India Pvt. Ltd	Disc filter

1 Introduction

Minerals are non-renewable natural resources that occurred in the Earth's crust. Minerals are an integral part of the nation's economy, meet human needs, associate with societal growth, contribute to industrialisation, and form important infrastructure for the development of the country.

Mineral Processing is a subject of interest for mining and metallurgical engineers and geologists to have knowledge of its techniques for application in judicial utilisation of mineral resources. Geological investigation of the Earth's crust for possible mineral resources forms the starting point for obtaining the required mineral product.

After a geologist estimates the mineral reserves, their quality, and assay value, it is the mining engineer who mines the reserves and makes it available for processing. A mineral engineer processes the mined minerals and makes them suitable either for metal extraction or direct use to produce various products such as cement, refractories, and fertilizers. Thus, following are the successive major steps involved to obtain the required end product.

Geological Investigation

⇩

Mining

⇩

Mineral Processing

⇩

Metal extraction or Industrial use

1.1. MINERALS

As defined by Dana, a mineral is a substance having definite chemical composition and internal atomic structure, and formed by the inorganic processes of nature (Read 1984).

All minerals contain one or more metals or metalloids. Minerals are broadly classified as metallic and non-metallic minerals.

Metallic minerals are the minerals from which a metal is extracted. Bauxite ($Al_2O_3.2H_2O$), Hematite (Fe_2O_3), Pyrolusite (MnO_2), Chromite (FeO Cr_2O_3), and Ilmenite (FeO TiO_2) are a few important metallic minerals.

Non-metallic minerals are used for industrial purposes such as manufacturing cement, refractories, glass and ceramics, insulators, and fertilizers. These minerals are also called as **industrial minerals**. Metals are not extracted from these minerals.

Some metallic minerals are also used for industrial purposes such as Bauxite, Chromite, and Zircon for the refractory industry, Pyrolusite for dry battery cells, and Ilmenite for pigment industry.

Following are important terms used in describing mineral deposits.

Rock is an aggregation of several minerals as occurred in the earth's crust. Granite, which is an aggregation of orthoclase, quartz, and mica, is an example of a rock.

Ore is also an aggregation of several minerals from which one or more minerals can be exploited or separated at profit. Today's rock, due to the unavailability of applicable technology, may become tomorrow's ore if the technology is available for utilisation. Similarly, at a point where the technology is available to profitably separate required minerals it is called as ore. The same ore with the same characteristics in all respects that cannot be exploited at profit is called a rock. Hence, it should be understood that the term ore is used to represent its economic viability so that it can be exploited or separated for the required minerals.

Ore Minerals or **Valuable Minerals** (also called **Economic Minerals**) are minerals of economic value and contain an economically exploitable quantity of some metal or non-metal.

Gangue Minerals are usually the non-metallic minerals associated with ore minerals that are worthless as a source for that metal or otherwise. These are usually unwanted, waste, and useless minerals. Gangue Minerals occasionally may find use as source of by-products. For example, pyrite present in Lead and Zinc ores is a Gangue Mineral, but it is separated as a by-product for extraction of Sulphur after Lead and Zinc minerals are separated.

Ore Deposits are the natural deposits of ore minerals. These are also called economic mineral deposits. The **ore** can be described as an aggregation of valuable minerals and gangue minerals.

Simple Ore is one from which a single metal can be extracted. For example, only Iron is extracted from Iron ore, Aluminium is extracted from Aluminium ore, and Chromium is extracted from Chrome ore. Such ores are called as simple ores.

Complex Ore is one from which two or more metals can be extracted. Lead and Zinc metals are extracted from lead Zinc ore; Lead, Zinc, and Copper metals are extracted from Lead–Zinc–Copper Ore. Such ores are called as complex ores.

Metal Content of a mineral is generally expressed in percent of metal present in the mineral. It is calculated by taking the atomic weights of the elements present in the mineral, as shown in example 1.1.1.

Example 1.1.1: *Calculate percent iron present in Magnetite (Fe_3O_4). Atomic weights of Iron and Oxygen are 55.85 and 16.00, respectively.*

Solution:

Given

Atomic weight of Iron $= 55.85$

Atomic weight of Oxygen $= 16.00$

Chemical formula of Hematite is Fe_3O_4

Molecular weight of Hematite $= 55.85 \times 3 + 16 \times 4 = 231.55$

Iron present in hematite $= \dfrac{55.85 \times 3}{231.55} \times 100 = 72.36\%$

This implies that 72.36% Fe (Iron metal) by weight is present in Magnetite.

Similarly, percent metal present in any metallic mineral can be calculated. Some of the metallic minerals, their chemical formulae, metal extracted from them, and the percent metal present in respective minerals by similar calculation are shown in Table 1.

Table 1: Metallic minerals

Mineral	Chemical Formula	Metal extracted	% metal
Hematite	Fe_2O_3	Iron	69.94
Magnetite	Fe_3O_4	Iron	72.36
Bauxite	$Al_2O_3.2H_2O$	Aluminium	39.11
Braunite	$3Mn_2O_3\ MnSiO_3$	Manganese	63.60
Pyrolusite	MnO_2	Manganese	63.19
Chromite	$FeO\ Cr_2O_3$	Chromium	46.46
Galena	PbS	Lead	86.60
Sphalerite	ZnS	Zinc	67.10
Chalcopyrite	$CuFeS_2$	Copper	34.63
Ilmenite	$FeO\ TiO_2$	Titanium	31.57
Rutile	TiO_2	Titanium	59.95
Zircon	$ZrSiO_4$	Zirconium	49.76
Monazite	$(Ce,La,Th)PO_4$	Thorium	---

Assay Value or **Tenor** is the percent metal, percent valuable mineral, or ounces precious metal per ton of the ore, depending upon the type of ore involved.

For ores of metallic minerals, percent metal present in the ore is the assay value. For an ore of precious metals such as gold, ounces or gram precious metal per ton of the ore is the assay value as the precious metal is present in minute quantities. Assay value of an ore is usually determined by chemical analysis of a sample collected from the ore. Different minerals present in an ore are qualitatively determined by microscopic analysis of a sample. When once the assay value of an ore and the minerals present in that ore are determined, valuable and gangue mineral contents can be estimated by calculation. This aspect is illustrated in example 1.1.2.

Example 1.1.2: *An Iron Ore of 60% Fe contains only one Iron Mineral Hematite. What is the weight of gangue per ton of Iron Ore? Atomic weights of Iron and Oxygen are 55.85 and 16.00, respectively.*

Solution:

Given

Atomic weight of Iron = 55.85

Atomic weight of Oxygen = 16.00

Chemical formula of Hematite is Fe_2O_3

Molecular weight of Hematite $= 55.85 \times 2 + 16 \times 3 = 159.7$

Iron present in hematite $= \dfrac{55.85 \times 2}{159.7} \times 100 = 69.94\%$

%Fe in Hematite Fe_2O_3 = 69.94%

%Fe in Iron Ore = 60%

$\therefore$ % Fe_2O_3 in Iron Ore $= \dfrac{60.0}{69.94} \times 100 = 85.79\ \%$

% gangue in Iron Ore $= 100 - 85.79 = 14.21\%$

$\therefore$ Weight of gangue per ton of Iron ore $= 14.21/100 = 0.1421$ ton

An assay value of an ore of non-metallic minerals is represented by the percent valuable or required constituent of the ore. For example, lime (CaO) is the required constituent in the limestone (predominantly $CaCO_3$) used for manufacturing cement. The assay value of the limestone is represented as %CaO. Example 1.1.3 illustrates this calculation.

Example 1.1.3: *Determine percent lime (CaO) present in limestone of $CaCO_3$ composition. Atomic weights of Calcium, Carbon, and Oxygen are 40.08, 12.01 and 16.00, respectively.*

Solution:

Given

Atomic weight of Calcium = 40.08

Atomic weight of Carbon = 12.01

Atomic weight of Oxygen = 16.00

Chemical formula of limestone is $CaCO_3$

Molecular weight of Limestone $= 40.08 + 12.01 + 16 \times 3 = 100.09$

Lime (CaO) present in limestone $= \dfrac{40.08 + 16.00}{100.09} \times 100 = 56\%$

It means that 56% lime (CaO) by weight is present in limestone.

Table 2 shows few non-metallic minerals, their chemical formulae and chief uses.

Table 2: Non-metallic minerals

Mineral	Chemical Formula	Main uses
Apatite	$Ca_4(CaF)(PO_4)_3$	Fertilizer, phosphorous chemicals, and gems jewellery.
Asbestos (Crysotile)	$Mg_3Si_2O_5(OH)_4$	Fire proof fabrics, asbestos cement, friction products, roofing, flooring, and pipe insulation.
Baryte	$BaSO_4$	Well drilling fluids, barium chemicals, glass, paint, and rubber.
Bentonite	$(Ca\ Mg)O\ SiO_2$ $(Al\ Fe)_2O_3$	Oil-well drilling, cosmetics, toothpaste, and binder for iron ore pellets.
Calcite	$CaCO_3$	Cement and lime, filler in paper, paint, adhesives, and sealants
Dolomite	$CaMg(CO_3)_2$	Furnace linings, building material, and flux in blast furnace.
Feldspar	$(Na,K,Ca)\ AlSi_3O_8$	Glass and ceramics.
Fluorite	CaF_2	Production of HF and flux in steel manufacturing.
Garnet (Almandine)	$3FeO\ Al_2O_3\ 3SiO_2$	Sand blasting, water filtration, wood polishing, as an abrasive, and gemstones.
Gypsum	$CaSO_4.2H_2O$	Cement and fertilizers.
Kaolinite (China clay)	$H_4Al_2Si_2O_9$	Paper, paint, cosmetics, ceramics, rubber, and fiberglass.
Kyanite	Al_2SiO_5	Refractories.
Limestone	Chiefly $CaCO_3$	Cement, sculptures, building material, and flux in blast furnace.
Magnesite	$MgCO_3$	Refractories, magnesium chemicals, and fertilizers.
Phosphate rock	$Ca_3(PO_4)_2$	Fertilizers, phosphorous chemicals, food additives, and detergents.
Quartz	SiO_2	Glass, ferrosilicon, silicon metal, silicon carbide, and metallurgical fluxes.
Sillimanite	$Al_2O_3\ SiO_2$	Refractories.
Talc	$H_2Mg_3(SiO_3)_4$	Ceramics, paint, paper, roofing, plastics, cosmetics, and pharmaceuticals.
Vermiculite	$3MgO(FeAl)_2O_3\ 3SiO_2$	Building industry, fireproofing, refractory insulation, acoustic panels, and brake linings.

Grade of an ore signifies the general quality of the ore. A high-assay-value ore is termed as **high-grade** or **rich ore**, whereas that of low assay value is termed as **low-grade** or **lean ore**. The terms low and high grade are relative to the specific ore of acceptable tenor. For example, Iron ore of about 60% Fe is acceptable for Iron extraction, whereas Lead ore of about 10% Pb is acceptable for Lead extraction. Therefore, Iron ore of more than 60% Fe and Lead ore of more than 10% Pb are rich ores.

Grade is expressed in terms of metal content in case of metallic ores such as iron, manganese, and base-metal ores in terms of oxide percentage for many metallic and non-metallic ores e.g. Cr_2O_3 in chrome ore, Al_2O_3 in Aluminium ore, CaO in limestone, and P_2O_5 in apatite and rock phosphate.

In some cases, impurities also determine grade: for example, the presence of Sulphur and Phosphorus in iron ore, manganese ore, and coal. Furthermore, material strength and colour is considered for grading purposes. Spinning and non-spinning types for asbestos; snow white, white, and off colour for barites; and friable, compact or massive, and crystalline for limestone are a few examples. Rubber manufacturers specify copper-free china clay because even a minute trace of copper results in early decay of rubber. However, in other cases, physical properties and size are the only two factors considered for grading, chemical composition is not considered at all; for example, mica and asbestos. Iron ore and coal are also graded on the basis of size as lumps, fines, etc.

Specification of an ore is intimately related with grade. It pinpoints the tolerance limits of all constituents present in it. Individual consumers may prescribe different specifications for the same grade of mineral, which is dependent upon two factors:

(1) manufacturing technique adopted by individual units and

(2) the grade of other raw material required to be used to obtain the end product.

1.2. MINERAL PROCESSING

Mineral Processing is a term that can be applied to a group of operations an ore is subjected to for obtaining a required end product. The ore is to be reduced in size to detach valuable mineral grains from the ore. Detached valuable mineral grains are then separated. Separation of valuable mineral from the aggregation of valuable and gangue minerals by physical methods is termed as Mineral Beneficiation. Thus, liberation and beneficiation are the two basic steps involved in Mineral Processing. The various operations in each step are as follows.

1. **Liberation:** Detachment or freeing of dissimilar particles from each other, i.e. valuable and gangue mineral particles.

   Operations: Crushing
   Grinding

2. **Beneficiation:** Actual separation of liberated dissimilar particles i.e. valuable and gangue mineral particles.

Operations: Gravity concentration

Heavy Medium Separation
Jigging
Spiralling
Tabling

Flotation
Magnetic separation
Electrical separation
Miscellaneous operations such as Hand Sorting

Before, during, and after these liberation and beneficiation operations, several operations are performed to obtain a required end product. These operations can be called as supporting operations. Supporting operations are essential operations of any plant without which no plant can exist. Following are the few supporting operations.

Preliminary washing,
Screening
Classification
Thickening
Filtration
Storage
Conveying
Feeding
Pumping
Pneumatic and Slurry transport

All these operations are known as unit operations that are defined as the operations conducted on any material and involve physical changes.

In practice, it is not possible to separate the pure valuable mineral product because it contains some gangue minerals. This product is called **concentrate**. Similarly, gangue-mineral products called **tailing** also contain minute quantities of valuable mineral. This is due to either complex locking nature of the ore, resulting in incomplete liberation, or inefficiency of the equipment separating the valuable mineral particles.

Considering all aspects, A.M. Gaudin has defined Mineral Processing as follows (Gaudin 2005):

> "Mineral Processing can be defined as processing of raw minerals to yield marketable products and waste by means of physical or mechanical methods in such a way that the physical and chemical identity of the minerals are not destroyed."

It follows that Mineral Processing is a process designed to meet the needs of the consumer of minerals. **Run-of-mine Ore** is an ore directly taken from the mine. It is a raw material for a Mineral Process Engineer from which much of the valuable mineral is separated during its processing. The primary object of Mineral Processing is to eliminate either unwanted chemical species or particles of unsuitable size or structure.

In a hypothetical situation, the concentrate from a beneficiation operation contains pure valuable mineral. Its assay value is equivalent to percent metal present in valuable mineral. Thus, the maximum grade of the concentrate obtainable by beneficiation is equivalent to percent metal present in valuable mineral. This is illustrated in example 1.2.1. Gangue content in the ore is also estimated in this example.

Example 1.2.1: *What is the maximum grade of the Zinc concentrate obtainable from an ore of 8% Zn containing sphalerite (ZnS)? Atomic weights of Zn and S are 65.37 and 32.06, respectively. Estimate the quantity of gangue per ton of ore.*

Solution:

Given

Atomic weight of Zinc $= 65.37$

Atomic weight of Sulphur $= 32.06$

Chemical formula of Sphalerite is ZnS

Molecular weight of Galena $= 65.37 + 32.06 = 97.43$

% Zinc $= \dfrac{65.37}{97.43} \times 100 = 67.10\%$

Maximum grade of Zinc concentrate obtainable $= 67.10\%$

% ZnS in zinc ore $= \dfrac{8.00}{67.10} \times 100 = 11.92\%$

% gangue in zinc ore $= 100 - 11.92 = 88.08\%$

Gangue in zinc ore $= 0.8808$ ton per ton of ore

During processing, much of the gangue minerals, which are usually present in large quantities in many ores, are removed so that lean ores can be made technically suitable for extraction of metal.

The benefits are

1. reduced freight and handling costs as no freight must be paid for the waste discarded by the processing operations,
2. reduced cost of extraction (smelting) owing to reduction in tonnage to be smelted, and
3. reduced metal loss in slag owing to reduction in amount of metal-bearing slag produced at the smelter.

Therefore, Mineral Processing is considered as major division in the science of extractive metallurgy. Mineral Processing is usually performed at the mine site and gaining importance owing to the following factors:

1. Depletion of good-quality (high-grade) ore reserves owing to much of such ore being continuously mined and utilised for extraction of metal; thus, it is unavoidable to use low-grade ores (which require beneficiation) for metal extraction to meet demands.
2. To use unmined reserves of a particular mine, switching over from **Selective mining** to a cheaper mining method, i.e. **Bulk mining**, is found economical wherein beneficiation is a must.
3. Compared to metallurgical processes, Mineral Processing is inexpensive.

1.3. MINERAL PROPERTIES USEFUL FOR BENEFICIATION

Mineral Processing employs many physical and physicochemical properties of minerals for their separation. Behavioural difference between valuable and gangue minerals affords methods for their separation. Following are the physical and physicochemical properties of interest in Mineral Processing.

Physical properties

- Hardness
- Tenacity, brittleness, and friability
- Structure and fracture
- Friction
- Aggregation
- Colour and lustre
- Specific gravity
- Magnetic property
- Electrical property

Physicochemical properties

- Surface properties
- Change of porosity by heat
- Decrepitation by heat
- Change of magnetism by heat
- Selective dissolution

Hardness: Mineral hardness is a characteristic property that can be defined as the ability of a mineral to resist scratching. Minerals differ greatly in their hardness, ranging from those of diamonds (10) to talc (1), as adopted by Dana (10-Diamond, 9-Corundum, 8-Topaz, 7-Quartz, 6-Feldspar, 5-Apatite, 4-Fluorite, 3-Calcite, 2-Gypsum, 1-Talc). Each mineral in the list can scratch all those below it. Hardness affects the wear of crushing machines-harder the mineral greater the wear. Although hardness does not necessarily affect the tendency of minerals to produce fines or slimes in crushing, it can be considered a property on the basis of which particles will be broken into different sizes when they are broken. This difference in size makes the minerals to separate.

Tenacity, Brittleness, and Friability: They refer to the breaking strength of minerals. Minerals such as mica, talc, and gypsum are extremely tough; however, they may at the same time be soft, making them difficult to break. Some forms of hornblende and feldspar exhibit extraordinary toughness, but are not very hard. Other minerals are brittle and break with comparative ease, e.g. some varieties of quartz. When in agitation, a hard and brittle mineral will make more fines or slimes than one that is soft and tough.

Hardness together with toughness, are the main determining factors in crushing and grinding. Hardness affects the brittleness and friability of a mineral, whereas toughness determines its resilience and elasticity. The interplay of these two qualities in a homogeneous piece of rock determines its response to the compressive forces (crushing) and abrasive forces (grinding) in size reduction.

Structure and Fracture: The shape into which a mineral will break is often determined by the presence (or absence) of crystalline structure. For example, galena tends to break into cubes, feldspar into elongated fragments, mica into flat scales, magnetite into rounded grains, and chalcopyrite into conchoidal fragments. All these shapes directly affect the ability of the grain to settle in water or move upon a plane surface, which is an important phenomenon used for separation of certain minerals.

Friction: The ability of particles to move or slide on a surface will be affected by their shapes and the coefficient of friction, as determined by the nature of the surfaces of the minerals. The automatic slate pickers used in cleaning coal utilize this principle.

Aggregation: Aggregation is the physical form in which the valuable mineral exists in the ore body. It is far easier to recover the valuable mineral in an ore when this valuable mineral occurs in pure and relatively large masses than when it occurs in extremely fine grains distributed throughout the rock mass or in a laminated or tabular form.

Colour and lustre: These qualities are of the utmost value in hand picking. Slight differences in colour or in lustre, e.g. the brass yellow of chalcopyrite, the pale yellow of pyrite, the white of arsenophyite, the vitreous lustre of quartz, the resinous of sphalerite, the adamantine of diamond and cerussite, the dull of chalk, and the pearly of talc furnish valuable aids in hand picking.

Specific gravity: The difference in specific gravity of minerals affords one of the surest means of separating them from each other. To accomplish satisfactory work in separation in water, there must be a difference in specific gravity of 1.0 between the valuable mineral and the gangue. Under favourable conditions with careful work the minimum difference must be 0.5. For separation in air, the difference should be at least 0.2. The minimum value of the difference is still lower as 0.1 for heavy solution separations. In water separations, unless some adverse condition is introduced, the denser the liquid (or suspension or pulp), the better will it serve for the separation of minerals.

Magnetic property: Magnetic attraction is strong in some minerals, notably magnetite and some forms of pyrrhotite. Minerals, such as franklinite, chromite, serpentine, iron-bearing sphalerite, and garnet, have low magnetic attraction. Minerals such as quartz, calcite, gypsum, and feldspar exhibit no attraction at all. Using appropriately constructed magnets, magnetic minerals can be separated from non-magnetic minerals; moreover, minerals of more magnetic property can be separated from minerals of less magnetic property. By careful adjustments of the instrument settings, most minerals can be separated even from complex mixtures of many minerals.

Electrical property: The fact that some minerals are relatively good conductors, whereas others are relatively poor conductors of electricity has rendered it possible to affect the separation of two or more minerals.

Surface properties: The surface properties of a mineral determine its behaviour at an interface or surface between two phases. A separation of minerals may be accomplished if one mineral or group of minerals adheres to or consolidates with a surface of some kind, presented equally to all the minerals. Air bubbles attach themselves to some minerals in flotation and not to others. Gold adheres to an amalgamated plate surface, whereas quartz and other common minerals do not. Diamonds adhere to a greasy surface, whereas quartz does not; thereby, affecting an economical separation.

In flotation, surface properties of minerals are varied by the addition of chemical reagents; thereby, the valuable minerals are given the ability to adhere to the air-water surfaces of air bubbles. The selective adhesion of oil for certain minerals has also been used in certain older flotation processes.

Change of porosity by heat: When heated, certain minerals lose a part of their volatile constituents and become porous or spongy. Air fills these pores and gives the mineral a lower apparent specific gravity, which sometimes aids in separation.

Decrepitation by heat: When laid upon a hot plate, some minerals decrepitate or fly to pieces through the unequal expansion that overcomes the cohesion of the molecules. Examples of such minerals are calcite, fluorite, and barite. A mineral that decrepitates may be separated from one that does not by screening. Concentration of the barite is one such example.

Change of magnetism by heat: Certain minerals, especially those of iron, when heated, lose Oxygen, carbonic acid, or Sulphur, and are changed from being nonmagnetic or only slightly magnetic to strongly magnetic. The magnet may then be employed for separating them from nonmagnetic minerals.

Selective dissolution: Valuable minerals from their ores can be recovered by selective dissolution of either desired or undesired constituents. For example, in cyanidation process, gold is selectively dissolved in cyanides.

Use of supplementary principles: The separation of the valuable mineral from the gangue is seldom accomplished in one step. Graded crushing and graded separation are more often necessary, with a division of work among the machines in a mill such that each machine is relatively closely limited to merely operating for which it is particularly suited.

Continuity of operation in all stages of treatment is desirable for uniform results and for most economical operation; hence, machines used for beneficiation are designed primarily to operate on a continuous flow of material. Batch methods are largely confined to laboratory testing or special small-scale work.

1.4. MINERAL CHARACTERISATION

Ore/mineral characterisation is the preliminary step before subjecting the ore to processing to obtain end product(s). It involves identification of various minerals present in the ore, their chemical composition, relative amounts, texture, size and physical properties such as density, magnetic susceptibility, electrical conductivity and surface property. Ore characterisation studies form an integral and often critical part of investigations. A comprehensive mineralogical investigation should encompass not only those minerals of economic value but also harmful minerals whose presence may negatively impact the processing or salability of end products.

Furthermore, the minerals' chemical composition, size, morphology and association are all factors influencing the products produced from it, and therefore its success as a commercial venture. Other factors such as local variations in composition, grain size, mineralogy or the mineralogical distribution of harmful trace elements can influence profitability and the selection of processing routines to be employed. In short, comprehensive ore characterisation is an indispensable advantage, which can guide its processing for maximum profit.

Following are some of the identifiable physical properties of minerals:

1. Light-dependent characters
    - (a) Colour
    - (b) Streak
    - (c) Lustre
    - (d) Transparency
    - (e) Phosphorescence
    - (f) Fluorescence
2. Aggregation state
    - (a) Form
    - (b) Habit
    - (c) Pseudomorphism, Polymorphism and Polytipism
    - (d) Cleavage
    - (e) Fracture
    - (g) Hardness
    - (h) Tenacity
3. Specific gravity (density)
4. Magnetic susceptibility
5. Electrical conductivity
6. Surface property

Table 3: Mineral characterisation methods

Method	Use
1. Conventional wet chemical analysis 2. X-Ray Fluorescence (XRF) Spectrometry	Major elemental analysis
1. Atomic Absorption Spectroscopy (AAS)	Minor elemental analysis
1. Microwave Plasma Atomic Emission Spectroscopy (MP-AES)	Minor and Trace elemental analysis
1. Inductively Coupled Plasma Optical Emission Spectrometry (ICP-OES) 2. Inductively Coupled Plasma Mass Spectrometry (ICP-MS)	Trace elemental analysis
Optical microscopy 1. Reflected Light Microscopy 2. Transmitted Light Microscopy 3. Phase Contrast Microscopy (PCM)	Mineral identification and quantification
Electron microscopy 1. Scanning Electron Microscopy (SEM) with Energy Dispersive Spectroscopy (EDS) 2. Field Emission-Scanning Electron Microscopy (FE-SEM) 3. Quantitative Evaluation of Minerals by Scanning Electron Microscopy (QEMSCAN) 4. X-ray Photoelectron Spectroscopy (XPS) 5. Electron Probe Micro-analyser (EPMA) with Wavelength Dispersive Spectroscopy (WDS) 6. Transmission Electron Microscope (TEM) **Proton microscopy** 1. Proton Induced X-ray Emission (PIXE)	Mineral structure, texture and elemental composition
1. Mineral Liberation analysis (MLA) with EDS/WDS	Mineral structure, texture and elemental composition with liberation characteristics
1. X-Ray Diffraction (XRD) 2. Raman spectroscopy 3. Infrared spectroscopy 4. Fourier transform infrared spectroscopy (FTIR)	Mineral phase identification
1. Atomic Force Microscopy (AFM)	Analysis of mineral surfaces
1. Differential Thermal Analysis (DTA)	Minerals identification, semi-quantitative mineral composition
1. Thermo Gravimetric Analysis (TGA)	Mineral purity, distinguishes mineral compositions

Mineral identification on the basis of their physical properties is termed as Megascopic Identification. Minerals are also identified on the basis of their optical properties under a microscope. Transparent minerals are identified under Petrological or Mineralogical microscopes, whereas opaque minerals are identified under an Ore microscope. Such microscopic examination, which is performed for thorough understanding of ore mineralogy, determines the mineral species present in the ore and its relative abundance.

Texture of mineral occurrences is an important property which is useful for separation of valuable minerals from their ores. Textures are mainly of three types:

Fine grained	< 1 mm
Medium grained	1 to 5 mm
Coarse grained	> 5 mm

Few mineral characterisation methods and their use are given in Table 3.

1.5. PROBLEMS FOR PRACTICE

1.5.1: *Calculate percent iron present in Hematite (Fe_2O_3). Atomic weights of Iron and Oxygen are 55.85 and 16.00, respectively.* *[69.94%]*

1.5.2: *Lead ore of 2% Pb contains Galena (PbS) as only lead mineral. All other minerals are gangue minerals. What is the weight of gangue per ton of lead ore? Atomic weights of lead and sulphur are 207.19 and 32.06, respectively.* *[0.9769 ton]*

1.5.3: *Calculate %Al_2O_3 and %SiO_2 present in china clay ($Al_2O_3.2SiO_2.2H_2O$). Atomic weights of Aluminium, Silica, Hydrogen and Oxygen are 26.98, 28.09, 16.00 and 1.00, respectively.* *[39.50%, 46.56%]*

1.5.4: *Determine %Cr_2O_3 and %Cr in Chromite $FeO.Cr_2O_3$. Atomic weights of Iron, chromium and Oxygen are 55.85, 52.00 and 16.00, respectively.* *[46.46%, 67.9%]*

1.5.5: *What maximum grade of the manganese concentrate can be obtained from an ore of 42% Mn containing braunite ($3Mn_2O_3.MnSiO_3$)? Atomic weights of Manganese, Silica and Oxygen are 54.94, 28.09 and 16.00, respectively. Estimate the quantity of gangue per ton of ore.* *[63.6%Mn, 0.3396 ton]*

2 Sampling

A Mineral Processing plant costs thousands of dollars to build and operate. The plant's success relies on the assays of a few small samples. Representing large ore bodies truly and accurately using a small sample which can be handled in a laboratory is a difficult task. The difficulties arise chiefly when collecting such small samples from the bulk of the material.

Sampling: The method or operation of taking the small amount of material from the bulk is called sampling. It is the art of cutting a small portion of material from a large lot.

Sample: The small amount of material taken is called a sample. It should be representative of the bulk in all respects in its physical and chemical properties.

Sampling was well defined by Taggart (1945) as "the operation of removing a part, convenient in quantity for analysis, from a whole which is of much greater in bulk, in such a way that the proportion and distribution of the quality to be tested are the same in both the whole and the part removed (SAMPLE)." The conditions of the more stringent definition, that the sample shall be completely representative of the whole as regards all aspects, are practically never fulfilled when heterogeneous mineral mixtures are sampled.

Sampling is a statistical technique based on the theory of probability. The first and most obvious reason for sampling is to acquire information about the ore entering the plant for treatment. Second is to inspect its condition at selected points during its progress through the plant so that a comparison can be made between the optimum requirements for efficient treatment and those existing, should these not coincide. Third is to disclose recovery and reduce losses.

The prerequisite for development of a satisfactory flowsheet is the acquisition of a fully representative ore sample; even though in respect of a new ore-body, this sample may have to be something of a compromise. A bad sample will result in wastage of all test work and can result in an ill-designed mill.

A sample can be taken from any type of material–dry, wet or pulp. However, sampling method and necessary apparatus are different for each type. Sampling process consists of the following three stages to obtain representative test sample.

1. **Gross sample from a huge lot**

   A gross sample, which is large in quantity, is collected from a huge lot by grab sampling in the field

2. **Reduction of gross sample to a conveniently handled quantity**

   Gross sample is reduced to a secondary and/or ternary sample, depending upon the quantity of gross sample in steps by fractional shovelling

3. **Sample preparation**

   Test sample is prepared by sample preparation

The sample process is illustrated in Fig. 1.

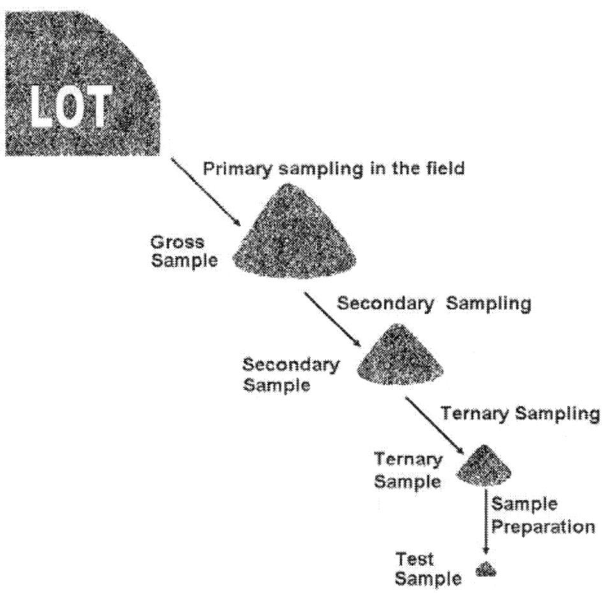

Fig.1 Sampling process

2.1. GROSS SAMPLE FROM A HUGE LOT

At first, a sample is collected in large quantities from a lot (known as **gross sample** or **primary sample** or **composite sample**) by means of various types of sampling equipment such as mechanical or hand-tool samplers using appropriate sampling methods and techniques.

The two broad sampling methods are hand sampling and machine sampling. Hand sampling is usually expensive: it is slow in batch sampling and labour wasting in stream sampling. The personal element enters so largely that accurate results are difficult to obtain. Machine sampling should be used whenever possible and hand sampling should be applied only when the material is unsuitable for treatment by machine, as with sticky ore, or when machinery is not available, or else the expense of its installation is not justified.

The two methods used to obtain gross sample are **Random sampling** and **Systematic** or **Stratified sampling**. Grab and stream sampling are the two methods used to collect the gross sample.

Grab sampling is used to collect samples from stationary bulk materials such as a stockpile. It is the simplest form of hand sampling. A small quantity of material is taken (or cut) from a place (or point) in the bulk. This quantity of material taken in one cut is called **increment**. Grab sampling consists in taking small, equal portions of material from several places by scoop or shovel selected at random (random sampling) or at regular intervals (systematic or stratified sampling) from the bulk of the material to be sampled. If large samples are desired, pits are dug at various points and the whole or part of the material excavated is taken for the sample.

Railroad cars or boats may be grab sampled in a manner similar to that of the stockpile. In this case, a net is frequently placed over the surface of the ore and the sample is taken at the knots in the net, or the sampler measures the sample intervals from the sides and ends of the cars with a shovel or measuring stick.

When an ore is being loaded or unloaded, grab samples may be taken at intervals along the working faces as excavation progresses. When the ore being transported in small lots by wheelbarrows, tram buckets, wagons or small cars, grab samples may be taken from each load and thrown into a box. The box is emptied at the end of each day or shift and taken as a sample of the ore transported for a certain period of time.

The material collected from various points is mixed together to form a primary or gross sample, which is the base for the final sample.

Advantages of grab sampling are that it is cheap and quick. The disadvantage is the difficultly involved when taking a small portion such as a shovelful or scoopful to get full representation of all sizes of particles, especially when lumps are present. The appropriate part of the lump and the amount of coarser and finer pieces to be taken at each grab is left to the judgment of the sampler. Moreover, either by over-zealousness or by carelessness, more of some size than is proper is almost certain to be taken. It is also practically impossible to take such an amount of material at each grab that every part of the lot shall have proportionate representation in the final sample.

The various hand-tool samplers used are Drill, Shovel, Scoop, and Pipe samplers. Pipe samplers are especially useful for collecting the sample from inside the stockpile, where it is inaccessible for other tools. Several varieties of pipe samplers such as slot, pocket and auger samplers are used.

The simplest form of a slot sampler consists of two pipes arranged one in another with longitudinal slots cut out: one end with a sharpened edge, the other fitted with a tee and two short nipples to form a handle. The pipes should be long enough to reach the bottom of the heap to be sampled. Following are the operational steps (Fig. 2).

1. Insert the sampler into the material to a required depth
2. Turn the handle to open the slot, the material will now be collected in the central tube
3. Turn the handle to close the slot
4. Remove the sampler from the material

5. Invert the sampler and discharge the sample

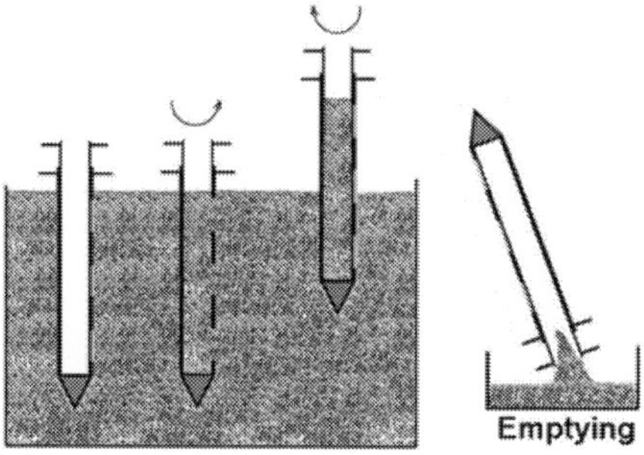

Fig. 2 Operational steps of a slot sampler

The Slot sampler is ideal for taking a large-volume, cross-sectional sample from a stockpile or container and is suitable for sampling free flowing powders, granules and even slightly cohesive powders.

Auger samplers are used in the same manner and on the same classes of materials as that of pipe samplers. They have the advantage of being useable on material that has packed so hard that it is difficult to force a pipe sampler into it.

Hand sample cutters and Automatic machine samplers are used for stream sampling in all plants. Automatic machine samplers are devices designed to substitute mechanical processes for the undesirable human element in hand sampling. Thereby, they lessen or eliminate the accidental and intentional errors introduced by the human factor, shorten the time required for cutting a sample, and reduce the amount of operating labour. Automatic machine samplers are typically used in any systematic program and consist of a cutter that goes across the material stream either by linear travel or rotating motion to obtain a slice of the flow. Following are criteria of these samplers,

- the sampling device must intersect the stream perpendicular to the flow
- entire width of the stream needs to be sampled
- the cutter should be sufficiently wide for the sample and should travel at constant speed.

Typically, the cutter's opening width should be over three times the top size.

The best known dry-ore samplers are Arc-path samplers. They are built in both revolving and swinging or oscillating forms. A common falling ore rotating-arc sampling device is the **Vezin sampler** (Fig. 3). It consists of a cutter attached to the vertical cylinder connected to the vertical shaft of an electric motor. The cylinder rotates as the motor shaft. The cutter rotates in a true circular path, passes under the end of the chute and discharges the ore to be sampled. The cutter cuts the falling ore as it passes through the stream of the ore. Thus, collected sample is discharged through a pipe connected to the cutter. The Vezin sampler is also used for sampling the pulps. The **Snyder sampler** is of rotating type, like the Vezin sampler. **Bocherdt** and **Vertical-swing rotary samplers** are the rotary-type wet-pulp samplers.

Oscillating and Swinging arc-path samplers constitute a considerable group in both dry- and wet-pulp sampling.

Straight-line cutter travel is theoretically ideal, since it is readily arranged to give equal time in each portion of the stream. A **Chain-bucket sampler** is representative of a type of straight-line sampler which is widely used for dry ore. In this sampler, buckets are of tilting type and the travel is perpendicular to the stream of ore falling from the belt conveyor. A **Geco sampler** is typical of a common form of drive for straight-line cutters, applicable both the dry- and wet-pulp sampling.

Automatic machine samplers have difficulty in handling sticky ore.

Sampling devices called poppet valves are used for pulp sampling in pipes. A **Poppet sampler** (Fig. 4) has a piston with wear resistant seals which pushes into the pressurised mainline, it pauses in the extended position to allow the chamber to fill and then it retracts to release the sample into a collection device at atmospheric pressure.

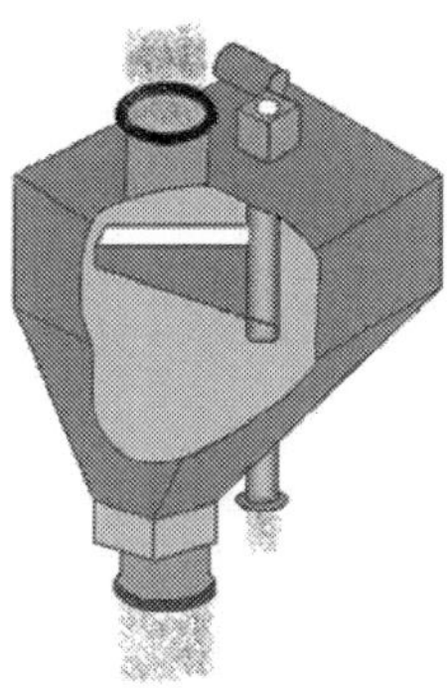

Fig. 3 Vezin Sampler

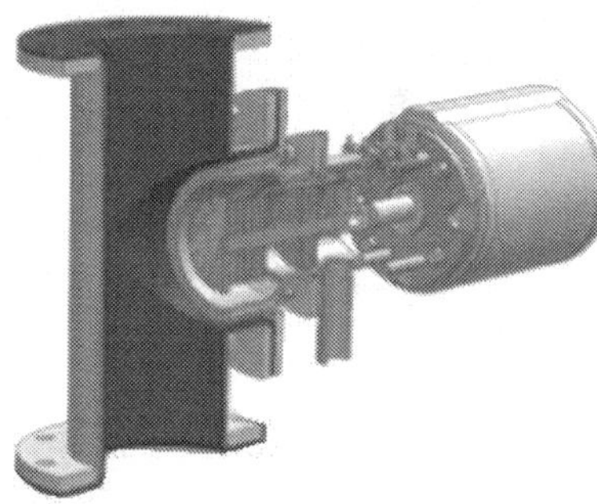

Fig. 4 Poppet sampler

(Courtesy of eDart, South Africa)

In case of wet ore or pulp, the gross sample is to be dried before it is subjected to the second stage of the sampling process.

2.2. REDUCTION OF GROSS SAMPLE

The gross samples, being large in quantity, must be appropriately 'reduced' so that the reduced sample is representative of the collected gross sample. **Shovel sampling** or **fractional shovelling** is one of the methods used to reduce the gross sample to a convenient size.

In shovel sampling, every alternate, third, fourth, fifth, or 10th shovelful is taken for the sample, depending upon the size of the permissible or desired sample. Common practice is to take the fifth (Fig. 5) or 10th shovels in unloading a car and to finish sampling by alternate shovels.

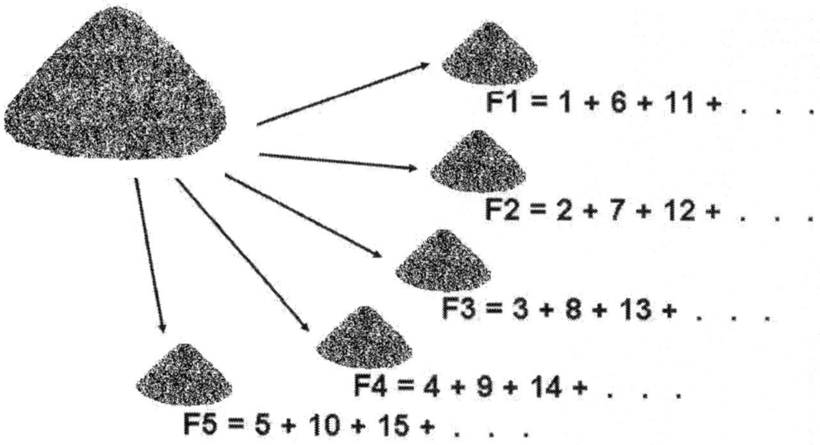

Fig. 5 Shovel sampling

Thus, the gross sample is reduced in stages, depending upon the quantity of gross sample. Such reduced samples are called **secondary** and **ternary samples,** depending upon the number of stages used.

2.3. SAMPLE PREPARATION

Reduction of reduced sample to a quantity necessary for analysis, known as a **final** or **test** or **laboratory sample**, is called **sample preparation**. It is the process of reducing the quantity by splitting. It is essential that the secondary or ternary sample be thoroughly mixed before reduction to obtain a representative sub-sample or laboratory sample.

The two broad sampling methods for sample preparation are hand and machine sampling. Coning and quartering is the oldest standard sample splitting method of hand sampling for sample preparation. For many years, the standard method of sampling for batches of ore value was to be determined between buyer and seller.

In this method, the given material is poured into a cone by shovelling all the material to one point on the quartering floor in a manner such that the particles roll down in all directions from the central point. Irregularities in the composition of the mass are thus distributed as concentric layers of a cone.

The top of the cone is then flattened with the edge of the shovel by equally spreading the material in all directions until a disc is formed. This disc is made into quadrants and the diagonally opposite quarters are cut out and all the material in the rejected quadrants are removed. The mass now contains half the original quantity and after reducing the large pieces proportionately, coning and quartering are repeated, securing one-fourth of the original mass. The cycle is repeated until the desired sample is secured. Fig. 6 illustrates the method of coning and quartering.

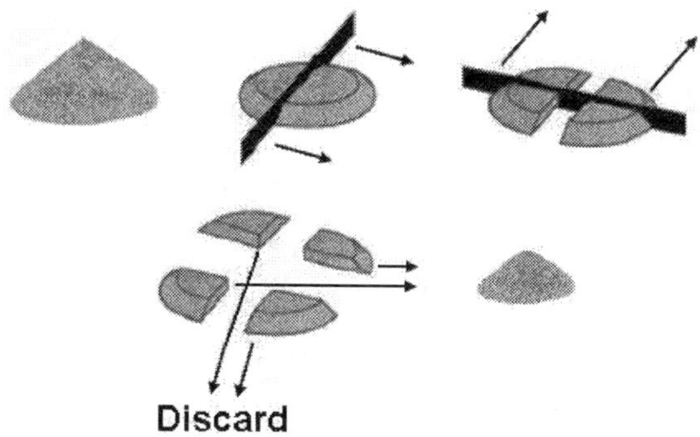

Fig. 6 Coning and quartering method

The Jones riffle and **Table samplers** are the two machine samplers used for reducing the secondary/ternary samples. Jones Riffle sampler (Fig. 7) is the stationary mechanical sampler and is most satisfactory device to obtain final sample. It consists of alternating rectangular slots, with adjacent slots discharging at opposite sides. Material is fed at the top of the riffle with a shovel. The Jones riffle sampler splits the sample and discharges them to two trays. The content of one of the trays is the sample. If required, this sample is re-fed to the same riffle for further reduction. It is mostly used in cutting down large samples by hand to assay weight, but may be used for initial sampling of a mill stream.

Table sampler (Fig. 8) is used for larger samples (>5 kg). It involves an inclined table that contains a series of holes and prisms. As the dropped sample goes down the incline, it meets rows of prisms that split and guide the sample. The material that falls through the holes gets discarded. The material that reaches the end of the table is the required sample.

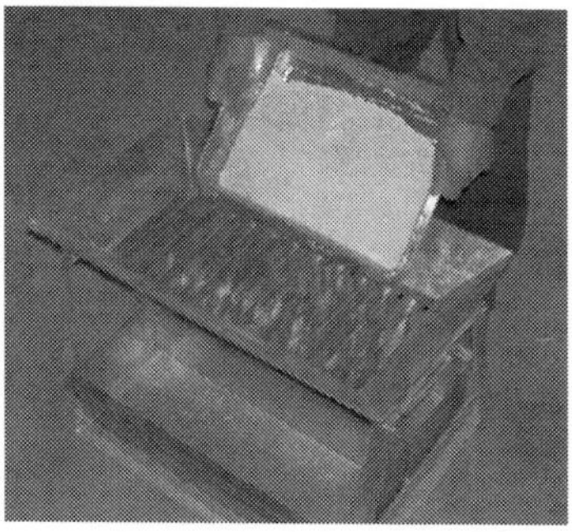

Fig. 7 Jones riffle sampler

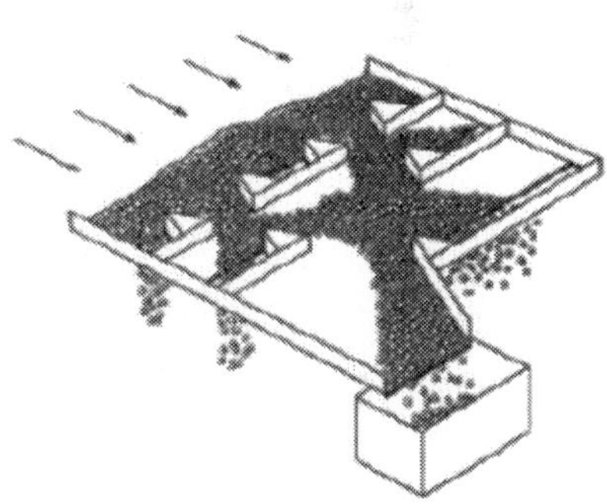

Fig. 8 Table sampler

Further splitting of the sample can be done by **Coning and quartering** or using **paper cone splitter**, a **riffle splitter** with smaller slot width, a **rotary cone splitter** (Fig. 9) or a **rotary table splitter** (Fig. 10).

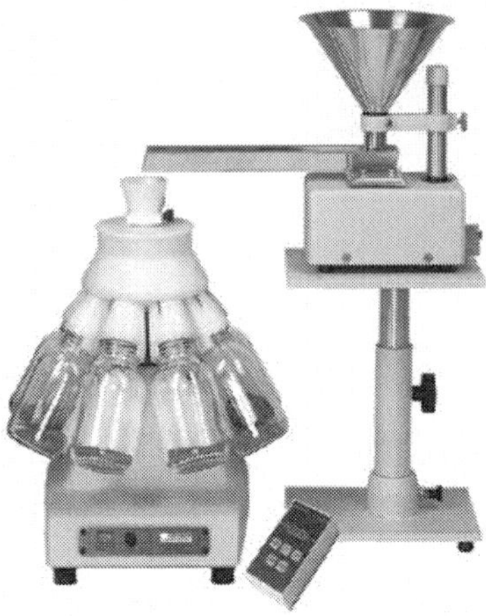

Fig. 9 Rotary cone splitter
(Courtesy of FRITSCH Milling and Sizing)

Fig. 10 Rotary table splitter
(Courtesy of www.lavallab.com)

The sampler's knowledge, experience, judgment and ability are of greater value because instructions cannot cover every point or combination of circumstances encountered on each preparation.

The procedure to be adopted for taking a sample and the amount of the sample depends on the size of the original material, particle size of the material, the method of sampling and the purpose for which the sample is taken. Table 4 shows one of the early sampling studies that proposed to relate the particle size of the material being sampled to the sample size required for a representative sample.

Table 4: Particle size and minimum weight of the sample

Particle size (inches)	Minimum weight of the sample (pounds)
0.04	0.0625
0.08	0.5
0.16	4
0.32	32
0.64	256
1.25	2048
2.50	16348

The basis for the lot to be sampled for this theory was 100 tons of ore. As one can observe, finer the material being sampled, smaller the size of sample required. The statistical fact is that the finer particles have many more individual particles per pound than do the coarser particles. Since the ore is made up of many different materials, the finer particles are much more likely to contain all of the individual elements of the entire sample.

When the sample is taken and subjected to analysis, the chances of error in a single sample will exist. One must take numerous samples to reduce error and keep the overall error within tolerable working limits.

3 Size Analysis

Size of the particle is most important consideration in mineral and coal processing as the energy consumed for reducing particle size depends on size. Particle size determines the type of equipment employed for size reduction, beneficiation, handling etc.

Particle size of standard configuration such as a sphere and cube can be easily specified. For example, the size of a spherical particle is its diameter and that of a cubical particle is the length of its side. As the mineral particles are of irregular in shape, it is difficult to define and determine their size. The size of a particle is usually defined by comparison with a standard configuration–normally, a spherical particle.

An **equivalent size** or **diameter** of an irregular particle is defined as the diameter of a spherical particle having the same controlling characteristics as that of the particle under consideration.

Volume diameter d_v is defined as the diameter of a spherical particle having the same volume as that of the irregular particle,

$$d_v = \left(\frac{6V_p}{\pi}\right)^{1/3} = 1.241\ V_p^{1/3} \qquad 3.1$$

where V_p is the volume of the irregular particle.

Surface diameter d_s is defined as the diameter of a spherical particle having the same surface area as that of the irregular particle,

$$d_s = \left(\frac{A_p}{\pi}\right)^{1/2} = 0.564\ A_p^{1/2} \qquad 3.2$$

where A_p is the surface area of the irregular particle.

Stokes' diameter is the diameter of a sphere having the same settling velocity as that of the particle under Stokisian conditions. **Newton's diameter** is the diameter of a sphere having the same settling velocity as that of the particle under Newtonian conditions.

It is evident that each definition has its limitations.

Size of the irregular particle is the decisive factor in determining its surface area. For example, let us consider one gram of pure crystalline silica spheres of diameter d cm having density ρ_p g/cm^3.

Surface area of an individual particle $= \pi d^2$ cm^2

Mass of an individual particle $= \frac{1}{6}\pi d^3 \rho_p$ g

Number of particles per gram $= \frac{6}{\pi d^3 \rho_p}$

$$\therefore \text{ Surface area of the particles per gram} = \frac{6}{\pi d^3 \rho_p}\pi d^2 = \frac{6}{d\rho_p} \text{ cm}^2 \qquad 3.3$$

This surface area per unit mass of the material is called as **Specific surface**; it is known as Blaine number in the powder industry.

Once particle size is defined, particle shape is also to be defined as the response of a particle during beneficiation is influenced to a large degree by its shape or the ratio of its surface to its volume. For a given volume, a sphere has minimum surface area, whereas a flat plate has maximum surface area. Larger the surface greater is the surface energy. The shape – the surface developed by the particle shape – has an impact on beneficiation.

The exact shape of an irregular particle is difficult to specify. One method of defining particle shape is using the term **Sphericity Ψ_S**. It is defined as the ratio of the surface area of the sphere having the same volume as that of the particle to the actual surface area of the particle. If d_v is the volume diameter of the sphere and A_p is the surface area of the particle, sphericity is defined as

$$\psi_S = \frac{\pi d_v^{\,2}}{A_p} \qquad 3.4$$

It must be noted that since sphericity compares the surface area of the particle to that of the equivalent spherical particle, it defines only the particle shape and is independent of particle size. Sphericity of a spherical particle is obviously 1.0. The sphericity of a cubical particle of size 'a' can be determined as follows:

Volumetric diameter of the cubical particle

$$d_v = \left(\frac{6V_p}{\pi}\right)^{1/3} = \left(\frac{6a^3}{\pi}\right)^{1/3} = 1.24a$$

$$\therefore \Psi_S = \frac{\pi(1.24a)^2}{6a^2} = 0.806$$

Similarly, the value of sphericity for a cylindrical particle of length equal to its diameter will be 0.874. Every mineral has a given crystallographic structure; thus, a given shape or configuration and therefore a specific sphericity. The value of sphericity for an irregular particle is always less than unity. For any crushed material, it is 0.6 – 0.7.

The reciprocal of sphericity is commonly called as the shape factor, or more precisely, the **surface shape factor λ_S**.

$$\text{Surface shape factor} = \lambda_S = \frac{1}{\psi_S} = \frac{A_p}{\pi d_v^2} \qquad 3.5$$

Sphericity Ψ_V is defined as the ratio of the volume of the sphere having the same surface area as that of the particle to the actual volume of the particle. If d_S is the surface diameter of the sphere and V_p is the volume of the particle, the sphericity is defined as

$$\psi_V = \frac{\frac{\pi}{6} d_S^3}{V_p} \qquad 3.6$$

$$\text{Then, \textbf{volume shape factor}} = \lambda_V = \frac{1}{\psi_V} = \frac{6V_p}{\pi d_S^3} \qquad 3.7$$

Volume shape factor λ_V is sometimes used for calculating the volume of an irregular particle.

In the mineral and coal industries, Standard test sieves are used to measure the size of the small and fine particles usually down to 74 microns. The side of a square aperture of a test sieve through which a particle just passes is taken as the size of the particle, though little or no importance is given to its shape.

Following are the definitions of important nomenclature used for size analysis.

Test Sieve is a circular shell of brass having an 8-inch diameter and a 2-inch height as shown in Fig.11.

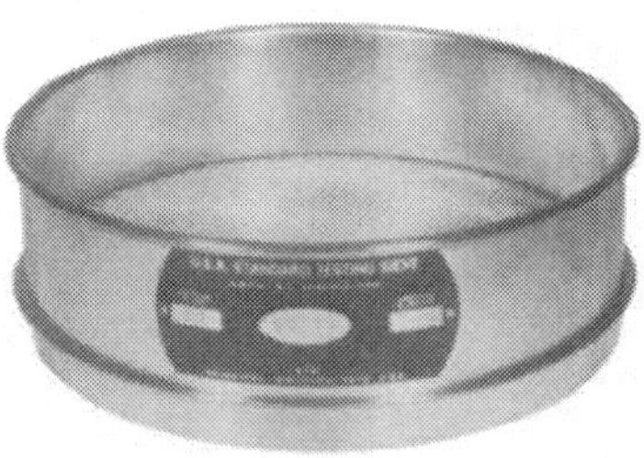

Fig. 11 Test sieve

Sieve cloth is made of wire and woven to produce nominally uniform cloth apertures (openings). It is placed at the bottom of the shell so that material can be held on the sieve.

Aperture (Opening) is the distance between two parallel wires.

Mesh number is the number of apertures per linear inch. Sieves are designated by Mesh number.

Mesh size is the size of an aperture, i.e. the distance between two parallel wires.

As mesh number increases, mesh size decreases.

Sieve Scale is the list of successive sieve sizes used in any laboratory, taken in order from coarsest to finest.

Standard Sieve Scale is the sieve scale adopted for size analyses and general testing work to facilitate the interchangeability of results and data. In this standard sieve scale, the sizes of successive sieves in series form a geometric progression.

For a standard sieve scale, the reference point is 74 microns, which is the aperture of a 200-mesh woven wire sieve. The ratio of the successive sizes of the sieves in the standard sieve scale is $\sqrt{2}$, which means that the area of the opening of any sieve in the series is two times that of the sieve just below it and half the area of the sieve next above it in the series.

In general, mesh number × mesh size in microns ≈ 15000.

For closer sizing work the sieve ratio of $\sqrt[4]{2}$ is common.

The different standards in use are

American Tyler Series

American Standards for Testing and Materials, ASTM E-11-01

British Standard Sieves, BSS 410-2000

French Series, AFNOR (Association Francaise de Normalisation) NFX 11-501

German Standard, DIN (Deutsches Institut fur Normung) 3310-1: 2000

The Indian Standard (IS) sieves, however, follow a different type of designation. For an IS sieve, the mesh number is equal to its aperture size expressed to the nearest decamicron (0.01 mm). Thus, an IS sieve of mesh number 50 will have an aperture width of approximately 500 microns. Such a method of designation has the simplicity that the aperture width is readily indicated from the mesh number.

For most size analyses it is usually impracticable and unnecessary to use all the sieves in a series. For most purposes, alternative sieves are adequate. For accurate work over certain size ranges of interest, consecutive sieves may be used. Intermediate sieves should never be randomly chosen as the data obtained will be difficult to interpret. In general, the sieve range should be chosen such that no more than about 5% of the sample material it retained on the coarsest sieve or passes the finest sieve. These limits may be lowered for more accurate work.

Table 5 shows the comparison of test sieves of different standards.

Table 5: Comparison of test sieves of different standards

TYLER			U.S.A ASTM E-11-01		BRITISH B.S 410-2000		INDIAN I.S. 460-1962		FRENCH AFNOR NFX-11-501		GERMAN DIN 3310-1: 2000	
Sieve designation Mesh No.	Width of aperture (mm)	Mesh double Tyler Series	Sieve designation Mesh No.	Width of aperture (mm)	Sieve designation Mesh No.	Width of aperture (mm)	Sieve designation Mesh No.	Width of aperture (mm)	Sieve designation Mesh No.	Width of aperture (mm)	Sieve designation Mesh No.	Width of aperture (mm)
-	-	-	-	-	-	-	-	-	38	5.00	-	5.00
4	4.75	-	4	4.75	3 ½	4.75	480	4.75	-	-	-	4.50
-	4.00	5	5	4.00	4	4.00	400	4.00	37	4.00	2E	4.00
6	3.35	-	6	3.35	5	3.35	340	3.35	-	-	-	-
-	-	-	-	-	-	3.15	320	3.18	36	3.15	-	3.15
-	2.80	7	7	2.80	6	2.80	280	2.80	-	-	-	2.80
8	2.36	-	8	2.36	7	2.36	240	2.39	35	2.50	-	2.50
-	2.00	9	10	2.00	8	2.00	200	2.00	34	2.00	3E	2.00
10	1.70	-	12	1.70	10	1.70	170	1.70	33	1.60	-	1.60
-	1.40	12	14	1.40	12	1.40	140	1.40	-	1.40	-	1.40
-	-	-	-	-	-	1.25	-	-	32	1.25	-	1.25
14	1.18	-	16	1.18	14	1.18	120	1.20	-	-	5	1.20
-	1.00	16	18	1.00	16	1.00	100	1.00	31	1.00	6	1.00
20	0.85	-	20	0.850	18	0.850	85	0.850	-	-	-	-
-	-	-	-	-	-	0.800	80	0.79	30	0.800	-	0.800
-	0.710	24	25	0.710	22	0.710	70	0.710	-	0.710	-	0.710
-	-	-	-	-	-	0.630	-	-	29	0.630	-	0.630
28	0.600	-	30	0.600	25	0.600	60	0.600	-	-	10	0.600
-	0.500	32	35	0.500	30	0.500	50	0.500	28	0.500	12	0.500
35	0.425	-	40	0.425	36	0.425	40	0.425	-	-	-	-
-	-	-	-	-	-	0.400	-	-	27	0.400	16	0.400
-	0.355	42	45	0.355	44	0.355	35	0.355	-	0.355	-	0.355
-	-	-	-	-	-	0.315	-	-	26	0.315	-	0.315
48	0.300	-	50	0.300	52	0.300	30	0.300	-	-	20	0.300
-	0.250	60	60	0.250	60	0.250	25	0.250	25	0.250	24	0.250
65	0.212	-	70	0.212	72	0.212	20	0..212	-	-	-	-
-	-	-	-	-	-	0.200	-	-	24	0.200	30	0.200
-	0.180	80	80	0.180	85	0.180	18	0.180	-	0.180	-	0.180
-	-	-	-	-	-	0.160	-	-	23	0.160	-	0.160
100	0.150	-	100	0.150	100	0.150	15	0.150	-	-	40	0.150
-	0.125	115	120	0.125	120	0.125	12	0.125	22	0.125	50	0.125
150	0.106	-	140	0.106	150	0.106	10	0.106	-	-	-	-
-	-	-	-	-	-	0.100	-	-	21	0.100	60	0.100
-	0.90	170	170	0.090	170	0.090	9	0.090	-	0.090	70	0.090
-	-	-	-	-	-	0.800	-	-	20	0.80	-	0.080
200	0.075	-	200	0.075	200	0.075	8	0.075	-	-	80	0.075
-	-	-	-	-	-	0.710	-	-	-	0.071	-	0.071

TYLER			U.S.A ASTM E-11-01		BRITISH B.S 410-2000		INDIAN I.S. 460-1962		FRENCH AFNOR NFX-11-501		GERMAN DIN 3310-1: 2000	
Sieve designation Mesh No.	Width of aperture (mm)	Mesh double Tyler Series	Sieve designation Mesh No.	Width of aperture (mm)	Sieve designation Mesh No.	Width of aperture (mm)	Sieve designation Mesh No.	Width of aperture (mm)	Sieve designation Mesh No.	Width of aperture (mm)	Sieve designation Mesh No.	Width of aperture (mm)
-	0.063	250	230	0.063	240	0.063	6	0.063	19	0.063	-	0.063
-	-	-	-	-	-	0.056	-	-	-	0.056	110	0.056
270	0.053	-	270	0.053	300	0.053	5	0.053	-	-	-	-
-	-	-	-	-	-	0.050	-	-	18	0.050	120	0.050
-	0.045	325	325	0.045	350	0.045	4	0.045	-	0.045	-	0.045
-	-	-	-	-	-	0.040	-	-	17	0.40	-	0.040
400	0.038	-	400	0.038	400	0.038	3	0.038	-	-	130	-

3.1. SIEVE ANALYSIS

Sieve analysis is the heart of mineral processing as separation is achieved primarily on the basis of particle size. Sieve analysis is a method of size analysis using test sieves at smaller and fine sizes. It is performed to separate the sample into numerous closely sized fractions by allowing the sample of material to pass through a series of test sieves and then determine the weight percentages of every fraction. **Closely sized material (closed size material)** is the material in which the difference between maximum and minimum sizes is less.

Sieving can be done by hand or machine. Hand sieving is considered more effective as it allows the particles to present in all possible orientations on to the sieve surface. However, machine sieving is preferred for routine analysis as hand sieving is long and tedious. Table model sieve shaker and Ro-tap sieve shaker (Figs. 12 and 13) are the two principal machines used in a laboratory for sieve analysis.

Fig. 12 Table model sieve shaker

Fig. 13 Ro-tap sieve shaker

(Courtesy: Jayant Scientific Industries, Mumbai)

Owing to irregular shapes, particles cannot pass through the sieve unless they are presented in a favourable orientation, particularly with the fine particles. Hence, there is no end point for sieving. For all practical purposes, the end point is considered to have been reached when there is little amount of material passing through after a certain length of sieving.

Sieving is generally done dry. Wet sieving is used when the material is in the form of slurry. When little moisture is present, a combination of wet and dry sieving is performed by initially adding water.

3.2 TESTING METHOD

The sieves chosen for the test are arranged in a **stack** (**nest**) starting from the coarsest sieve at the top and to finest at the bottom. A pan (receiver) is placed below the bottom sieve to receive final undersize and a lid is placed on top of the coarsest sieve to prevent escape of the sample.

The material to be tested is placed on the uppermost coarsest sieve and closed with a lid. The nest is then placed in a sieve shaker and sieved for certain time. After sieving by sieve shaker, hand sieving is usually performed for better accuracy by taking individual sieves. Fig. 14 shows the sieve analysis at the end of the sieving.

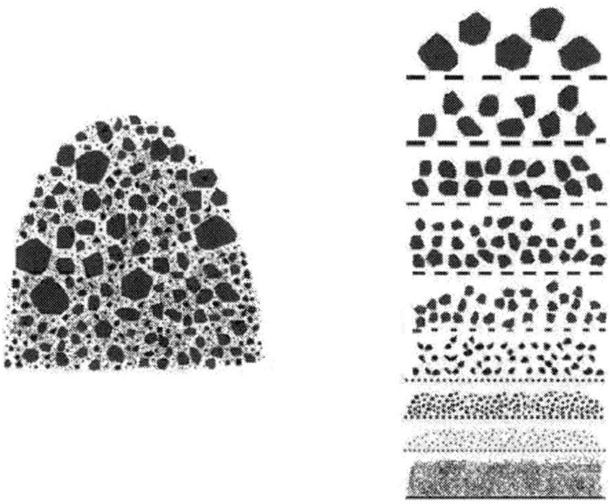

Fig. 14 Sieve analysis at the end of sieving

The nest of sieves is taken out from the sieve shaker. The material collected on each sieve is removed and weighed. For better accuracy, after the nest is taken out from the sieve shaker, each individual sieve is taken by hand with a lid and receiver and sieved for greater length of time starting from the top sieve. The material collected in the receiver is added to the next bottom sieve. The procedure is continued till the bottom-most sieve. The complete set of values is known as **Particle Size Distribution** data. This data for an ore is further elaborated in tabular form, as shown in Table 6.

Table 6: Particle Size Distribution data from size analysis test

Mesh number	Retained mesh size in microns	Weight of material g
+20	840	7.7
−20 + 28	595	13.2
−28 + 35	420	29.7
−35 + 48	297	40.7
−48 + 65	210	28.6
−65 + 100	149	22.0
−100 + 150	105	16.5
−150 + 200	74	8.8
−200		52.8
		220.00

'+' designates particles retained on that sieve
'–' designates particles passed through that sieve

Another way of representing the material with mesh numbers is 20/28, implying that the material passes through the 20 mesh and is retained on the 28 mesh.

The weight percentages of the material retained on each sieve are to be calculated to form differential analysis. Cumulative weight percentage retained is obtained from differential analysis by adding cumulatively the individual differential weight percentages from the top of the table. Cumulative weight percentage passing is obtained by adding the individual weight percentages from the bottom of the table.

All fractions except for the first are closely sized. Hence, the size of the particles in each fraction may be calculated as arithmetic mean of the limiting sizes.

For example, the size of −20+28 mesh fraction is $\frac{840+595}{2} = 717.5$ microns.

It means, the particles which pass through 20 mesh and retain on 28 mesh are having the mean size of 717.5 microns. Similarly, the mean sizes of each fraction are to be calculated. Table 7 shows all values.

Average size of the material is determined using the following simple arithmetic formula

$$\therefore \text{ Average size} = \frac{100}{\sum \frac{w_i}{d_i}}, \qquad 3.8$$

where w is the weight percent of the material retained by the sieve and d is the mean size of the material retained by the same sieve.

$$\text{Specific surface of the sample of particles} = A = \frac{6\lambda_S}{\rho_p}\sum\frac{w_i}{d_i} \qquad 3.9$$

where w is the weight fraction of the material of size d.

Calculation of average size and specific surface is illustrated in example 3.2.1.

Table 7: Calculated values for Particle Size Distribution

Mesh number	Retained mesh size microns	Mean size d_i microns	Weight (g)	Wt% retained (w_i)	Cum wt% retained	Cum wt% passing (W)
						100.0
+20	840		7.7	3.5	3.5	96.5
−20 + 28	595	717.5	13.2	6.0	9.5	90.5
−28 + 35	420	507.5	29.7	13.5	23.0	77.0
−35 + 48	297	358.5	40.7	18.5	41.5	58.5
−48 + 65	210	253.5	28.6	13.0	54.5	45.5
−65 + 100	149	179.5	22.0	10.0	64.5	35.5
−100 + 150	105	127.0	16.5	7.5	72.0	28.0
− 150 + 200	74	89.5	8.8	4.0	76.0	24.0
−200		37.5	52.8	24.0	100.0	
			220.00			

Example 3.2.1: *A sample of pyrite particles (Sp.gr. = 5) was screened. The screen analysis is given in Table 8. Calculate the average size of the sample and specific surface.*

Table 8: Screen analysis for example 3.2.1

Mesh	% Retained	Size (cm)
8/10	0	1.651
10/14	21.2	1.168
14/20	19.6	0.833
20/28	17.4	0.589
28/35	14.0	0.417
35/48	15.8	0.295
48/65	12.0	0.208

	100.0	

Solution:

$$\text{Average size of the sample} = d_a = \frac{100}{\sum \frac{w_i}{d_i}}$$

$$= \frac{100}{\frac{21.2}{1.168} + \frac{19.6}{0.833} + \frac{17.4}{0.589} + \frac{14.0}{0.417} + \frac{15.8}{0.295} + \frac{12.0}{0.208}}$$

$$= \frac{100}{216.05} = 0.4629 \text{ mm}$$

$$\text{Specific surface A} = \frac{6\lambda_S}{\rho_p} \sum \frac{w_i}{d_i}$$

Assuming particles are spheres, $\lambda_S = 1$

$$\therefore \text{ A} = \frac{6\lambda_S}{\rho_p} \sum \frac{w_i}{d_i} = \frac{6 \times 1}{5} \times \frac{1}{0.4629} = 2.592 \text{ cm}^2/\text{g}$$

3.3 PARTICLE SIZE DISTRIBUTION DATA PRESENTATION

Particle size distribution refers to the manner in which particles are quantitatively distributed among various sizes in a sample – a statistical relation between quantity and size. This data is best presented for use in the form of graphs. The simplest method is to plot a histogram of the weight percent of the material in the size interval against the size interval. When the size intervals are sufficiently small, the histogram can be presented as a continuous curve taking the middle points of histogram. In other words, a graph is plotted between the weight percent of the material as ordinate and the arithmetic mean size as abscissa. This is called a **linear scale frequency plot**. It gives the quantitative picture of the relative distribution of the material over the entire size range. In this plot, the size scale at the fine end is compressed and the data points are congested because the size intervals are in geometric series. Hence, a logarithmic scale is used for abscissa of mean sizes. As the distances separating the points corresponding to consecutive mean sizes are equal, the data points at the finer end will spread. Such a plot is called a **semi-log frequency plot**. In this plot, the area under the curve is equal to unity or 100 percent in all cases. This is often useful.

In studies of size distribution in crushed products, a graph known as a **log–log frequency plot** is often obtained by plotting logarithm of the weight percent of the material against the logarithm of the mean size. All three graphs for the tabulated data of Table 7 are shown in Fig. 15.

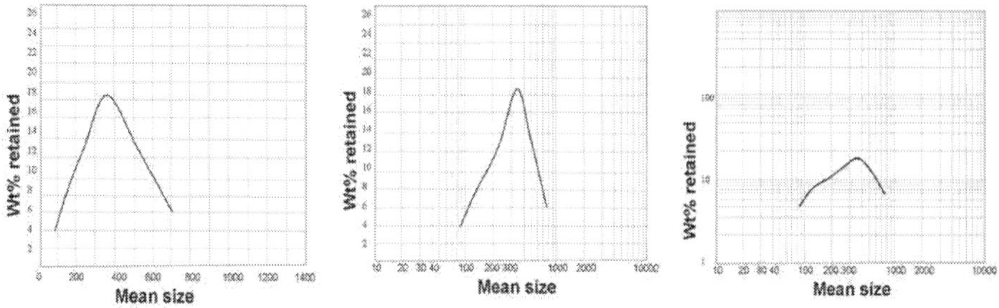

Linear scale frequency plot Semi-log frequency plot Log–log frequency plot

Fig. 15 Graphical presentation of data tabulated in Table 7

In many cases, the data is more commonly plotted as cumulative wt% passing versus actual size of opening. It is called **linear scale cumulative plot**. The size of the sieve at which 80% of the particles pass through (d_{80} or simply called 80% passing size) may be found from this plot. Crowding of data points at finer end is not a drawback in this plot since plotting of all points is not necessary. However, the logarithmic scale of abscissa for actual size of opening is recommended, in which case it is called a **semi-log cumulative plot**. These two plots are shown in Fig.16, wherein determination of 80% passing size is also shown.

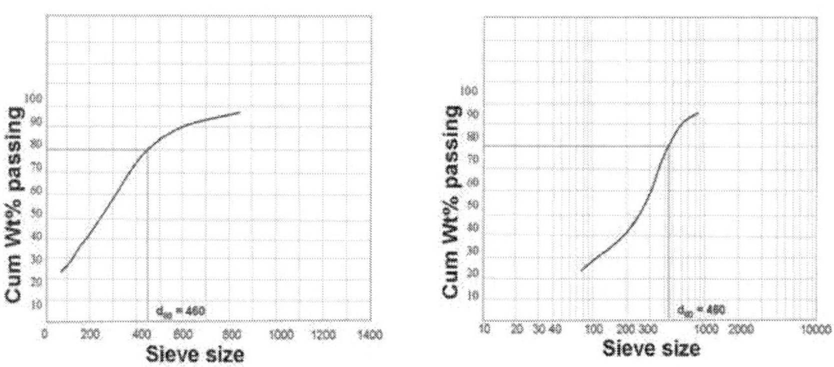

Linear scale cumulative plot Semi-log cumulative plot

Fig. 16 Linear scale and semi-log cumulative plots

80% passing size (d_{80}) is the size of the sieve at which 80% of the particles pass through that sieve. 80% passing size can be determined from the plot of cumulative weight percent passing versus sieve size.

F80 is the 80% passing size of the feed material that is fed to any operation.

P80 is the 80% passing size of the product material obtained from any operation.

80% passing size is used in all calculations to determine energy requirements for reducing the size of the particles by comminution equipment. In example 3.3.1, frequency and cumulative plots are drawn and 80% passing size is determined.

Example 3.3.1: *For the sieve analysis test data of a sample shown in Table 9, draw linear scale frequency plot, semi-log frequency plot, log–log frequency plot, linear scale cumulative plot and semi-log cumulative plot. Determine 80% passing size.*

Table 9: Sieve analysis test data of a sample for example 3.3.1

Mesh number	Retained mesh size in microns	Weight of material g
+12	1410	6.53
−12 + 14	1190	23.25
−14 + 20	840	29.50
−20 + 28	595	32.23
−28 + 35	420	24.52
−35 + 48	297	20.48
−48 + 65	210	19.35
−65 + 100	149	14.00
−100 + 150	105	15.22
−150 + 200	74	9.70
−200		55.22
		250.00

Solution:

All necessary calculations for drawing frequency and cumulative plots are performed and shown in Table 10.

Table 10: Calculated values for example 3.3.1

Mesh number	Retained mesh size (microns)	Mean size (microns)	Weight (g)	wt% retained	Cum wt% retained	Cum wt% passing W
						100.00
+12	1410		6.53	2.61	2.61	97.39
−12 + 14	1190	1300.0	23.25	9.30	11.91	88.09
−14 + 20	840	1015.0	29.50	11.80	23.71	76.29
−20 + 28	595	717.5	32.23	12.89	36.60	63.40
−28 + 35	420	507.5	24.52	9.81	46.41	55.59
−35 + 48	297	358.5	20.48	8.19	54.60	43.40
−48 + 65	210	253.5	19.35	7.74	62.34	37.66
−65 + 100	149	179.5	14.00	5.60	67.94	32.06
−100 + 150	105	127.0	15.22	6.09	74.03	25.97
−150 + 200	74	89.5	9.70	3.88	77.91	22.09
−200		37.0	55.22	22.09	100.00	
			250.00			

Linear scale frequency plot, semi-log frequency plot and log–log frequency plot are drawn with the values of Table 10 as shown in Fig 17.

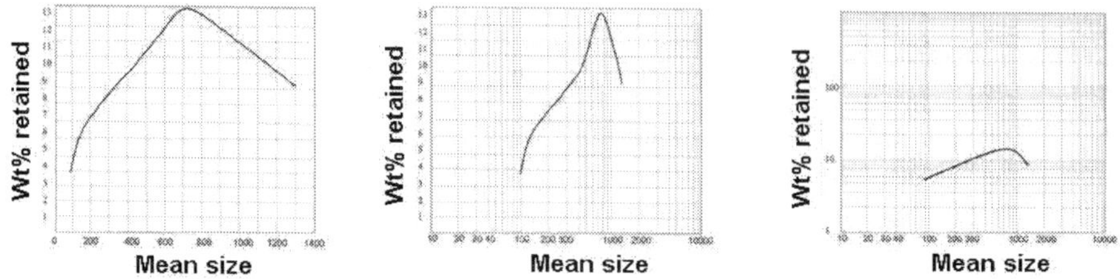

Linear scale frequency plot   Semi-log frequency plot   Log–log frequency plot

Fig. 17 Graphical presentation of data tabulated in table 10

Linear scale cumulative plot and semi-log cumulative plot are drawn with the values of Table 10, as shown in Fig. 18.

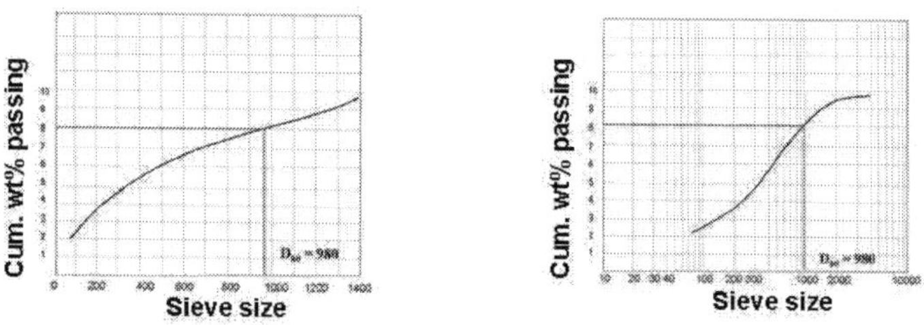

Linear scale cumulative plot   Semi-log cumulative plot

Fig. 18 Linear scale and semi-log cumulative plots for example 3.3.1

80% passing size is determined from these two plots, which is 980 microns.

3.4 SIZE ASSAY ANALYSIS

Assay value of the sample of material is determined by chemical analysis. Similarly, assay value of each size fraction of the material is also determined. This is usually called size assay analysis. Size assay analysis of an ore and their products is of prime importance through which much information can be obtained. For example, distribution of the values among various sizes will be determined by size assay analysis. Size assay analysis with or without certain physical properties of particles in each size fraction, such as specific gravity, is the basis for selection of the processes to be employed to obtain required product particles. It also gives the average assay value of the total material through which price of the material is decided on the basis of the required product's market value.

Size assay analysis of run-of mine (ROM) Iron ore of −150-mm size is given in Table 11. Size assay analysis also includes percentages of alumina and silica in addition to iron. This is because a minimum ratio of alumina to silica is required in the ore for extraction of iron.

Table 11: Size assay analysis of ROM Iron ore sample

Size	Wt%	% Fe	% Al_2O_3	% SiO_2
−150 + 50 mm	14.5	65.81	0.68	1.13
−50 + 40 mm	10.1	64.62	0.71	1.27
−40 + 30 mm	7.6	63.90	0.84	1.81
−30 + 20 mm	5.8	63.27	1.09	2.15
− 20 + 10 mm	4.0	62.84	1.28	2.68
−10 mm + 100 μm	37.2	63.15	1.82	3.84
−100 μm	20.8	61.63	2.16	4.23
	100.0			

From this data, the assay of the sample (head sample or simply head) can be calculated. Table 12 shows calculated values for determining head assay. Iron distribution in terms of percent of total iron among different sizes is also shown in the last column.

Table 12: Calculated values for Table 11

Size A	Wt% B	%Fe C	Fe D = (B × C)	Cum. Fe E	Cum wt% F	Cum Fe% G	% Fe distribution H = (D/6341.957)100
−150 + 50 mm	14.5	65.81	954.245	954.245	14.5	65.81	15.05
−50 + 40 mm	10.1	64.62	652.662	1606.907	24.6	65.32	10.29
−40 + 30 mm	7.6	63.90	485.640	2092.547	32.2	64.99	7.66
−30 + 20 mm	5.8	63.27	366.966	2459.513	38.0	64.72	5.79
−20 + 10 mm	4.0	62.84	251.360	2710.873	42.0	64.54	3.96
−10 + 100 μm	37.2	63.15	2349.180	5060.053	79.2	63.88	37.04
−100 μm	20.8	61.63	1281.904	6341.957	100.0	63.42	20.21
	100.0		6341.957				100.00

From Table 12, Head assay = 6341.957/100 = 63.42% Fe

Size assay analysis of Table 12 shows percent iron [column C] decreases with decrease in size of the particles except in the − 10 mm + 100 μm fraction. The −10 mm iron ore is not suitable for extraction of iron in blast furnace. Therefore, this fraction is to be removed. If so, the +10-mm fraction obtained is 42% [column F] containing 64.54%Fe [column G]. The −10 mm fraction removed is 58% [100 − 42] containing 62.6%Fe $\left[\frac{2349.180+1281.904}{37.2+20.8}\right]$. Percent iron distributed in the −10-mm fraction is 57.25% [37.04+20.21] and 42.75% [100 − 37.04 − 20.21] in the +10-mm fraction.

3.5 SUB-SIEVE SIZING

Sizing of the particles that are less than 40 microns is known as **Sub-sieve sizing**. The particles at fine sizes are termed as slimes and colloids. Following are the approximate size ranges of different particles:

Particles	**size ranges**
Sands	2 mm–74 microns
Slimes	74–0.1 microns
Colloids	0.1 micron - 0.001 micron

Size analysis methods used for particles less than 40 microns are shown in the Table 13.

Table 13: Size analysis methods for sub-sieve sizing

Method	Approximate range (μm)
Sedimentation (gravity)	40–1
(a) Beaker decantation	
(b) Andresen pipette	
Elutriation	40–5
Sedimentation (centrifugal)	5–0.05
Cyclosiser	
Microscope (optical)	150–1
Electron microscope	1– 0.005
Electrical Resistance method	400–0.5
Coulter counter	

Sedimentation and elutriation techniques are based on the settling behaviour of the particles of various sizes. The analysis is performed by separating the particles into various size fractions. Microscopic sizing is similar to measurement with a yardstick. The particles are sprinkled on a glass slide or mounted in some way on a slide and the size of individual particles is measured under the microscope. Laser beam particle size analysis is of recent origin. The Particle Size Measurement (PSM) system has been installed in several mineral processing plants for continuous measurement of particle size.

In industry, classifiers and hydrocyclones are used for sizing sub-sieve size particles. Fine particles are more difficult to handle and beneficiate. Greater stress is to be given on the process of fine particles not only to recover the values but also to control the pollution.

3.6 PROBLEMS FOR PRACTICE

3.6.1: *From the sieve analysis data of a sample shown in Table 14, determine 80% passing size.* *[580 μm]*

Table 14: Sieve analysis data for problem 3.6.1

Mesh number	Mesh size microns	Direct wt% retained
+18	853	7.0
−18 + 25	599	10.4
−25 + 36	422	14.2
−36 + 52	295	13.6
−52 + 72	211	9.2
−72 + 100	152	8.1
−100 + 150	104	8.2
−150 + 200	74	5.1
−200		24.2

3.6.2: *Sieve analysis test data of a sample of sand of specific gravity 2.65 is shown in Table 15. Calculate the average size and specific surface of the sample.*

Table 15: Sieve analysis test data of a sample for problem 3.6.2

Mesh number	Mesh size microns	Direct wt% retained
+18	853	7.0
−18 + 25	599	10.4
−25 + 36	422	14.2
−36 + 52	295	13.6
−52 + 72	211	9.2
−72 + 100	152	8.1
−100 + 150	104	8.2
−150 + 200	74	5.1
−200		24.2

[106 microns, 213.9 cm^2/g]

3.6.3: *For the sieve analysis test data of a sample shown in Table 16, draw linear scale frequency plot, semi-log frequency plot, log–log frequency plot, linear scale cumulative plot and semi-log cumulative plot. Determine 80% passing size.*

Table 16: Sieve analysis test data of a sample for problem 3.6.3

Mesh number	Retained mesh size in microns	Weight of material g
+14	1200	02.5
−14 + 22	710	18.0
−22 + 30	500	18.5
−30 + 44	355	21.0
−44 + 60	250	27.5
−60 + 72	210	36.0
−72 + 100	150	31.5
−100 + 150	105	26.0
−150 + 200	74	18.5
−200		50.5
		250.0

[400 microns]

3.6.4: *Size and assay analysis of ground copper ore sample is given in Table 17. Calculate grade of the sample and distribution of copper among various sizes.*

Table 17: Size and assay analysis of ground copper ore

Size range (μm)	Wt%	Assay (%Cu)
+500	2.0	0.05
−500 + 355	5.5	0.08
−355 + 250	10.5	0.35
−250 + 180	12.0	0.80
−180 + 125	15.5	1.20
−125 + 90	20.5	1.55
−90 + 63	22.0	1.80
−63	12.0	2.89

[1.38% Cu]

■■■■

4 Screening

The method of separating the particles according to their sizes using a suitable surface having holes (apertures) of certain dimensions is called screening. The minimum clear space between the edges of the opening in a screen surface is called the **aperture size** or **screen size**. In Industrial screening, the particles of various sizes are fed to the screen surface. The material passing through the screen aperture is **underflow** and contains **undersize** or **fines** or **minus** material, whereas the material retained on the screen surface is **overflow** and contains **oversize** or **coarse** or **plus** material.

Screening separates a mixture of various sizes of particles into two or more fractions. Each fraction is more uniform in particle size than in the original mixture. If the maximum and minimum size of the particles in any screened fraction is very close to each other, then such particles are called **closed size particles** or **closely sized particles**.

Screening and sieving are very similar in principle. The particles are repeatedly presented to the openings or the apertures of the screening surface until passage occurs or the process is terminated. The essential difference between screening and sieving is that screening is a continuous process and is used mainly on an industrial scale, whereas sieving is a batch process used almost exclusively for test purposes.

The screen can be of any size in dimension and size of opening. It is manufactured according to industrial requirements. The sieve cannot be of any size. It has a specific dimension and aperture size as standardised by different standards in use. It is restricted for laboratory use only for test purposes.

Screens are primarily used in mineral industries to control the size of feeds to crushers and grinding mills. The costs of crushing are usually lower when undersize material is not present in feed because fine material restricts crusher capacity while the capital and operating costs of screens are low.

The specific purposes of screening in mineral industries are as follows:

1. To remove oversize material before it is sent to next unit operation as in closed circuit crushing operations
2. To remove undersize material before it is sent to next unit operation which is set to treat material larger than this size

3. To grade materials into a specific series of sized (finished) products
4. To prepare a closely sized feed to any other unit operation

Screening is generally used for dry treatment of coarse material. Dry screening refers to the screening of a material containing a natural amount of moisture or a material that has been dried before screening. Dry screening can be performed down to 10 mesh with reasonable efficiency.

Wet screening refers to a screening operation in which water is added to the material being screened. Wet screening is usually applied to materials from 10 down to 30 mesh (0.5 mm); however, recent developments of the Sieve Bend Screen have made possible wet screening at the 50-micron size.

4.1 SCREEN

Screen is a machine used to separate particles by size. It consists of a main frame which generally rectangular in shape, wherein the screen surface that contains apertures for the passage of undersize particles is fixed. A support frame is installed directly below the screen surface to provide the necessary strength. The main frame is fixed on to the stationary base frame or structure. All other mechanical parts necessary to provide the required motion to the screen is attached to the main frame. The screen must be strong enough to support the weight of the material being screened, flexible enough to yield to the vibrating forces applied to it yet light enough to provide a reasonable percentage of open area to attain a practical through-put. The **percentage of opening** is the ratio of the net area of the apertures to the entire area of the screen surface. Large percentage of opening is beneficial for an industrial screen for improved rate of screening.

4.2 SCREEN SURFACES

Screen surface is the medium containing the apertures for the passage of the undersize material. The three general types of screen surfaces in common use are profile bars or parallel rods, punched or perforated plates and woven wire cloth.

4.2.1 Profile bars or Parallel rods

The profile bar surface is also known as the formed semi-permanent, or rod-deck surface. These surfaces are made up of parallel bars or rods anchored to heavier cross members. The bars may be circular, triangular or tapered (wedge) in cross-section. More recent types of bars taper longitudinally as well as vertically, facilitating the passage of slightly oversize lumps. For very coarse and heavy duty work involving heavy wear, parallel iron or steel bars or rails is used. These surfaces are used for both static and dynamic screens. Fig. 19 shows the common type of parallel rod surface.

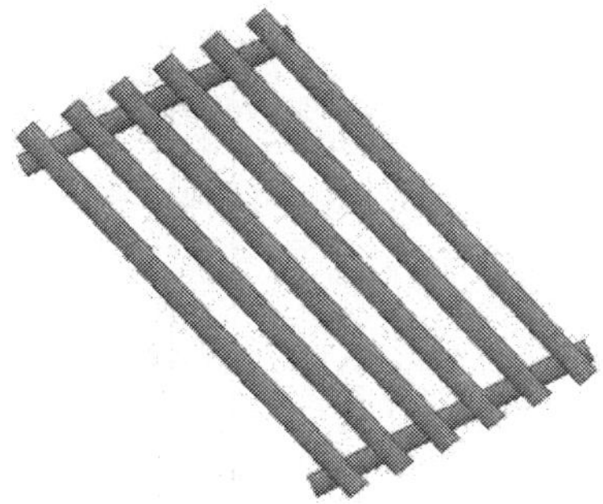

Fig. 19 Parallel rod screen surface

4.2.2 Punched or Perforated plates

The perforated plate surface is made by punching perforations or apertures in a plate. Perforated plate is available in various different materials such as Steel, Stainless steel, Brass, Copper, Aluminium, Monel and Munz metal. Perforations are available in various shapes–circular, square, hexagonal, slotted and other special forms. Plates are punched in the following manner.

1. Round, in-line and staggered
2. Square, inline and staggered
3. Hexagonal, inline and staggered
4. Slots, end staggered and side staggered
5. Slots, diagonal and straight parallel rows

Staggered arrangements of perforations (60° pattern) are the standard. Normally, perforations should be staggered in the direction of flow of the material being screened. Fig. 20 shows the perforated plates with round inline and round staggered perforations:

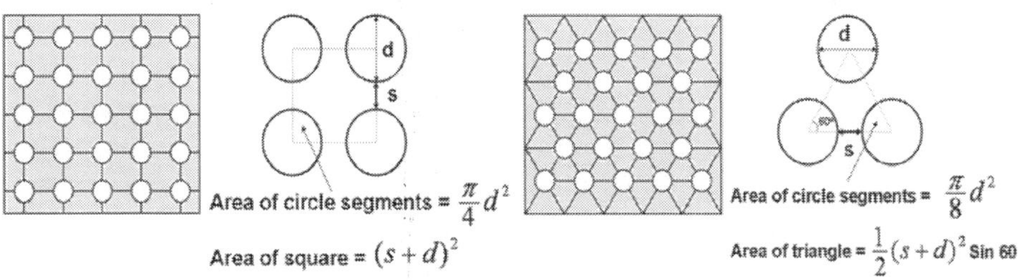

Fig. 20 Round inline and staggered punched plate screen surfaces

The amount of open area for square and diagonal spacing can be estimated by simple geometry as shown in Fig. 20. If d is the diameter of the aperture and s is the minimum spacing between them, then the percent open area for square and diagonal spacing are

$$A = \frac{\left(\frac{\pi}{4}d^2\right)}{(s+d)^2} = \frac{0.785d^2}{(s+d)^2} \text{ for square spacing and}$$

$$A = \frac{\left(\frac{\pi}{8}d^2\right)}{\frac{1}{2}(s+d)^2 \sin 60} = \frac{0.907d^2}{(s+d)^2} \text{ for diagonal spacing.}$$

Fig. 21 shows perforated plates with square inline and staggered perforations, hexagon inline and staggered perforations, slots end staggered and slots side staggered perforations and slots diagonal and straight parallel rows perforations.

Fig. 21 Perforated plates with various perforations

A perforated plate generally wears longer than woven wire cloth, and is stronger and more rigid. It is heavier and has a lower capacity than woven wire. Therefore, a perforated plate is generally restricted to coarser separations where surface loadings are high.

The perforated plates are often rubber clad. The rubber sheets have apertures slightly larger than that of the base plate. The holes in the rubber conform to the product size. The rubber cladding helps absorb the force of impact of feed material falling onto the screen. They also retard abrasion of the steel and promote a longer screen life. The elasticity of the rubber helps to reduce blinding of the screens. An added advantage of rubber-clad screens is a considerable reduction of noise level.

Nowadays, perforated plate surfaces made of rubber are employed as they have several advantages such as longer life, lower blinding and significant noise abatement. Disadvantages are limited resistance to high temperatures and a smaller percentage of open area in the finer size ranges.

If the screening surface is to make a separation at a coarse size, circular openings are used. If the separation is fine in size, slotted openings are used owing to lesser tendency to bind.

4.2.3 Woven wire cloth

Woven wire cloth consists of wires of usually a circular cross-section woven into a definite pattern or weave to provide square or rectangular openings. The wires are usually held in position by a crimp or bent in various wires at their intersection point. The wires running along the screen are termed the **warp** wires and those across the screen are the **shute** or **fill** wires. Large sheets of woven metal wire are available in thousands of specifications covering various aperture-to-wire ratios, weaves, aperture shapes, wire shapes and wire materials.

Circular wires with square openings make the cloth more rigid and are used for fairly coarse screening. Rectangular openings are more suitable for fine screening. The percentage of openings in a cloth of a rectangular opening is more than that of a square opening. Wedge wires are employed in many fine screens such as sieve bends. Such screens are strong and have relatively large open areas. The wedge profile minimises the particle blinding. Cross bars or a coarse mesh screen underneath the fine wire cloth are used as additional support. With any screen surface, it is good practice to have the surface divided into sections so that only part of the surface has to be replaced when failure occurs.

Steel, stainless steel, Monel metal, copper and bronze are chief metals from which woven wire cloths are made. Cloths are also available in natural and synthetic fabrics, but these are normally incapable of withstanding the rigors of mineral processing operations. Polyurethane or rubber offer exceptional resistance to abrasion and impact. Although initially more expensive, they are guaranteed to incur lower overall operating costs than conventional cloths due to longer life and reduced blinding. Noise levels are also lower. Fig. 22 shows woven wire screen surface with three different weaves.

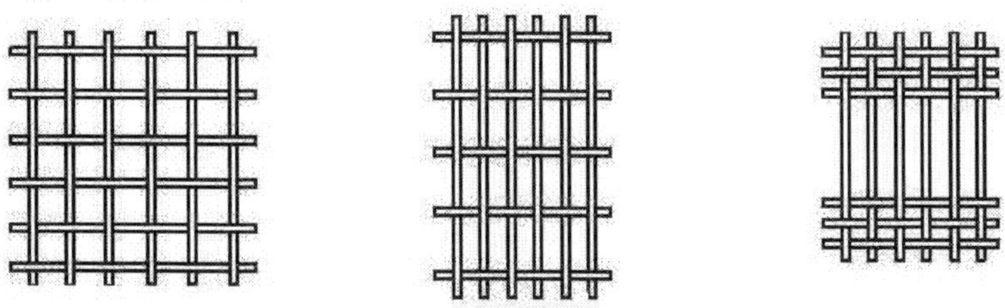

Square Rectangular Triple shute elongated

Fig. 22 Woven wire screen surfaces

If a is the size of the aperture and d is the diameter of the wire, then the percentage of opening $P = \frac{a^2}{(a+d)^2} \times 100$ applicable to square-mesh woven wire screen.

Screen surfaces can be classified according to how they are fixed to the screen as bolt-in, tensioned and modular fixing systems. Bolt-in screen surfaces consist of large sheets of punched steel plate often sandwiched with polyurethane or rubber wear surface to maximize wear life. These sheets are rigid and are bolted to the screen. Tensioned screen surfaces consist of cloths stretched taut either between the sides of the screen or along the length of the screen. Tensioned screens are available in various wire weaves as well as polyurethane and rubber mats. Self-cleaning wires are crimped to form apertures but individual wires are free to vibrate and, therefore, have a high resistance to blinding. Modular screening surfaces are most popular in harsh screening duties. Polyurethane and rubber screen decks are assembled in modules or panels and are fixed on to a sub-frame. Modular screens do not require tensioning and damaged sections can be replaced in situ.

Selection of a screening surface for a particular duty will depend on the aperture required and the nature of the work. Selection of size and shape of the apertures, proportion of open area, material properties of the screening surface and flexibility of the screen surface are important for the performance of the industrial screen.

4.3 SCREENING PROCESS

Consider a simplified screen as in Fig. 23.

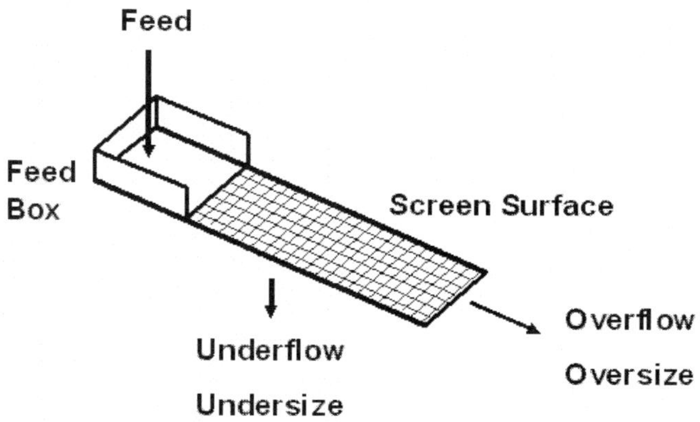

Fig. 23 Simplified screen

The material is fed at one end of the screen. Screening is affected by continuously presenting the material to be sized (the feed) to the screen surface which provides a relative motion with respect to the feed. The screen surface can be fixed or moveable. Agitation of the bed of material must be sufficient to expose all particles to the screen apertures several times during the travel of the material from feed end to the discharge end of the screen. At the same time, the screen must act as a transporter for moving retained particles from the feed end to the discharge end. Particles of size more than the aperture size of the screen are retained and less size particles are passed through the apertures.

The particles must be brought to the aperture and presented at such a velocity and direction that passage of undersize will not be hindered or prevented by bouncing from the edges or walls of the aperture. If every undersize particle is individually delivered to an aperture, with zero velocity, on a line normal to the surface, with least dimension centred on aperture, as shown in Fig. 24a; moreover, if screen surface had no thickness, every such particle will immediately pass through. However, practically, due to tonnage requirements, particles arrive in a crowd (Fig. 24b) with appreciable velocity on a line parallel to the surface with the greatest section presented to the aperture and screen surface has some relative thickness. Many undersize particles are prevented to pass through the aperture for considerable time. Others fail to enter the aperture; some others which enter are delayed in passage or stopped by friction against the walls of the aperture.

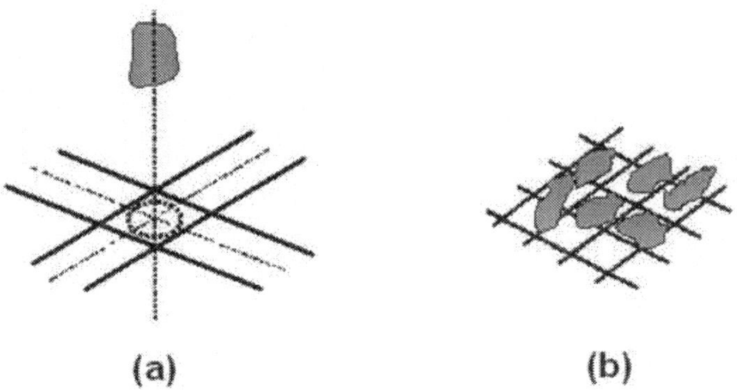

Fig. 24 (a) Ideal and (b) actual screening conditions

Undersize particles much smaller than the aperture pass through very rapidly almost like water. Oversize particles much coarser than the aperture slide over the screen surface freely. A fraction of the particles which are relatively small or close to the size of the aperture do not initially pass through as they fall on the screen surface across the apertures. In subsequent approaches, there is a high degree of probability that it will pass through one of the apertures before it reaches the end of the screen. If the particles are relatively large or close to the same size as the aperture, there is a high degree of probability that it will travel over the entire screen and get discharged as oversize particles. Whether the given near size (near mesh) particle is oversize or undersize is very largely dependent on merely how the particle is presented to the aperture. For instance, the three near size particles can be classified as oversize or undersize depending upon the probability of their reaching the screen surface in the right orientation, as shown in Fig. 25. The first particle will pass through as the two dimensions are smaller than the size of the aperture. The latter two particles will pass through only when they approach the aperture in that orientation. If they approach the aperture in any other orientation, they will not pass through the apertures and become oversize particles. As the particle's breadth of (b) is smaller than size of the screen aperture (a), they can also pass through when they approach normally provided the third dimension is smaller than either screen aperture or length of the particle (l).

Particles slightly larger than the screen aperture wedge in and cause blinding, whereas those slightly smaller than the screen aperture pass through with difficulty, i.e. slowly.

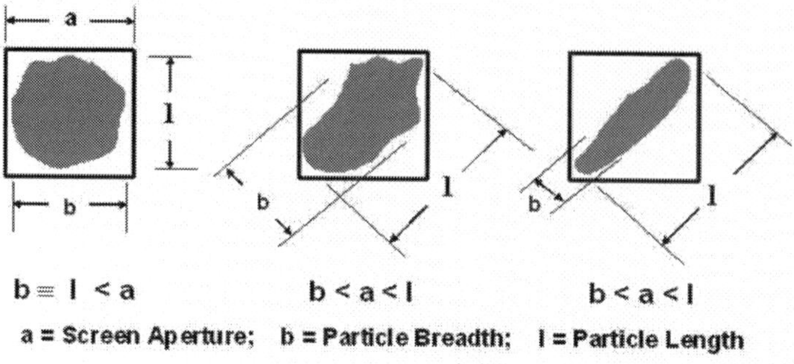

Fig. 25 Three near size particles

The velocity and direction of approach of particles to the aperture have a marked effect on rate of screening. If a spherical particle is traveling at a given rate along a screening surface whose apertures are several times the diameter of the particle, it strikes almost at the bottom of the opposite wall of the aperture and particle will pass through the aperture, as shown in Fig. 26A. With smaller aperture, the same particle will strike the far edge of the opening and pass through, as shown in Fig. 26B. However, an irregular particle may fail to pass. If the approach and the angle at which the particle is approaching is represented as **a** in Fig. 26C, it will pass through. If it is represented as **b**, it rebounds as shown and fail to pass through.

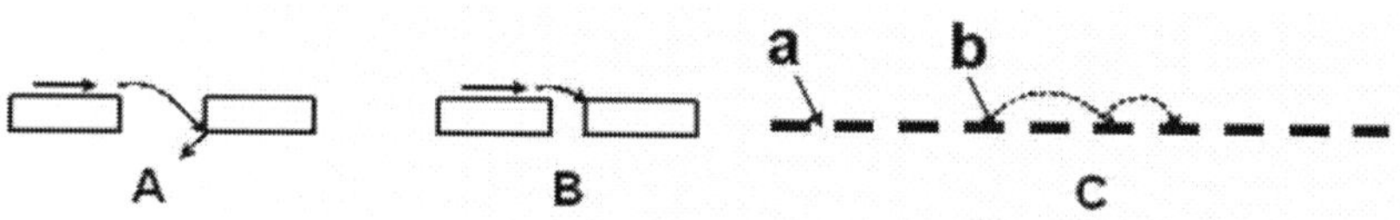

Fig. 26 Direction of approach of a particle

To attain the required angle, a particle may require several presentations. If the screen is sufficiently long, it can eventually approach the aperture at the appropriate angle and pass through. If the length is insufficient then, despite the particle size being smaller than the aperture, it may report as oversize. Thus, the length of the screen is important as it provides the screening efficiency and the width indicates the throughput rate. Ultimately, the probability of passing will depend on the initial mass of particles in the feed stream that would pass after several approaches to the aperture and the mass fraction remained over the screen surface.

When the screen is overloaded such that the top layer does not come in contact with the screen surface, the top layer will be discharged as oversize, which also contains fine particles. As the screening surface is vibrated, the bed of material tends to develop fluid-like characteristics. Moreover, the larger particles rise to the top, whereas the smaller particles sift through the voids and find their way to the bottom of the bed. This phenomenon of differentiation is called stratification.

Without stratification of the material, there would be no opportunity for separation to take place. There is an optimum depth of bed at which stratification occurs and produces the maximum rate of screening. An excessively thin bed reduces efficiency, whereas an excessively thick bed delays stratification, dampens bed movement and decreases accuracy of separation. When the bed is excessively thin, failure of stratification occurs and fines are reported to overflow. When a satisfactory bed depth is available, stratification is nearing completion and aids for better separation.

As the material progresses further on the screen surface, the material will spread. The depth of the bed of material is thick at the feed end, becomes thinner towards discharge end of the screen. Hence, three distinct flow regions can be identified. The lowest flow rate occurs at the beginning of the screen (region I of Fig. 27). At this region, the bed is excessively deep and the presentation for passage is severely restricted owing to a high quantity of oversize material and insufficient stratification. The overlying bed of material probably helps reduce the amount of bouncing of the lower particles, encourages greater frequency of presentation and may actually force some particles through the apertures. Undersize particles, occupying upper layers, need time and agitation to work their way down to the screen surface. Agitation of the screen surface imparts fluid properties to the bed of particles. The material on the screen expands and the larger particles tend to travel up. The smaller particles tend to gravitate down through the voids created by the expansion of the bed. This way, the bed stratifies as it travels forward. The process is successively repeated until all fines are passed. It can be summarised that two processes—stratification and separation—are involved in effecting the size separation.

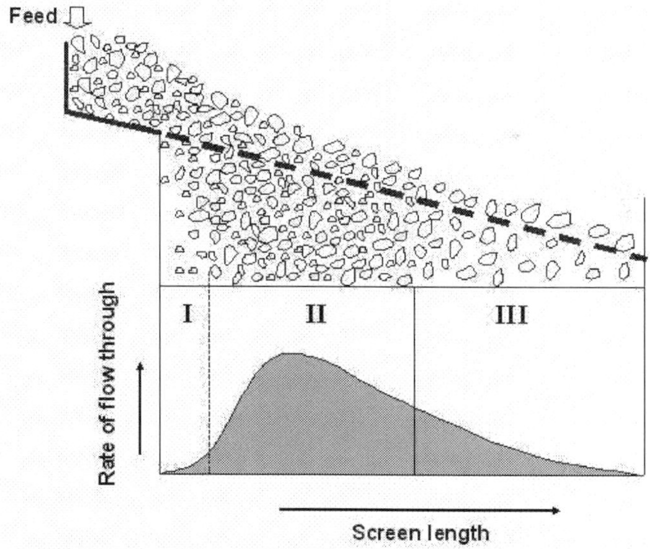

Fig. 27 Relative screening rates at different regions

Maximum flow rate occurs at the region II of Fig. 27. Closeness of the particles tends to restrain the bouncing action; moreover, the likely passing material has maximum exposure to the screen aperture. This region exists for a considerable length during which many particles pass through the apertures. Subsequently, the remaining particles over the screen surface are insufficient to form a monolayer on the screen. As the particles in the region III tend to have excessive unrestrained motion and the screen surface is not fully occupied by the particles, the flow rate is low. All these regions are schematically represented in Fig. 27.

As a general indication, the feed thickness should not exceed four times the size of the aperture for material having a bulk density of 1600 kg/m^3, 2½–3 times the aperture for material with a bulk density of 800 kg/m^3.

As the most efficient screening occurs in region II, it is desirable to maintain this region over as much of the surface as is possible. One way of achieving this is to decrease the slope along screen surface. This reduces the flow rate along the screen and maintains, or even increases, the bed depth. Such an approach takes advantage of stratification that has already occurred and prevents the formation of region III. It is claimed that this method can reduce area requirements by up to 50%.

The bed thickness at the discharge end (region III) is more important than the feed end (region I) and depends on the length of the screen. It is suggested that for the screens of length 1.8 m, the bed depth at the discharge end should be at least 1.5–2.0 times the average particle size, and 2.5–3.0 times for screens of length 7.2 m.

The size of the aperture has an effect on rate of screening. The passage of number of particles is inversely proportional to the cube of the screen aperture. The number of openings per unit surface area of the screen is inversely proportional to the square of the screen aperture. The number of passages per opening will then be inversely proportional to the size of the aperture. This leads to a conclusion that the capacity of a screen is directly proportional to the screen aperture.

The size of the apertures of the screen is more than the required size in case of inclined screens. As shown in Fig. 28, the apparent or effective size of the aperture is reduced to cos θ times the actual size of the aperture, where θ is the angle of inclination of the screen. As the slope increases, material moves across the screen faster, throughput increases and stratification takes place more rapidly. Most of the screens operate best at 12–18° angles.

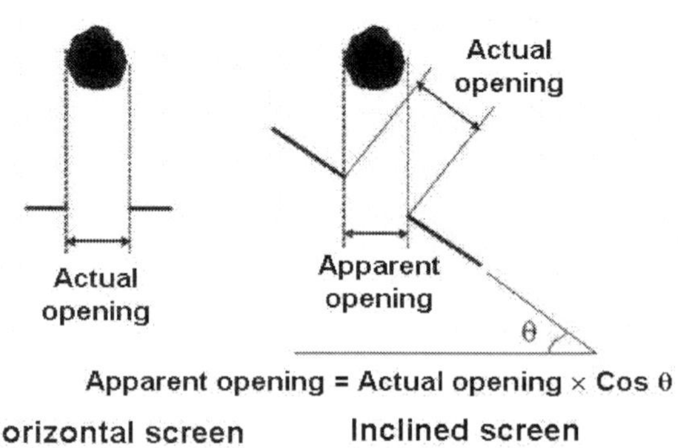

Fig. 28 Apparent opening in inclined screen

The shape of the aperture has a remarkable effect on screening. Screening with square openings or round holes is only two-dimensional screening. If the particle exceeds the size of the aperture in just one dimension, the probability of its being rejected to oversize is fairly high. The particle will pass through the aperture if at least two dimensions of the particle are smaller than the size of the aperture.

Square openings are preferred when accurate separation is required. If accurate sizing is not important, slotted rectangular openings or parallel rods or bars accomplish the task of screening at a higher rate than square openings. In a screen of rectangular apertures, open area is more and the probability of passage through a rectangular aperture is higher. It is less susceptible to blinding and easy passage of elongated particles. Rectangular apertures running parallel to the axis show the greatest wear but maximize the capacity and the passage of elongated particles.

The efficiency of a screen may be improved by increasing the number of times that particle is presented at an aperture. Following are few of the several different ways.

1. The length of the screen may be increased
2. The rate of flow may be decreased so that the material spends more time on the screen surface i.e. screening time is increased

   Rate of flow can be decreased by

   (a) changing the vibrator rotation from with-flow to against-flow (counter-flow)

   (b) decreasing the angle of inclination of the screen

   (c) decreasing the amplitude of vibration

In with-flow rotation, the motion of the vibrator is with the flow of material along the length of the screen, whereas counter-flow rotation is against the flow of the material. Generally, the combination of counter-flow rotation with steeper angle of inclination is preferred because it produces higher efficiencies than flow rotation at a flatter angle, but the same rate of travel. A desirable rate of travel for both horizontal and inclined units is usually 1 fps. Tests have shown that on circle-throw screens, direction of rotation may have a more critical influence on efficiency than either stroke or speed.

Amplitude of vibration, speed and slope of the screen determine particle bounce, advance and the duration and number of presentations. Generally, the higher the throw at constant speed, the less likely a particle is to cleanly fall through the aperture, the higher the probability of blockage and the lower the residence time. Therefore, amplitude is increased only when necessary and is more likely to be beneficial on overloaded screens. Increase in amplitude may generate free presentations on overloaded screens than on reasonably loaded units. On the other hand, an increase in the speed of a screen at constant amplitude has roughly the same effect as an increase in amplitude at constant speed. Generally, as the speed is increased, amplitude is decreased to compensate for the higher power drafts at high speeds. As a general rule, higher speeds with lower amplitudes are normal for finer separations.

With regard to the properties of the materials, moisture content and the presence of clayey material adversely affect the rate of screening. Either bone-dry or wet pulp is relatively easy to screen, but small percentage of water increases the difficulty of screening. When clayey material is present, sprays of water are employed to drain out the clayey material.

As a conclusion of the discussion of the principles of screening, numerous factors affecting the rate of screening can be listed out in the following two groups:

Material factors

1. Bulk density of the material
2. Size and size distribution of the particles
3. Size of the particle relative to the aperture
4. Shape of the particle
5. Presence of clay material
6. Moisture content of the material

Machine factors

1. Size of the aperture
2. Shape of the aperture
3. Size of the screen surface
4. Percent opening area
5. Amount of near size material in the feed
6. Method of feeding the particles
7. Angle of incidence of the particle on the screen surface
8. Rate of feed, i.e. rate of flow over the screen
9. Speed at which the particle approaches the screen surface
10. Thickness of the material on the screen surface
11. Blinding of the screen surface
12. Type of screening, i.e. wet or dry screening
13. Type of motion given to the screen surface
14. Slope of the screen deck
15. Mechanical design for supporting and tightening the screen deck

4.4 TYPES OF INDUSTRIAL SCREENS

Screens employed in industries are mainly classified as **stationary** and **dynamic** screens on the basis of their method of support. Stationary screens are placed at any angle between 0° and 45° from the horizontal and are used primarily for coarse screening. Dynamic screens are classified by the method used to produce relative motion between the feed and screen surface as revolving, shaking, vibrating, reciprocating, gyrating and probability screens. Fixed screens are most invariably run dry, whereas moving screens are run either dry or wet. Neither type will operate satisfactorily on damp (moist) material.

Most commonly, screens are used for size separations in conjunction with crushing operations. In mineral industry, screens are rarely used for separations below 0.2 mm because they have inadequate capacity. However, sieve bends are used for separations as low as 50 μm as these devices give sharper separations than wet classifiers. Screens are classified as stationary and dynamic screens as shown below.

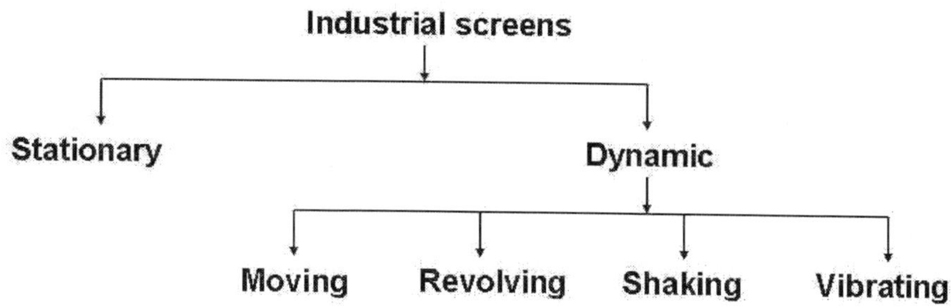

4.4.1 Stationary Screens

Grizzly, Divergator and Sieve bend are the three important stationary screens. **Grizzly** (Fig. 29) consists of a series of equally spaced heavy parallel bars set in a frame. Some grizzlies are shaken or mechanically vibrated to help the sizing and to aid in the removal of the oversize. The bars of a grizzly are inclined usually at an angle between 20° and 50°. Feed flows down the grizzly in the direction of the bars to assist flow and to reduce clogging. The bars are usually tapered in cross-section to minimize clogging once particles have entered between them. Inverted railroad rails are frequently employed. Grizzlies are used to handle dry materials. They are not suitable for moist or sticky materials. Grizzlies are normally employed for very coarse material; most commonly for sizing the feed to primary crushers. Grizzlies are used for sizing the particles in the range of 20–300 mm. Grizzlies are simple and strong, requiring no external power supply and needs little maintenance, but are wasteful of headroom.

In **divergator** (Fig. 29), the bars are longitudinally fitted in the direction of flow of the process material and fixed at one end only, i.e. the end at which the material arrives. The bars are parallel from above but alternate bars diverge at an angle of about 5° or 6° from the basic angle of installation. The gap between the bars increases from the fixed to the free end of the machine. Any pieces of material which become lodged in the bars are knocked out by other material flowing on. Therefore, the machines are self-cleaning and blockage free. The bar profile, the distance between the bars and angle of divergence as well as the slope of the total bar system determine the flow of material and size of separation. The simple robust design ensures maintenance-free operation even with difficult, sticky materials contaminated with impurities. The range of separation is normally between 25 and 400 mm.

Grizzly

Divergator
(Courtesy of Mogensen)

DSM Sieve Bend
(Courtesy of Deister Concentrator, LLC)

Fig. 29 Stationary screens

The **sieve bend** (Fig. 29), also called as the DSM sieve bend (Dutch State Mines Sieve Bend), is a stationary curved screen composed of horizontal wedge bars oriented at right angles to the line of flow. The screen deck is clamped to the screen housing, the lower section of which allows undersize to be collected. A complete sieve bend is comprised of a feed distribution chamber, a structure for holding the screen surface and collection chambers or chutes for oversize and undersize materials. The screen surface is curved in a 60° arc at a radius between 20 and 60 inches. The feed end of the sieve is tangent to the vertical so that discharge end is elevated 30° from the horizontal for a 60° sieve bend. Oversize product tends to slow down at this end, contributing to dewatering.

In operation, the feed slurry is introduced through a feed chamber mounted at the top of the screen and is distributed evenly over the top end of the screen surface. Most sieve bends are fed with gravity feed in such a manner that a falling height of approximately 20 inches is provided before the feed enters upon the screen surface. Feed slurry tangentially enters the upper surface in a direction perpendicular to the openings between wedge bars. Liquid and undersize particles pass through the openings while dewatered oversize particles flow on down the screen. The concave shape of the screen surface serves to develop centrifugal forces which press the slurry against the surface. Each wedge bar appears to act as a knife on the underside of the passing slurry. A layer of liquid and undersize particles is sliced off and directed downwards through the slot. This slicing action produces an undersize particle which is approximately half the dimension of the space between the bars. This feature is particularly advantageous because it practically eliminates blinding (clogging the aperture with particle) of the screen surface.

Sieve bend is gaining wide popularity in the mineral industry for very fine wet screening purposes. Sieve bends are successfully used for fine particles of size down to 50 μm. Sieve bends are especially important for particles in which density effects make the classifier unsuitable. They are relatively high in efficiency and capacity. In the DSM sieve bend, the particles tend to adhere to the back of the screen cloth after passing through the apertures. This adherence of the particles is minimised by providing mechanical devices to periodically vibrate or rap the screen surface to remove blinded particles. Fig. 29 shows Grizzly, Divergator and DSM sieve bends.

4.4.2 Moving Screens

In **Linear** and **Pancep** screens, the screen surface moves linearly. Linear screen (Fig. 30) consists of a synthetic monofilament moving screen cloth. It is supported on rollers and driven by a head pulley coupled to a variable speed drive unit. Dilute slurry introduced in a feed box enters on to the moving cloth. The undersize material passes through the cloth by gravity and is collected in the underpan. The oversize material is retained on the screen and is discharged at the drive pulley, and any adhering material is washed from the screen cloth using water sprays. As this screen operates with no vibration, it prevents realignment of oversize material with the cloth aperture, ensuring retention of all oversize material. Screening efficiencies are as high as 98%. It has advantages of low installation and maintenance costs, low power consumption, low noise level and cloth replacement is easier and quicker. The linear screen is primarily suited to scalping coarse material from process slurries and effective removal of thrash material. The widest use of the linear screen is in the processing of gold, iron ore and base metals.

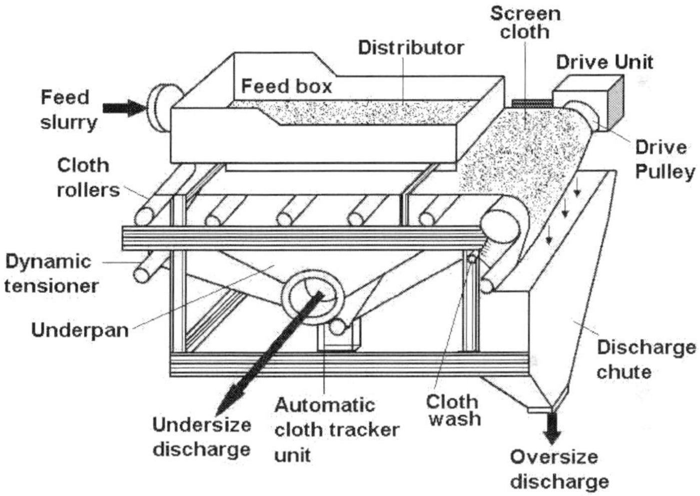

Fig. 30 Linear screen

The principle of the Pansep screen is similar to that of linear screen; however, the deck is divided into a series of pans that move in a manner similar to that of a conveyor. The base of each pan consists of a tensioned wire screen mesh permitting finer cut points than on linear screens. Cut points are in the range of 45–600 microns. As the screens do not have density effect, Pancep screens can be used to separate coarse material from hydrocylone overflow in grinding circuits to increase recovery.

4.4.3 Revolving Screens

Trommel is the most important revolving screen. It consists of punched plate or wire mesh, bent to form a cylinder, which is rotated about its axis at a speed between 35%

and 45% of critical speed. The axis is set at a slight inclination to the horizontal to make the oversize material to travel longitudinally within the cylinder. Trommels can be made to deliver several sized products by having smaller apertures at the feed end and coarser apertures at the discharge end. The main problem is that the fine screen wears quickly, as the whole of the feed must be fed on to it. The problem of undue wear on the fine-screen trommel is overcome in the Compound trommel wherein a number of trommels are mounted—one inside the other on a common shaft. The innermost will be the coarsest and the outermost the finest, the feed being introduced into the innermost screen. It is tedious to inspect the inner screens for plugging and damage. Trommel screens are used for screening particles of sizes 55–60 mm. Fig. 31 shows a simple and compound trommel screens.

Trommel screens are cheaper, mechanically robust and vibration free. They typically have lower capacities due to usage of part of the screen surface at any one time and can be more prone to blinding. They are widely used in screening plants and the screening of mill discharge streams. Trommel screens are fixed at the discharge end of the ball mill, FAG and SAG mills to prevent ball scats from reaching subsequent processing equipment and to prevent a build-up of pebbles in the mill. Trommel screens are also used for wet-scrubbing of ores such as bauxite.

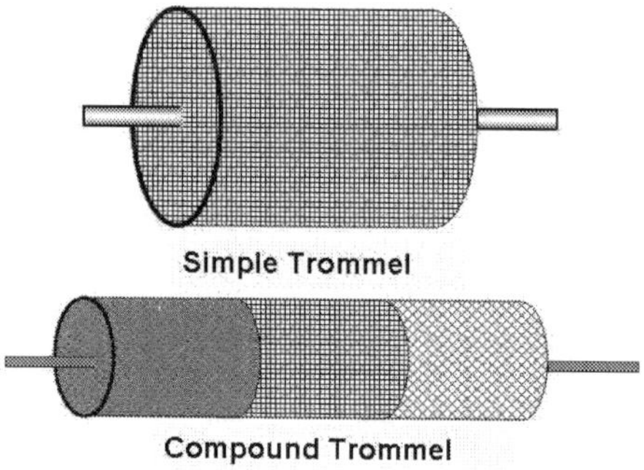

Fig. 31 Trommel screens

Fig. 32 Roller screen

Roller screen (Fig. 32) consists of rotating rolls or discs which are driven in parallel. The rolls are arrayed at a set distance forming square or rectangular screen apertures, allowing the passage of undersize material. Oversize material is transported across the series of rolls to the downstream crusher. Due to the even rotation of the rolls, the feed material is subjected to the harmonic rolling motion and gently separated.

In case of very moist, sticky or clayey material, the rolls are fitted with scrapers which clean the bottom of the rolls to ensure efficient separation. Roller screens can be used for screening applications from 3 to 300 mm. Advantages are high capacity, low noise levels, require little head-room, subject the material to less impact and permit screening of very sticky materials. **Rotaspiral** is a trammel-like device designed for ultra-fine screening between 1000 and 75 microns. The drum contains an internal spiral to move the material through the screen. Water sprays are used to fluidise the screen bed and wash the screen surface. Rotaspiral can also be used in a dewatering duty.

4.4.4 Shaking screens

Shaking screens, also called **jigging screens**, have a reciprocating movement mechanically induced in the horizontal direction and are mounted either horizontally or with a gentle slope. Shaking screens consist of a rectangular plate perforated with holes of the required size and shape, and fixed in a suitable frame. The screen is mounted on flexible supports and is connected to an eccentric or other driving mechanism which moves the screen backwards and forwards a distance of between 2 and 6 inches about 100 times per minute. The supports are usually arranged so that during the forward stroke the motion carries the screening surface and the particles on it slightly upwards. On the return stroke, the screen moves backwards and slightly downwards, withdrawing itself from under the particles. This motion, together with the slope of the screen, causes the particles to slide jerkily in contact with the screen. Screens may be superimposed or may be arranged end to end to prepare a number of different grades. Shaking screens are used for coarser particles down to 12 mm size. These are widely used for coarse coal sizing. Fig. 33 shows the shaking screen.

Fig. 33 Shaking screen (Courtesy of JXSC Mining, China)

4.4.5 Vibrating screens

Vibrating screens are most nowadays widely used and have a large variety of designs. They can be classified as horizontal or nearly horizontal and inclined screens. They are arranged as single- and multi-deck screens. The screen having one screening surface is called a single-deck screen; moreover, if a screen has two or more screen surfaces, it is called a multi-deck screen.

Fig. 34 shows single- and two-deck horizontal vibrating screens and Fig. 35 shows single- and two-deck inclined vibrating screens.

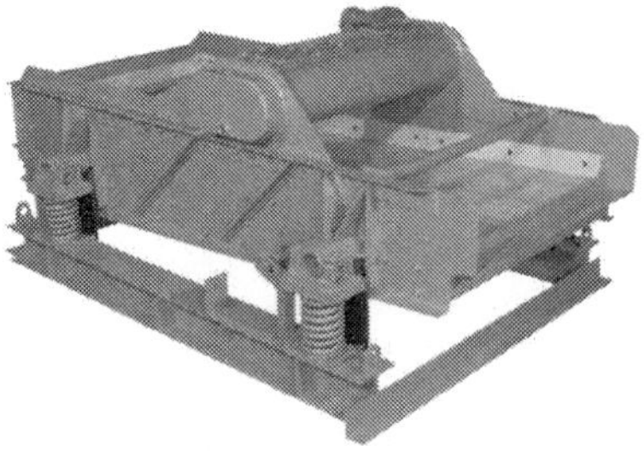

Single deck horizontal vibrating screen

Two deck horizontal vibrating screen

Fig. 34 Horizontal vibrating screens (Courtesy of Deister Machine Co. Inc)

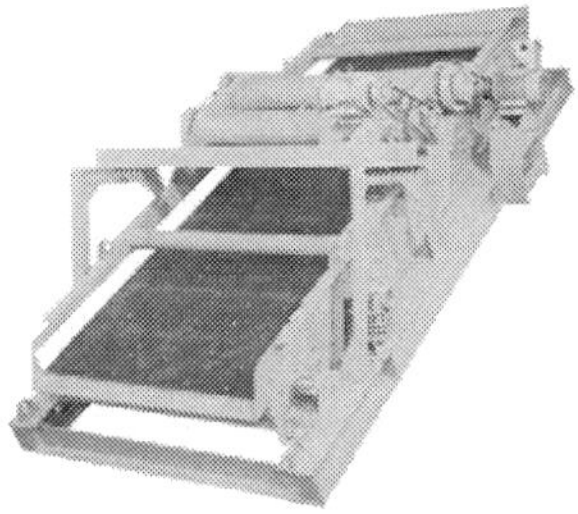

Single-deck inclined vibrating screen

Two-deck inclined vibrating screen

Fig. 35 Inclined vibrating screens (Courtesy of Deister Machine Co. Inc)

Vibration is induced vertically either by the rotation of mechanical reciprocating device applied to the casing or by electric devices operating directly on the screen. In mechanically vibrated screens, vibrations are produced by Cam and spring, Eccentric or Unbalanced flywheels. Mechanically vibrated screens are most widely used for coarse sizing.

Vibration in the finer screening ranges is often produced by unbalanced weights or flywheels attached to the drive shaft. The vibrator generates an elliptical motion, slanting forward at the feed end; a circular motion at the centre; and an elliptical motion, slanting backwards at the discharge end. Forward motion at the feed end serves to move oversize material rapidly out of the feed zone to keep the bed as thin as possible. This action facilitates passage of fines which should be completely removed in the first one-third of the screen length. As the oversize bed thins down, the motion gradually changes near the centre of the screen to the circular pattern to slow down the rate of travel of the solid. At the discharge end, the oversize and remaining near-size materials are subjected to the increasingly retarded effect of the backward elliptical motion. This allows the near size material more time to find openings in the screen surface.

Electrically vibrated screens operate with a high-frequency motion of very small throw, created by a moving magnet activated by alternating current. The electro-magnetic vibrator is mounted above and connected directly to the screening surface. The high-speed vibrating screen finds favour for separating fine sizes on account of its high throughput of difficult material. Vibrating screens are principally used in making separations of 25 cm to 250 μm.

Screening is performed either dry or wet. Usually, dry screening is used for coarse material and wet screening is used for fine material. Wet screening is superior because adhering fines are easily washed off and the dust problem is avoided. Wet screens tend to have lower capacities than dry screens perhaps as much as 50% less. The cost of dewatering and drying of the products from wet screening becomes more.

A **multi-slope vibrating screen** is made up of two or three decks that are installed at different angles from around 35° at the feed end reducing to around 0°-15° at the bottom end which provides the banana shape hence called as **Banana screen**. The steep section at the feed end cause the feed material to flow rapidly resulting thin bed of particles stratifies more quickly and, therefore, has a faster screening rate. The slope decreases towards the discharge end to slow down the remaining material enabling more efficient screening of the near-size material.

The **Mogensen Sizer** is made up of several oscillating and sloping screen decks of decreasing aperture size which are arranged on top of one another and have different inclinations, as shown in Fig. 36. On a Mogensen Sizer, the feed material is first loosened up and then flows through the different screen decks almost vertically in free fall, the coarse particles being removed from the material flow early so that the load on the lower screen decks is considerably reduced. Depending on the number of the cut-points required, the coarse fractions are collected in the discharge of the Sizer and removed via co-vibrating outlets. Compared to traditional screens, the feed material spends much less time in a Mogensen Sizer so that it can be operated at high throughputs.

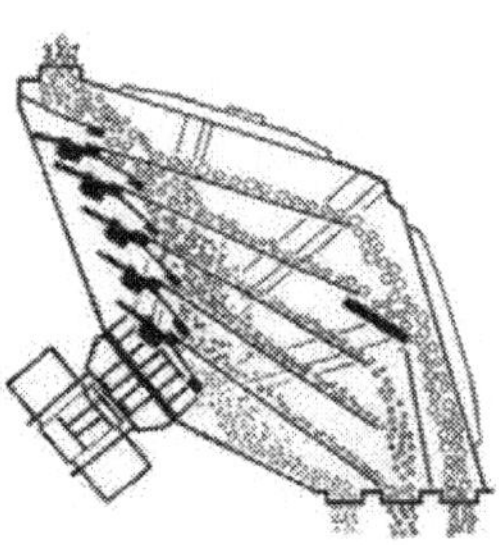

Fig. 36 Mogensen sizer (Courtesy of Mogensen)

Fig. 37 Vibrating grizzly (Courtesy of Mogensen)

Very coarse material is usually screened on a **vibrating grizzly** screen (Fig. 37). Vibrating grizzlies are usually inclined at an angle of around 20° and have a circular throw mechanism. The gap between grizzly bars is usually greater than 50 mm and can be as large as 300 mm with feed top size as large as 1 m. The most common use of grizzlies in mineral processing is for sizing the dry feed to primary and secondary crushers. If a crusher has a 100 mm setting, then the feed can be passed over a grizzly with a 100 mm gap so that the load on the crusher reduces and the grizzly also serves as a feeder.

The concept used in the **flip-flow screen** is a system of flexible screen panels that are alternately stretched and relaxed to impart motion to the screen bed instead of relying only on mechanical vibration of the screen body. The throwing action can generate forces of up to 50 g on the screen surface, preventing material from blinding. Flip-flow screens can be used for separation of difficult, sticky or wet materials ranging from 0.5 to 50 mm. They are particularly suited for fine separations of damp material that cannot be efficiently screened on conventional vibrating screens.

Reciprocating, Gyratory, Rotating-probability and Resonance screens are some of the other dynamic screens. They are similar to vibrating screens but differ in the type of motion given to the screen deck.

Following are some of the screens named in mineral industry according to the purpose for which they are used:

1. **Sizing Screen**: used to separate the particles according to their sizes
2. **Feed screen**: used to prepare the feed to any unit operation
3. **Trash Screen**: used to remove the trash material
4. **Scalping screen**: used to remove small amounts of either oversizes or undersizes
5. **De-watering screen**: used to remove water from mixture of solids and water
6. **De-sliming screen**: used to remove slimes from the coarse material
7. **Dedusting screen**: used to remove extremely fine particles from dry material
8. **Medium recovery screen**: used to remove medium solids from coarse material

4.5 SCREEN EFFICIENCY

The efficiency (effectiveness) of the screen and the capacity of the screen depend on the rate of screening. However, efficiency and capacity are opposite to each other in the sense that capacity can be increased at the expense of efficiency and vice versa.

Screen efficiency (often called the effectiveness of a screen), designated as η, is a measure of the success of a screen in closely separating oversize and undersize particles. It is nothing but the recovery of required size particles. There is no standard method for defining the screen efficiency. Depending on whether one is interested in removing oversize or undersize particles, screening efficiencies may be defined in numerous ways. In an industrial screen, if there are no broken or deformed apertures and screen is perfectly made, no single coarse particle coarser than the size of aperture pass through. In example 4.5.1, the screening is of this type.

Example 4.5.1: *An ore containing 55% of −1 mm material is screened on a 1 mm screen at the rate of 500 tons per day. If 250 tons per day of undersize is obtained from the screen and it contains no oversize, determine the efficiency of the screen.*

Solution:

Given

Quantity of the feed material	= 500 tons/day
Quantity of the undersize material obtained	= 250 tons/day
Percent −1 mm material in the feed	= 55%

Quantity of undersize material present in the feed = $500 \times \frac{55}{100}$ =275 tons per day

Screen efficiency = Effectiveness of the screen = Recovery of undersize material

$$\eta = \frac{\text{Undersize mterial obtained}}{\text{Undersize material present in feed ore}} \times 100 = \frac{250}{275} \times 100 = 90.9\%$$

In reality, some coarse particles, may be less in quantity, will report to the underflow fraction and few undersize particles will remain in overflow fraction. Screen efficiency is defined with respect to the required particles to separate.

In processing plants, samples from feed ore, overflow and underflow fractions are collected and subjected to size analysis to know the different sizes present in feed and two product fractions to evaluate the performance or efficiency of the screen. Under such cases, equations for efficiency can be derived by writing material balance equations on the screen as follows:

Total material balance	$F = P + U$	4.1
Oversize material balance	$Ff = Pp + Uu$	4.2
Undersize material balance	$F(1 - f) = P(1 - p) + U(1 - u)$	4.3

where F = weight flowrate of feed material

P = weight flowrate of overflow material obtained from the screen

U = weight flowrate of underflow material obtained from the screen

f = fraction of oversize material in the feed

p = fraction of oversize material in the overflow obtained from the screen

u = fraction of oversize material in the underflow obtained from the screen

On computation of equations 4.1, 4.2 and 4.3,

$$\frac{P}{F} = \frac{f-u}{p-u} \quad \text{and} \quad \frac{U}{F} = \frac{p-f}{p-u}$$

The recovery of oversize material into the screen overflow is referred as Screen Efficiency (or Screen Effectiveness) η_p based on the oversize material

$$\eta_p = \frac{\text{Oversize particles obtained in overflow}}{\text{Oversize particles present in feed ore}} = \frac{Pp}{Ff} = \frac{p(f-u)}{f(p-u)} \qquad 4.4$$

The recovery of undersize material into the screen underflow is referred as Screen Efficiency (Screen Effectiveness) η_u based on the undersize material

$$\eta_u = \frac{\text{Undersize particles obtained in underflow}}{\text{Undersize particles present in feed ore}} = \frac{U(1-u)}{F(1-f)} = \frac{(1-u)(p-f)}{(1-f)(p-u)} \qquad 4.5$$

A combined overall efficiency, or overall effectiveness, η is then obtained by multiplying the above two equations together

$$\eta = \frac{p(f-u)}{f(p-u)} \times \frac{(1-u)(p-f)}{(1-f)(p-u)} = \frac{p(f-u)(1-u)(p-f)}{f(p-u)^2(1-f)} \qquad 4.6$$

If there are no broken or deformed apertures and screen is perfectly made, no single coarse particle will pass through the screen, i.e. u = 0. Then, the formula for fines recovery η_u and the formula for overall efficiency η both reduce to

$$\eta = \frac{p-f}{p(1-f)} \qquad 4.7$$

This formula is widely used and implies that recovery of the coarse material in the overflow is 100%. Example 4.5.2 is illustrated with screen efficiency calculations.

Example 4.5.2: *Ore particles contains 80% by weight of fine material. To remove these fines, it is screened at the rate of 250 tons/h using 1.8 mm screen. Overflow and underflow products are sampled and analysed. It is found that +1.8 mm material in overflow and underflow products are 40% and 10%, respectively. Estimate the tonnage of fines removed in underflow and the tonnage of fines left in overflow. Calculate the efficiency of the screen in separating undersize material and oversize material. Also calculate overall efficiency.*

Solution:

Given

Fraction of +1.8 mm material in the feed	= f = 1 − 0.8 = 0.2
Fraction of +1.8 mm material in the overflow product	= p = 0.4
Fraction of +1.8 mm material in the underflow product	= u = 0.1
Weight flowrate of feed material	= F = 250 tons/h
Total material balance	F = P + U
+1.8 mm material balance	Ff = Pp + Uu

On substitution of the given values, the above equations become

$$250 = P + U$$

$$250 \times 0.2 = P \times 0.4 + U \times 0.1$$

Solving the above two equations ⇒ U = 166.67 tons/h

Tonnage of – 1.8 mm fines removed in underflow

$$= U(1-u) = 166.67 \times (1.0 - 0.1) = 150 \text{ tons/h}$$

Tonnage of overflow = F − U = 250 − 166.67 = 83.33 tons/h

Tonnage of – 1.8 mm fines left in overflow

= P(1 − p) = 83.33 × (1.0 − 0.4) = 50 tons/h

Efficiency of the screen in separating undersize material can be calculated using equation 4.5.

$$\eta_u = \frac{\text{Undersize particles obtained in underflow}}{\text{Undersize particles present in feed ore}} = \frac{U(1-u)}{F(1-f)} = \frac{(1-u)(p-f)}{(1-f)(p-u)}$$

$$= \frac{(1-0.1)(0.4-0.2)}{(1-0.2)(0.4-0.1)}$$

$$= 0.75 \Rightarrow 75\%$$

Efficiency of the screen in separating oversize material can be calculated using equation 4.4

$$\eta_p = \frac{\text{Oversize particles obtained in overflow}}{\text{Oversize particles present in feed ore}} = \frac{Pp}{Ff} = \frac{p(f-u)}{f(p-u)}$$

$$= \frac{0.4(0.2-0.1)}{0.2(0.4-0.1)}$$

$$= 0.667 \Rightarrow 66.7\%$$

$$\text{Overall efficiency} = \eta = \frac{p(f-u)(1-u)(p-f)}{f(p-u)^2(1-f)} = \frac{0.4(0.2-0.1)(1-0.1)(0.4-0.2)}{0.2(0.4-0.1)^2(1-0.2)}$$

$$= 0.50 \Rightarrow 50\%$$

(or) Overall efficiency = η = 0.75 × 0.667 = 0.50 ⇒ 50%

If f, p and u are respectively expressed in terms of the fractions of undersize material in feed, overflow and underflow by forming the mass balance equations, three expressions for efficiency of the screen can be obtained as

$$\eta_p = \frac{P(1-p)}{F(1-f)} = \frac{(1-p)(f-u)}{(1-f)(p-u)} \quad 4.8$$

$$\eta_u = \frac{Uu}{Ff} = \frac{u(p-f)}{f(p-u)} \quad 4.9$$

$$\eta = \frac{u(u-f)(1-p)(f-p)}{f(u-p)^2(1-f)} \quad 4.10$$

In example 4.5.2, if these formulae are used by taking fraction of undersize for f, p and u, the same result will be obtained as follows:

Fraction of −1.8 mm material in the feed = f = 0.8

Fraction of −1.8 mm material in the overflow product = p = 1− 0.4 = 0.6

Fraction of −1.8 mm material in the underflow product = u = 1 − 0.1 = 0.9

$$\eta_p = \frac{(1-p)(f-u)}{(1-f)(p-u)} = \frac{(1-0.6)(0.8-0.9)}{(1-0.8)(0.6-0.9)} = 0.667 \Rightarrow 66.7\%$$

$$\eta_u = \frac{u(p-f)}{f(p-u)} = \frac{0.9(0.6-0.8)}{0.8(0.6-0.9)} = 0.75 \Rightarrow 75\%$$

$$\eta = \frac{u(u-f)(1-p)(f-p)}{f(u-p)^2(1-f)}$$

$$= \frac{0.9(0.9-0.8)(1-0.6)(0.8-0.6)}{0.8(0.9-0.6)^2(1-0.8)} = 0.50 \Rightarrow 50\%$$

The formulae 4.4–4.10 do not give an absolute value of the efficiency because no allowance is made for the difficulty of the separation. A feed composed mainly of particles of a size near to that of the screen aperture — **near size material (near mesh material)** — presents a more difficult separation than a feed composed mainly of very coarse and very fine particles with a screen aperture intermediate between them. In such cases, it is proposed to define the efficiency as the ratio of the near size material taken out by the screen to the near-size material present in the feed.

4.6 PROBLEMS FOR PRACTICE

4.6.1: *An ore containing 45% of −0.5 mm material is screened on 0.5 mm screen at the rate of 700 tons per day. If 280 tons per day of undersize is obtained and it contains no oversize, determine the efficiency of the screen.*

[88.9%]

4.6.2: *Road metal product obtained from a crusher consists of 25% of −¼" size is to be used for preparation of concrete. To get +¼" fraction required for the preparation of concrete, it is screened by using ¼" screen. The overflow and underflow products of the screen contain 80% of +¼" size and 95% of −¼" size respectively. If the feed to the screen is 55 tons, what is the +¼" size metal lost with underflow of the screen? Estimate the effectiveness of the screen.* *[0.18 tons; 99.56%]*

4.6.3: *−10 mesh material is fed to a double deck vibrating screen for separation. The desired product is −40+60 mesh material. Therefore, 40- and 60-mesh screens are used. The feed is introduced on the 40-mesh screen. From the sieve analysis of the feed and the three products given in Table 18, calculate the effectiveness of the double deck screen in separating −40+60 mesh material.*

Table 18: Sieve analysis of feed and three products for problem 4.6.3

	Mass fraction			
Mesh	**Feed**	**Oversize from 40 mesh screen**	**Oversize from 60 mesh screen**	**Undersize from 60 mesh screen**
−10 + 20	0.097	0.197	0.026	0.0005
−20 + 30	0.186	0.389	0.039	0.0009
−30 + 40	0.258	0.337	0.322	0.0036
−40 + 60	0.281	0.066	0.526	0.3490
−60 + 85	0.091	0.005	0.061	0.2990
−85 + 100	0.087	0.006	0.026	0.3470

[23.08%]

4.6.4: *Calculate screen efficiency when quartz mixture is screened through a 1.5-mm screen to obtain +1.5-mm fraction. The size analyses of feed, overflow and underflow is obtained by sampling and sieve analysis. The results are given in Table 19. If the feed rate to the screen is 100 tons/h, calculate the tonnage of fines remained in overflow.*

Table 19: Size analyses of feed, overflow and underflow for problem 4.6.4

Screen size mm	Weight percent retained this size		
	Feed	**Overflow**	**Underflow**
3.3	3.5	7.0	----
2.3	13.5	36.0	----
1.5	33.0	37.0	15.0
1.0	22.7	13.0	43.0
0.8	16.0	4.0	25.0
0.6	5.4	3.0	8.0
0.4	2.1	----	3.0
0.2	1.8	----	2.0
−0.2	2.0	----	4.0

[82.6%, 10.76 tons/h]

■■■■

5 Density

Density is important next to the size of the particle. Density of the particle can be of mass density and weight density. **Mass density** is the mass of the particle per unit volume, whereas **Weight density** is the weight of the particle per unit volume. **Specific gravity** is defined as the ratio of the density of the particle to the density of water at 4°C. In the C.G.S system of units, the numerical value of density and specific gravity is same. Hence, it is normal practice to indicate specific gravity for density.

In all beneficiation operations, particularly in Gravity Concentration operations, density has an important role together with the size and shape of the particles. As the ore contains different minerals, the density of an ore varies depending on the minerals it contains. Before the ore is to be beneficiated, the density of the ore and the density of the different minerals present in the ore are to be determined.

A bulk solid (bulk material) is combination of particles and space. For a bulk material, the average particle density can be determined by dividing the weight of the material (solids) by the true volume occupied by the particles (not including the voids). This can be determined using the density bottle.

The stepwise procedure for determination of density of an ore is as follows:

1. Wash, dry and weigh the density bottle with stopper. Let this weight be w_1.
2. Thoroughly dry the ore sample.
3. Keep 5–10 grams of ore sample in the bottle and reweigh. Let this weight be w_2.
4. Now fill the bottle with a liquid of known density. The liquid used should not react with the ore.
5. Insert the stopper, allow the liquid to fall out of the bottle, wipe of excess liquid and weigh the bottle. Let this weight be w_3.
6. Remove ore and liquid from the bottle and fill the bottle with liquid alone and repeat step.5. Let this weight be w_4.

$w_2 - w_1$ is the weight of the ore sample

$w_4 - w_1$ is the weight of the liquid occupying whole volume of the bottle

$w_3 - w_2$ is the weight of the liquid having the volume equal to the volume of density bottle less volume of ore sample taken

$(w_4 - w_1) - (w_3 - w_2)$ is the weight of the liquid of volume equal to that of the ore sample

If ρ_l is the density of the liquid

$$\text{Density of the ore sample} = \frac{w_2 - w_1}{(w_4 - w_1) - (w_3 - w_2)} \times \rho_l \qquad 5.1$$

The bulk material is a combination of particles and space, the percentage of the total volume not occupied by the particles is referred as the **voidage** or **void fraction**. Sometimes the term **porosity** is applied to bulk material to mean the same as voidage.

The **bulk density** is the overall density of a material kept in large quantities, which can be defined as the mass of the material divided by its total volume (particles and voids) and depends upon the true density of the material and the pore space between the particles. It is a measure of the storage capacity.

Three kinds of bulk density that apply to materials handling calculations are

1. **Aerated density**

   When the sample of the bulk material is carefully poured into a measuring cylinder to measure its volume, then the computed density is called as 'Aerated', 'loose' or 'poured' bulk density (ρ_a).

2. **Packed density**

   If the sample is packed by dropping the cylinder vertically several times from a height of one or two centimetres on to a table, then the computed density is called 'packed' or 'tapped' bulk density (ρ_c).

3. **Dynamic** or **Working density**

   The dynamic or working density (ρ_d) is a function of Aerated and Packed densities. Its value is intermediate to that of Aerated and Packed densities. The dynamic or working density (ρ_d) is expressed as

$$\rho_d = (\rho_c - \rho_a)\,\mathbf{C} + \rho_a \qquad 5.2$$

where **C** is the Compressibility expressed as

$$\mathbf{C} = \frac{(\rho_c - \rho_a)}{\rho_c} \qquad 5.3$$

The lower the compressibility, the material is more free flowing. The dividing line between free flowing (granular) and non-free flowing (powder) is 0.20–0.21 compressibility. A higher value indicates a powder that is non-free flowing and will be likely to bridge in a hopper. The compressibility of a material often helps to indicate uniformity in size and shape of the material, its deformability, surface area, cohesion and moisture content.

5.1 SOLIDS AND PULP

Most of the mineral processing operations are wet. Water is added to the ore particles to aid beneficiation. The mixture of water and solid particles is known as **Pulp**.

Other terms commonly used are

Suspension: When the solid particles are held up in the water, the pulp is called suspension. In other words, in suspension, the solid particles are well dispersed throughout.

Slurry: A mixture of fine solids (called slimes) and water

Sludge: Thick pulp i.e. pulp with less quantity of water

Pulp or slurry density is most easily measured in terms of weight of the slurry per unit volume (g/cm^3 or kg/m^3). A sample of slurry taken in container of known volume is weighed to give slurry density directly. Marcy Scale available in the market gives direct reading for the density of the slurry and % solids in the slurry.

The composition of a slurry is often represented as the fraction (or percent) of solids by weight. It is determined by sampling the slurry, weighing, drying and reweighing.

$$C_w = \text{fraction of solids by weight} = \frac{\text{Weight of the particles}}{\text{Weight of the slurry}}$$

$$C_v = \text{fraction of solids by volume} = \frac{\text{Volume of the particles}}{\text{Volume of the slurry}}$$

Knowing the densities of the slurry (ρ_{sl}), water (ρ_w) and dry solids (ρ_p), the fraction of solids (C_w) by weight can be calculated. A material balance equation in terms of the volume can be written as

Volume of the solids + Volume of the water = Volume of the slurry

$$\frac{C_w}{\rho_p} + \frac{1-C_w}{\rho_w} = \frac{1}{\rho_{sl}} \qquad 5.4$$

$$\Rightarrow \quad C_w = \frac{\rho_p(\rho_{sl} - 1)}{\rho_{sl}(\rho_p - 1)} \qquad [\because \rho_w = 1 \text{ g/cm}^3] \qquad 5.5$$

$$\Rightarrow \quad \rho_{sl} = \frac{\rho_p}{\rho_p + C_w(1-\rho_p)} \qquad 5.6$$

Similarly, a material balance equation in terms of the weight can be written as

Weight of the solids + Weight of the water = Weight of the slurry

$$C_v\,\rho_p + (1 - C_v)\,\rho_w = \rho_{sl} \qquad 5.7$$

$$\Rightarrow \quad C_v = \frac{(\rho_{sl} - 1)}{\rho_p - 1} \qquad [\because \rho_w = 1 \text{ g/cm}^3] \qquad 5.8$$

$$\Rightarrow \quad \rho_{sl} = 1 + C_v\,(\rho_p - 1) \qquad 5.9$$

Dilution ratio is the ratio of the weight of the water to the weight of the solids in the slurry.

$$\text{Dilution ratio} = \text{DR} = \frac{1 - C_w}{C_w} \qquad 5.10$$

Substituting the value of C_w of equation 5.5

$$\text{Dilution ratio} = \text{DR} = \frac{\rho_p - \rho_{sl}}{\rho_p(\rho_{sl} - 1)} \qquad 5.11$$

Dilution ratio is particularly important as the product of dilution ratio and weight of the solids in the slurry is equal to the weight of the water in the slurry.

The following examples illustrate the applications of density calculations in different situations.

Example 5.1.1: *450 g of solids are taken and prepared a pulp of 500 cc by adding water to it. If the weight of 500 cc pulp is 800 g, what is the specific gravity of solids? Determine the density of solids in kg/m³. Calculate percent solids in the pulp by weight and by volume.*

Solution:

Given

Weight of the solids	= 450 g
Weight of the pulp	= 800 g
Volume of the pulp	= 500 cc
Weight of water	= 800 – 450 = 350 g
Volume of water	= 350/1 = 350 cc
Volume of solids	= 500 – 350 = 150 cc
Weight density of solids	$= \frac{\text{Weight of the solids}}{\text{Volume of the solids}} = \frac{450}{150} = 3$ g/cc
Specific gravity of solids	$= \frac{\text{Density of the solids}}{\text{Density of the water}} = \frac{3}{1} = 3$
Density of solids	$= 3 \times 1000 = 3000$ kg/m^3
% solids by weight	= 450/800 = 0.5625 ⇒ 56.25%
% solids by volume	= 150/500 = 0.30 ⇒ 30%

Example 5.1.2: *When a 2 l slurry sample of hematite of specific gravity of 4.8 is filtered, 250 g of wet filter cake is obtained. Subsequently the filter cake is perfectly dried. The weight of dried hematite is 180 g.*

Calculate (a) % solids by volume in the slurry

(b) % solids by weight in the slurry

(c) Liquid solid ratio by volume and weight
(d) Density of the slurry
(e) % moisture in the filter cake
(f) Bulk density of the filter cake

Solution:

This problem is worked out by using density definition alone

Weight of the wet filter cake	= 250 g
Volume of the slurry sample	= 2 l = 2000 cm^3
Weight of dried hematite	= 180 g
Density of hematite	= 4.8 g/cm^3
Volume of hematite	= 180/4.8 = 37.5 cm^3
Volume of water in the slurry	= 2000 − 37.5 = 1962.5 cm^3
% hematite by volume	= 37.5 × 100/2000 = **1.875%**
Weight of water	= 1962.5 g
Weight of Hematite	= 180 g
% hematite by weight	= 180 × 100/2142.5 = **8.4%**
Liquid solid ratio by volume	= 1962.5/37.5 = **52.3**
Liquid solid ratio by weight	= 1962.5/180 = **10.9**
Density of the slurry	= 2142.5/2000 = **1.07125**
Weight of moisture in filter cake	= 250 − 180 = 70 g
% moisture by weight in filter cake	= 70 × 100/250 = **28%**
Volume of filter cake	= 37.5 + 70 = 107.5 cm^3
Bulk density of the filter cake	= 250/107.5 = **2.33**

Example 5.1.3: *Calculate the pulp density if the % quartz by weight is 40 in the pulp and the specific gravity of particles is 2.65.*

Solution:

Given

% quartz by weight in pulp = 40% $\Rightarrow C_w = 0.4$

Density of the particle $= \rho_p = 2.65$ g/cm^3

Substituting these values in equation 5.4

$$\frac{C_w}{\rho_p} + \frac{1-C_w}{\rho_w} = \frac{1}{\rho_{sl}} \Rightarrow \frac{0.4}{2.65} + \frac{1-0.4}{1.0} = \frac{1}{\rho_{sl}} \Rightarrow \rho_{sl} = 1.33 \text{ g/cm}^3$$

Pulp density = ρ_{sl} = 1.33 g/cm^3

Alternately by using equation 5.6

$$\text{Pulp density} = \rho_{sl} = \frac{\rho_p}{\rho_p + C_w(1-\rho_p)} = \frac{2.65}{2.65+0.40(1-2.65)} = 1.33 \text{ g/cm}^3$$

Example 5.1.4: *Determine % solids by weight in the pulp if the sp.gr. of solids is 4.95 and the pulp density is 1.9 g/cm³. Also calculate % water by volume in the pulp.*

Solution:

Given

Density of solids $= \rho_p = 4.95$ g/cm³

Density of the pulp $= \rho_{sl} = 1.90$ g/cm³

Substituting ρ_p and ρ_{sl} values in equation 5.4

$$\frac{C_w}{\rho_p} + \frac{1-C_w}{\rho_w} = \frac{1}{\rho_{sl}} \Rightarrow \frac{C_w}{4.95} + \frac{1-C_w}{1.0} = \frac{1}{1.9} \Rightarrow C_w = 0.5936 \Rightarrow 59.36\%$$

$$\text{Alternately, } C_w = \frac{\rho_p(\rho_{sl}-1)}{\rho_{sl}(\rho_p-1)} = \frac{4.95\,(1.9-1)}{1.9(4.95-1)} = 0.5936 \Rightarrow 59.36\%$$

% Solids by weight in the pulp = 59.36%

Substituting the values of density of solids and pulp in equation 5.7

$$C_v\rho_p + (1-C_v)\rho_w = \rho_{sl} \Rightarrow C_v \times 4.95 + (1-C_v) \times 1.0 = 1.9$$

$$\Rightarrow C_v = 0.2278 \Rightarrow 22.78\%$$

$$\text{Alternately, } C_v = \frac{(\rho_{sl}-1)}{\rho_p-1} = \frac{(1.9-1)}{4.95-1} = 0.2278 \Rightarrow 22.78\%$$

% Solids by volume in the pulp = 22.78%

% water by volume = 100.00 – 22.78 = 77.22%

Example 5.1.5: *How much galena of density of 7.5 g/cm³ must be added to water to obtain 1 l of pulp at a pulp density of 2.0 g/cm³?*

Solution:

Given

Density of galena $= \rho_p = 7.5$ g/cm³

Density of the pulp $= \rho_{sl} = 2.0$ g/cm³

Volume of the pulp = 1 l = 1000 cm³

Substituting these values in equation 5.4

$$\frac{C_w}{\rho_p} + \frac{1-C_w}{\rho_w} = \frac{1}{\rho_{sl}} \Rightarrow \frac{C_w}{7.5} + \frac{1-C_w}{1} = \frac{1}{2} \qquad \Rightarrow C_w = 0.5769$$

Fraction of galena by weight in the pulp = 0.5769

Weight of the pulp = 1000 × 2.0 = 2000 g = 2.0 kg

Weight of galena to be added = 2.0 × 0.5769 = 1.154 kg

Alternately,

substituting the values of density of galena and the pulp in equation 5.7

$$C_v\rho_g + (1 - C_v)\rho_w = \rho_{sl} \Rightarrow C_v \times 7.5 + (1 - C_v) = 2 \Rightarrow C_v = 0.1538$$

Fraction of galena by volume in the pulp = 0.1538

Weight of galena to be added for 1 cm^3 of pulp = 0.1538 × 7.5 = 1.154 g

Weight of galena to be added for 1 l of pulp = 1.154 × 1000 = 1154 g = 1.154 kg

Example 5.1.6: *Magnetite slurry is flowing at the rate of 2 l/s in a beneficiation plant. The density of magnetite and slurry are 5000 kg/m^3 and 1500 kg/m^3, respectively. Calculate the tonnage of the magnetite treated in the plant in an hour.*

Solution:

Given

Flow rate of magnetite = 2 l/s = 2000 cm^3/s

Density of magnetite = ρ_p = 5000 kg/m^3 = 5.0 g/cm^3

Density of the slurry = ρ_{sl} = 1500 kg/m^3 = 1.5 g/cm^3

Substituting these values in equation 5.4

$$\frac{C_w}{\rho_p} + \frac{1 - C_w}{\rho_w} = \frac{1}{\rho_{sl}} \Rightarrow \frac{C_w}{5.0} + \frac{1 - C_w}{1.0} = \frac{1}{1.5} \Rightarrow C_w = 0.417$$

Fraction of magnetite in the pulp = C_w = 0.417

Weight of magnetite per cc of slurry = 1.5 × 0.417 = 0.6255 g/cm^3

Weight flow rate of magnetite in the slurry

= 0.6255 × 2000 = 1251 g/s = 1.251 kg/s = 4.5036 tons/h

Example 5.1.7: *Determine the density of the pulp having dilution ratio of 3 and solids density of 2650 kg/m^3.*

Solution:

Given

Density of the solids = 2650 kg/m^3 = 2.65 g/cm^3

$$\text{Dilution ratio} = 3 = \frac{1 - C_w}{C_w} \Rightarrow C_w = 0.25$$

Fraction of solids in the pulp = $C_w = 0.25$

Substituting density of solids and fraction of solids in equation 5.4

$$\frac{C_w}{\rho_p} + \frac{1-C_w}{\rho_w} = \frac{1}{\rho_{sl}} \Rightarrow \frac{0.25}{2.65} + \frac{1-0.25}{1.0} = \frac{1}{\rho_{sl}} \Rightarrow \rho_{sl} = 1.18 \text{ g/cm}^3$$

Density of the pulp = $\rho_{sl} = 1.18$ g/cm^3

5.2 MISCIBLE LIQUIDS

When it is required to prepare a liquid of definite density by mixing two miscible liquids of known densities as in case of float and sink analysis, the equation is written as

$$C_v \rho_1 + (1 - C_v) \rho_2 = \rho_{12} \qquad 5.12$$

where ρ_1 = density of liquid 1

ρ_2 = density of liquid 2

ρ_{12} = density of the resultant liquid after mixing two liquids

C_v = fraction of liquid 1 by volume

$$C_v = \frac{\rho_{12} - \rho_2}{\rho_1 - \rho_2} \qquad 5.13$$

If C_w is the fraction of liquid 1 by weight, the equation is

$$\frac{C_W}{\rho_1} + \frac{1-C_w}{\rho_2} = \frac{1}{\rho_{12}} \qquad 5.14$$

$$\Rightarrow C_w = \frac{\rho_1(\rho_{12} - \rho_2)}{\rho_{12}(\rho_1 - \rho_2)} \qquad 5.15$$

Example 5.2.1 shows the calculation of quantity of two miscible liquids required to be added to yield a solution of desired density.

Example 5.2.1: *Calculate the volume of benzene and carbon tetrachloride required to prepare 100 cc of solution of heavy liquid of specific gravity 1.2. Specific gravities of benzene and carbon tetrachloride are 0.8 and 1.6, respectively.*

Solution:

Given

Density of benzene = $\rho_1 = 0.8$ g/cm^3

Density of Carbon tetrachloride = $\rho_2 = 1.6$ g/cm^3

Density of resulting solution = $\rho_{12} = 1.2$ g/cm^3

Volume of the solution required = 100 cc

Substituting density values in equation 5.12

$$C_v\rho_1 + (1 - C_v)\rho_2 = \rho_{12}$$

$$\Rightarrow \quad C_v \times 0.8 + (1 - C_v) \times 1.6 = 1.2$$

$$\Rightarrow \quad C_v = 0.50 \Rightarrow 50\%$$

i.e. quantity of the benzene required is 50% by volume.

As the total quantity of the solution is 100 cc, 50 cc of benzene and 50 cc of carbon tetrachloride are required to prepare 100 cc of solution.

5.3 PROBLEMS FOR PRACTICE

5.3.1: *Determine % ferrosilicon by weight to be added to prepare the suspension of specific gravity of 1.80 if the specific gravity of the ferrosilicon is 7.5.* [51.3%]

5.3.2: *What is the density of a slurry containing 25% by weight SiO_2 (density = 2.65 g/cm^3), 15% by weight Magnetite (density = 5.18 g/cm^3) and 15% by weight Galena (density = 7.5 g/cm^3)?* [1.69 g/cm^3]

5.3.3: *It is required to prepare a pulp of 30% solids by volume with 4.2 l of water. Determine the weight of solids to be added in kilograms if the specific gravity of solids is 2.65.* [4.77 kg]

5.3.4: *A quartz slurry flowing at the rate of 720 kg/h is sampled and found its density as 1400 kg/m^3. If the specific gravity of the quartz is 2.65, calculate the mass flowrate of quartz within the slurry.* [330.5 kg/h]

5.3.5: *A pump is fed by a slurry having a flowrate of 5 m^3/h containing 40% solids by weight. Specific gravity of solids in the slurry is 3.0. Calculate the tonnage of dry solids pumped per hour.* [2.728 tons/h]

5.3.6: *Calculate the pulp density and pulp dilution if the pulp consists of 30% galena and 40% quartz by weight. Specific gravities of galena and quartz are 7.5 and 2.65, respectively.* [2.04 g/cm^3, 0.43]

5.3.7: *Two liquids of specific gravities 1.26 and 1.6 are to be mixed to obtain 300 cc solution of specific gravity 1.4. Calculate the quantities of two liquids required by volume and by weight.* [176.4 cc, 123.6 cc, 222.3 g, 197.7 g]

■■■■

6 Liberation

Liberation is the first and the most important step in processing of minerals. The second step—beneficiation—is impracticable if the liberation is not successful.

Liberation : Liberation can be defined as freeing or detachment of dissimilar mineral grains. The operation employed to liberate the dissimilar mineral grains is Size reduction or Comminution.

Free particles : If the particles of ore consist of a single mineral, they are termed as Free particles.

Locked particles : If the ore particles consist of two or more minerals, they are termed as locked particles. If the locked particles contain valuable mineral, they are termed as middling particles.

Grain size : It is the size of a mineral as it occurs in the ore.

Particle size : It is the size of any particle whether free or locked particle.

Grain and grain size pertain to uncrushed ore and particle and particle size pertain to crushed or ground ore.

Liberation size : It is the size of a mineral particle at which that mineral is completely liberated. It is the size of a free particle of required (valuable) mineral.

Various mineral grains, present in the ore, exist in physical combination with each other. To detach the valuable mineral grains from all other gangue mineral grains, it is essential to reduce the size of the ore particles.

If one mineral species in an ore is to be physically separated from all other species in the ore, all grains of the desired species must be physically detached from all remaining species in the ore. In an ore containing minerals A, B and C (Fig. 38) if all grains of mineral A (considered as valuable mineral) is to be separated from the ore, all grains of A must be detached from other minerals B and C (gangue minerals). When such detachment is complete, mineral A is said to be liberated. However, liberation of mineral A does not require liberation of other minerals B and C in the ore.

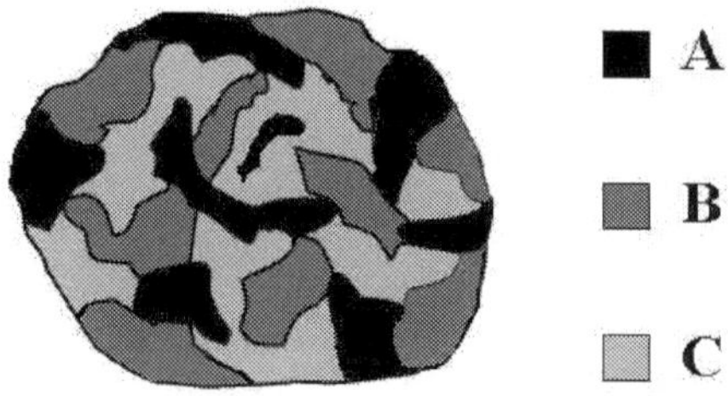

Fig. 38 A particle of an ore containing A, B and C minerals

Let us consider a particle in two dimensions (Fig. 39). Let ABCD be an 8 × 8 cm size particle consisting of valuable mineral particles (shaded portion) and gangue mineral particles (white portion) of 2 × 2 cm size each in equal proportion. This particle is reduced by size reduction to 2 × 2 cm size. If the fracture takes place along the boundary lines as shown in Fig. 39a, it yields 16 particles of 2 × 2 cm size each. All 16 particles are free particles out of which eight are of valuable mineral and eight are of gangue mineral. Here, all the valuable mineral particles are liberated. This type of liberation is called liberation by detachment.

In this case, **Grain, Particle** and **Liberation sizes** are equal.

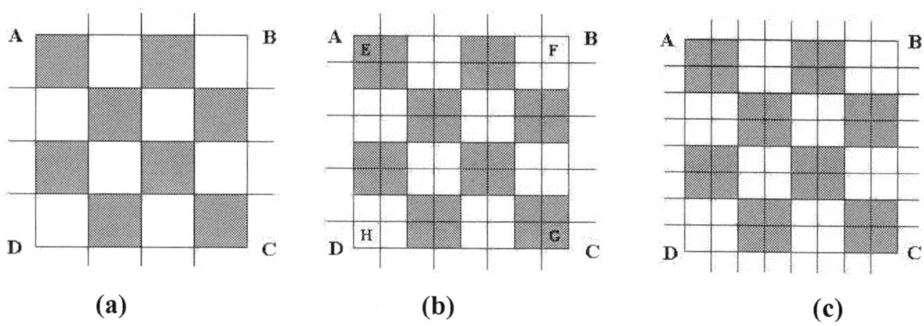

Fig. 39 Liberation methods

If the fracture takes place across the grains, as shown in Fig. 39b, it yields 25 particles out of which four particles E, F, G and H are of 1 × 1 cm size, 12 particles are of 2 × 1 cm size and the remaining nine particles are of 2 × 2 cm size. Here,

E and G are free particles of valuable mineral

F and H are free particles of gangue mineral

All others are locked particles

Hence, valuable mineral has been liberated partially to a lesser extent.

If the resulting particles are again crushed and fracture takes place along the boundary lines, as shown in Fig. 39c, it yields 64 particles; all are free particles of 1 × 1 cm size out of which 32 are valuable mineral particles and 32 are gangue mineral particles. Here, all valuable mineral particles are liberated. This happens when the particles are reduced to 1 × 1 cm size which is less than grain size. 1 × 1 cm is the particle size and also the liberation size.

If the fracture during second time also does not take place along the boundary lines, it yields still locked particles which need further reduction in size. This reduction is continued till all valuable mineral particles occur as free particles. Then, the particle size, hence liberation size, is less than 1 × 1 cm. This happens because the fracture takes place across the mineral grain. This type of liberation is called liberation by size reduction. In this case, **Liberation size** (also **Particle size)** is less than the **Grain size**.

It is noteworthy that the particles are reduced by size reduction in both the cases, but it matters whether the fracture takes place along the boundaries or across the grains. Locked particles will be produced when the fracture takes place across the grains.

Fig. 40 shows an example of typical comminution product wherein black refers valuable mineral and white refers gangue mineral. Only four are liberated free valuable mineral particles, eight are free gangue mineral particles, six are locked particles containing valuable and gangue minerals. These locked particles are known as **middling particles**.

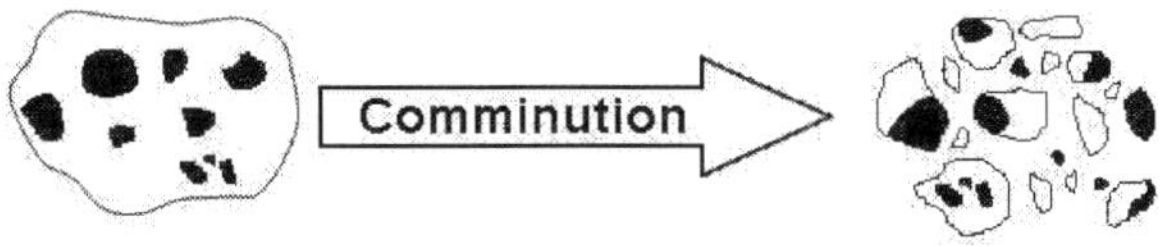

Fig. 40 Typical comminution product

Degree of liberation—quantitatively referred to as **percent liberation**—of a certain mineral, is defined as the percent of that mineral liberated and occurring as free particles in relation to the total amount of the same mineral present in the ore.

For example, let the amount of valuable mineral present in one ton of ore is 300 kg. When it is comminuted to certain size, the amount of valuable mineral exists as free particles is 294 kg. The remaining 6 kg of valuable mineral exists as locked particles with gangue minerals.

$$\text{Then, the percent liberation} = \frac{294}{300} \times 100 = 98\%$$

Beneficiation is performed at this size to separate 98% valuable minerals so that 2% of the valuable minerals are lost. To separate 100% valuable minerals, the ore is to be further reduced in size to obtain 100% liberation, which consumes additional power. If the cost of this additional power is more than the cost of 2% valuable minerals, more than 98% liberation is not at all desirable.

Chrome ore of different size fractions are analysed for liberation of chromite particles by optical microscopic examination. Data showing liberation of chromite at various size fractions is presented in Table 20. The percentage chromite liberation is calculated using the relationship

$$\text{Chromite liberation \%} = \frac{\text{Free chromite \%}}{\text{Free chromite \%} + \text{Locked chromite \%}} \times 100$$

Table 20: Liberation of the chrome ore at different particle size fractions (Atiq et al 2005)

Particle size (mesh #)	Free chromite particles (%)	Locked particles (%)	Free gangue (%)	Chromite liberation (%)
−8 + 10	15.38	58.91	25.70	20.70
−10 + 25	30.4	34.16	35.43	47.09
−25 + 50	39.2	22.38	38.41	63.66
−50 + 60	49.0	10.23	40.71	82.72
−60 + 80	50.1	9.0	40.8	84.77
−80	52.2	6.4	41.3	89.08

The data reveals that the chrome ore required grinding up to a −80 # for adequate liberation of the chromite particles.

Analysis of comminution product of Manganese ore is carried out on individual sieve fraction by mounting the mineral grains on square glass plate and observed under reflected light microscope to know the degree of liberation. Data showing liberation of manganese mineral at various size fractions is presented in Table 21.

Table 21: Liberation of the manganese ore at different particle size fractions (Chinnaiah 2004)

Particle size (mesh #)	Free particles	Locked particles	Degree of liberation %
−22 + 44	12	40	23.08
−44 + 60	32	51	38.55
−60 + 100	34	30	53.12
−100 + 150	38	10	79.17
−150 + 200	58	8	87.88
−200	68	8	89.47

The data reveals that the manganese ore under investigation required grinding up to −200 # for adequate liberation of the manganese mineral particles.

Beneficiation method to be used depends on liberation size of the ore which in turn depends on type of the ore. Ore types can be conveniently classified as follows.

Massive ores : In these ores, reasonable amount of crushing makes much of the valuables liberated.

Example: Coal, bedded iron ores

Intergrown Ores : In these ores, valuables can be freed only partially by crushing and require certain amount of grinding to complete the liberation.

Example: Chrome ore

Disseminated ore : In these ores, valuables are sparely distributed through a waste rock matrix and require fine grinding to liberate the valuables.

Example: Gold ore

To liberate the valuables, the ore particles are to be reduced in size by the application of the forces. When the forces are applied on the ore particle, fracture takes place depending upon the method of application of the forces.

7 Comminution

The operation of applying a force on the particle to break it is called **size reduction**. **Comminution** is a general term for size reduction that may be applied without regard to the actual breakage mechanism involved.

In any industrial comminution operation, the breakage of any individual particles simultaneously occurs with that of many other particles. The breakage product of any particle is intimately mixed with those of other particles. Thus, an industrial comminution operation can be analysed only in terms of a distribution of feed and product particles. However, each individual particle breaks as a result of the stresses applied to it and it alone.

7.1 FRACTURE

Fracture in the particle occurs as a result of application of a force. When a force is applied on a particle, stress will develop within the particle. When this stress exceeds ultimate stress, the particle will break. Let us consider a particle subjected to two opposing forces by a concentrated load, as shown in Fig. 41.

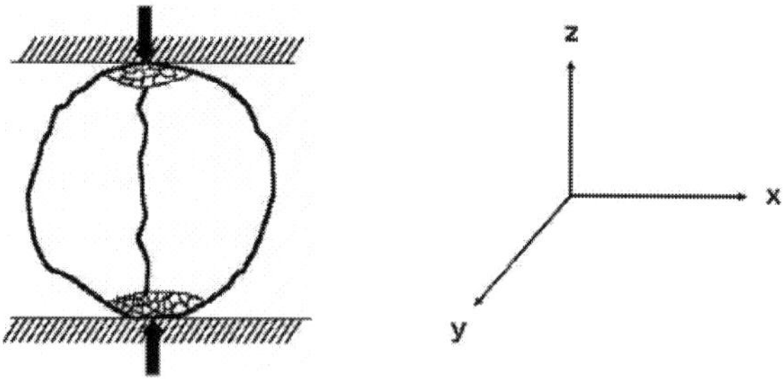

Fig. 41 Compressive forces

The principal stress in the z direction is a compressive stress throughout the particle. The principal stress in the x and y directions is a compressive stress adjacent to the load points, but a tensile stress within the particle. This tensile stress is lower than compressive stress. As the tensile strength is as little as one-tenth of the compressive strength, the fracture primarily occurs because of the tensile stress which results in breakage into a small number of large pieces. Due to the compression adjacent to the loading points, it results a large number of small pieces.

Compressive force can be applied at either a fast or a slow rate. Under the conditions of slow compression, energy applied is just sufficient to load comparatively few regions of the particle to the fracture point and only a few particles result. Their size is comparatively close to the original particle size. Under the conditions of rapid loading such as in high-velocity impact, applied energy is well in excess of that required for fracture. Many areas in the particle are overloaded, and the result is a comparatively large number of particles with a wide size distribution. Impact causes immediate fracture with no residual stresses.

Attrition or abrasion fracture occurs when a force (shear force) acts parallel to the surface of the particle. Due to insufficient energy applied on the particle, localized stress occurs and a small area is fractured to give very fine particles.

Another type of fracture is chipping, wherein the edges or corners of a particle will break due to the application of oblique forces—generally, impact forces—on the particle.

In practice, these events do not occur in isolation. For example, when the particles are crushed by compression as in the case of a crusher, coarse particles will produce resulting from the induced tensile stress, fine particles will produce resulting from compressive stress near the points of loading and by attrition due to particle interaction. All these types of forces and fractures, and sizes of the particles after fracture are shown in Fig. 42.

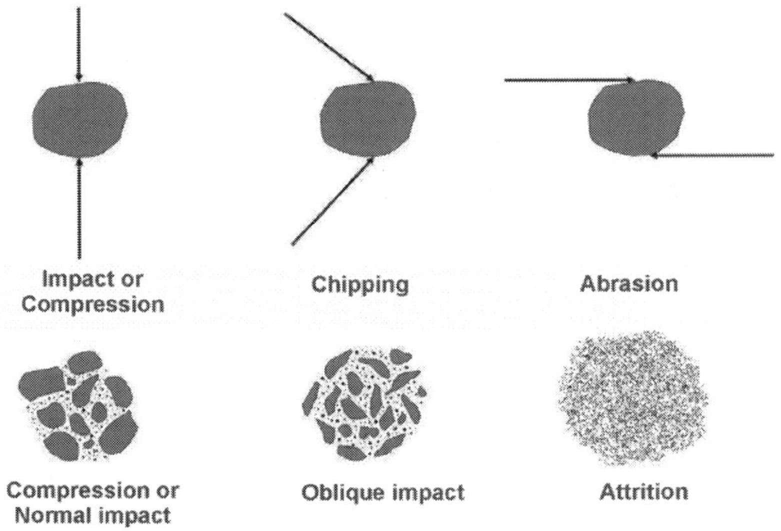

Fig. 42 Mechanism of fracture

It can be summarised that all types of forces exist in any size reduction operation even though individual size reduction units are predominantly designed for application of one type of force.

7.2 OBJECTIVES OF COMMINUTION

Following are some of the objectives of comminution:

1. Reduction of large lumps into small pieces
2. Production of solids of desired size range
3. Liberation of valuable minerals from gangue minerals
4. Preparation of feed material for different beneficiation operations
5. Convenience in handling and transportation

Energy consumed for the comminution operation is high when compared to other operations such as screening, beneficiation, dewatering and conveying in mineral and mining industries. Therefore, much attention is to be paid to minimise the production of fines (finer than required) which will consume additional power for reducing to fines.

7.3 LAWS OF COMMINUTION

Laws of comminution are concerned with the relationship between energy input and the sizes of feed and product particles. Three laws of comminution for energy requirements have been put forward by Rittinger, Kick and Bond. None of these three laws are applicable over a wide range of sizes. Rittinger's and Kick's laws, while tenable in some cases, were never of much use as practical tools. The third law proposed by F.C. Bond in 1952 is based on a detailed compilation and study of numerous laboratory and plant comminution data and provides the technician with a reasonably accurate measure of power requirements.

Rittinger's Law: Rittinger (1867) postulated that the energy required for size reduction of a solid is directly proportional to the new surface area created. Mathematically, this can be expressed as

$$\frac{E}{m} = W = K_r\left[\frac{1}{d_P} - \frac{1}{d_F}\right] \qquad 7.1$$

where

E = Energy or Power in kW

m = Feed rate to machine in ton/h

W = Energy or power kWh/ton

d_P = Size of the product particles in microns

d_F = Size of the feed particles in microns

K_r = Rittinger's constant

The reciprocal of K_r is called the **Rittinger's number**. Rittinger's law is best applicable to coarse and intermediate size reduction.

Kick's Law: Kick (1885) proposed that the energy required to produce analogous changes of configuration of geometrically similar bodies of equal technologic state varies with the volumes or weights of these bodies. If a particle is only deformed so that only its configuration or surface area is altered, then the above statement remains to be true. However, in comminution where the number of particles increases due to breakage of large particles and new surfaces are created, the above statement ceases to be applicable.

The modified Kick's law which can be applied to crushing states that the work required for crushing a given quantity of material is constant for a given reduction ratio irrespective of original size. Mathematically, this law can be expressed as

$$\frac{E}{m} = W = K_k \log \frac{d_F}{d_P} \qquad 7.2$$

where K_k is Kick's constant

According to this law, comminution energy depends only on the reduction ratio and is independent of the original size of the feed. It is evidently absurd to say that the energy required for reducing a 100-mm particle to 50 mm will be the same as that for reducing a 1-mm particle to 0.5 mm. In fact, higher amount of energy is required for reducing fine particles to still finer size than for breaking down large pieces of rock. Kick's law is reasonably accurate in the crushing range above about 1 cm in diameter.

Bond's Law: Bond (1952) states that the total work useful in breakage that has been applied to a given weight of homogeneous broken material is inversely proportional to the square root of the average size of the product particles, directly proportional to the length of crack tips formed and directly proportional to the square root of new surface created. Mathematically, Bond's law is expressed as

$$\frac{E}{m} = W = K_b \left[\frac{1}{\sqrt{d_P}} - \frac{1}{\sqrt{d_F}} \right] \qquad 7.3$$

where K_b is Bond's constant

To apply Bond's law, Bond's constant must be evaluated. Bond's constant is evaluated by defining what is called work index, W_i. It is defined as the gross energy in kWh/short ton of feed necessary to reduce a very large feed to such a size that 80% of product particles passes a 100-micron screen.

On the basis of this definition, it can be written that

$$\text{if } d_F = \infty, \text{ and } d_P = 100 \text{ microns, } \frac{E}{m} = W_i \text{ kWh/short ton}$$

On substitution in 7.3,

$$\frac{E}{m} = W_i = K_b \left[\frac{1}{\sqrt{100}} - \frac{1}{\sqrt{\infty}} \right] \Rightarrow \text{Wi} = \frac{K_b}{10} \Rightarrow K_b = 10\ W_i$$

Thus, if 80% of feed particles passes d_F micron screen and 80% of product passes d_P micron screen, then

$$\frac{E}{m} = W = 10W_i \left[\frac{1}{\sqrt{d_P}} - \frac{1}{\sqrt{d_F}} \right] \qquad 7.4$$

The work index includes the friction in the crusher and the power W is gross power. An experimental procedure for determination of Work Index of a given material as suggested by Bond has been given in Annexure. Bond's law is applicable reasonably in the range of conventional rod mill and ball mill grinding.

It is a fact that most of the energy supplied to a comminution machine is absorbed by the machine itself to move various parts of the machine, and only a small fraction of the total energy is available for breaking the material. For example, in a ball mill, less than 1% of the total energy input is used for actual size reduction, the bulk of the energy is utilized in running the mill and in the production of heat.

7.4 TYPES OF COMMINUTION OPERATIONS

The run-of-mine ore is quite coarse and cannot be reduced to fine size in one stage. It may require three or more stages. Each stage requires separate equipment. The comminution operations are divided into two broad groups as **Crushing,** wherein large lumps are reduced to fragments or smaller particles, and **Grinding** wherein relatively coarse particles are reduced to the ultimate fineness.

The machines used for crushing and grinding are entirely different. It is to be noted that the energy required for comminution of unit mass of smaller particles is more than the energy required for unit mass of coarser particles. However, the energy required to reduce coarser particle is more than that of smaller particle. Hence, the machines used for crushing (crushers) must be massive and rugged and the machines used for grinding (mills) must be capable of dispersing energy over a large area. In crushers, the breakage forces are applied either by compression or impact whereas in grinding mills shear forces are predominantly applied. The comminution laws are crucial in calculating the power consumption in comminution operations. Examples 7.4.1–7.4.4 illustrate the calculations of power consumption.

Example 7.4.1: *A crusher fed at the rate of 1 ton/h is crushing 12 mm cubes to a product size of 80% retained on a 3-mm screen, 10% retained on a 2-mm screen and 10% retained on a 1-mm screen. If the crusher requires 4HP for this operation, how much power is required to crush the same material at the same rate from a feed size of 8 mm to product size of 1 mm. Estimate the power using (a) Rittinger's law (b) Kick's law.*

Solution:

Given

Feed size during first operation	$= d_{F1} = 12$ mm;
Feed size during second operation	$= d_{F2} = 8$ mm;
Product size during second operation	$= d_{P2} = 1$ mm;

Energy required during first operation $= E_1 = 4$ HP;
Energy required during second operation $= E_2 = ?$
Mass flow rate during both operations $= m_1 = m_2 = 1$ ton/h

Average size of the product during first operation $= d_{P1} = \dfrac{100}{\Sigma \dfrac{w}{d}}$

$$= \frac{100}{\frac{80}{3} + \frac{10}{2} + \frac{10}{1}} = 2.4 \text{ mm}$$

(a) As per Rittinger's law, $\dfrac{E}{m} = K_r\left(\dfrac{1}{d_P} - \dfrac{1}{d_F}\right)$

Substituting the values in this equation gives

$$\frac{4}{1} = K_r\left(\frac{1}{2.4} - \frac{1}{12}\right) \quad \text{......I}$$

$$\frac{E_2}{1} = K_r\left(\frac{1}{1} - \frac{1}{8}\right) \quad \text{......II}$$

Eq.II/Eq.I gives $E_2 = 10.5$ HP

Power required to crush from a feed size of 8 mm to a 1-mm product = 10.5 HP

(b) As per Kick's law, $\dfrac{E}{m} = K_k \log\dfrac{d_F}{d_P}$

Substituting the values in this equation, gives

$$\frac{4}{1} = K_k \log\frac{12}{2.4} \quad \text{.........III}$$

$$\frac{E_2}{1} = K_k \log\frac{8}{1} \quad \text{.........IV}$$

Eq.IV/Eq.III gives $E_2 = 5.169$ HP

Power required to crush from a feed size of 8 mm to 1 mm product = 5.169 HP

Example 7.4.2: *What is the power required to crush 100 tons/h of limestone if 80% of the feed passes a 2" screen and 80% of the product passes a 1/8" screen? The work Index for limestone may be taken as 12.74 kWh.*

Solution:

Given

80% passing size of the product $= d_P = 1/8'' = 3175$ microns;
80% passing size of the feed $= d_F = 2'' = 50800$ microns;

Work Index of the limestone = W_i = 12.74 kWh

As per the Bond's law $W = 10W_i\left(\frac{1}{\sqrt{d_P}} - \frac{1}{\sqrt{d_F}}\right)$

$$= 10 \times 12.74\left(\frac{1}{\sqrt{3175}} - \frac{1}{\sqrt{50800}}\right) = 1.696 \text{ kWh/ton}$$

Power per 100 tons/h = 1.696 × 100 = 169.6 kW

Example 7.4.3: *If a ball mill requires 3 kWh/ton of feed to reduce a feed of 1600 microns to a product of 400 microns, calculate the work index.*

Solution:

Given

80% passing size of the product = d_P = 400 microns;

80% passing size of the feed = d_F = 1600 microns;

Power required = W = 3 kWh

As per the Bond's law $W = 10W_i\left(\frac{1}{\sqrt{d_P}} - \frac{1}{\sqrt{d_F}}\right)$

$$3 = 10W_i\left(\frac{1}{\sqrt{400}} - \frac{1}{\sqrt{1600}}\right)$$

$\Rightarrow W_i$ = 12 kWh/ton

Example 7.4.4: *A comminution process consumed 460 HP/ton to crush a rock from 7.5 to 0.75 cm. If the power consumed was 697 HP/ton to reduce the feed of the same size, calculate the product size in the event.*

Solution:

Given

Power consumption = W = 460 HP/ton

80% passing size of the product = d_P = 0.75 cm = 7500 microns

80% passing size of the feed = d_F = 7.5 cm = 75000 microns

Power consumption in second operation = W = 697 HP/ton

As per the Bond's Law $W = K_b\left(\frac{1}{\sqrt{d_P}} - \frac{1}{\sqrt{d_F}}\right)$

For first operation

$$460 = K_b\left(\frac{1}{\sqrt{7500}} - \frac{1}{\sqrt{75000}}\right) \quad \ldots\text{I}$$

For second operation

$$697 = K_b\left(\frac{1}{\sqrt{d_P}} - \frac{1}{\sqrt{75000}}\right) \quad \ldots\text{II}$$

Evaluating I and II gives d_P = 4101.2 microns

Product size in second operation = d_P = 4101.2 microns

7.5 PROBLEMS FOR PRACTICE

7.5.1: *A rock of average particle size of 25 mm is crushed to a product of average particle size of 6 mm at a rate of 10 tons/h. At this rate, the mill consumes 18 kW of power. It requires 0.5 kW to run the mill empty. How much power will be consumed if the same feed was crushed to a particle diameter to 10 mm? Assume that Rittinger's Law is valid.* *[8.79 kW]*

7.5.2: *Determine the power necessary to reduce 110 tons of an ore from 6 mm to 150 microns, if the work index is 10 kWh/ton.* *[755.92 kW]*

7.5.3: *Copper ore having 80% passing size of 1530 microns was grounded to a product size of 80% passing 79 microns. If power consumed per hour is 11.26 kW, calculate the work index of copper ore.* *[12.95 kWh/ton]*

7.5.4: *Calculate the power required to crush the phosphate rock from a feed size of 80% passing 4" to a product size of 80% passing 1/8" at a rate of 100 tons/h. The work index of phosphate rock is 2.74. Also calculate the power required to crush the product further to 80% passing 1000 microns*

[40.027 kW, 78.037 kW]

■■■■

8 Crushing

As defined by Gaudin (2005), crushing is that operation or group of operations in a mineral processing plant whose object is to reduce large lumps to fragments, the coarsest particles in the crushed product being one-twentieth of an inch or more in size. The size of coarsest particles in the crushed product is half an inch in many cases.

The crushing action in all crushing machines (crushers) results from forces applied to the particles by a moving part working against a stationary or another moving part. It is the first stage of size reduction. It must crush run-of-mine ore which contains large-size particles. It requires greater force to be applied on the particles. Crushing is generally a dry operation and is usually performed in two or three stages.

Crushers are designed to reduce rock in such a manner that all pieces are less than a stated size. However, no crusher has been devised which produces only fragments greater than a specified size. A crusher always produces various sizes of particles with a substantial number of fines. As investigated by Gaudin and Hukki (1944), when the products of crushing are separated into a series of closely sized fractions, the total surface of each fraction will be more or less equal.

The extent of size reduction achieved by any crushing operation is described by the **reduction ratio**. The reduction ratio is broadly defined as the ratio of the feed size to the product size in any crushing operation. It is beneficial in determining what a crusher can do, or is doing, in terms of size reduction. It can also be used as a partial indicator of the stresses the crusher will be subjected to during operation, an element in determining crusher capacity and as an indicator of crusher efficiency. However, there is no one method of calculation which will provide a useful figure for the aforementioned criteria. Therefore, there are various types of reduction ratios in use; three most important ratios are defined as follows:

$$\text{Limiting reduction ratio} = \frac{\text{Size passing all feed particles}}{\text{Size passing all product particles}}$$

$$\text{Average reduction ratio} = \frac{\text{Average size of the feed particles}}{\text{Average size of the product particles}}$$

$$\text{80\% passing reduction ratio} = \frac{\text{80\% passing size of the feed particles}}{\text{80\% passing size of the product particles}}$$

Reduction ratio is a convenient measure for comparing the performance of different crushers.

As crushing is performed in stages, it may be divided into primary, secondary and tertiary stages on the basis of particle size. Crushers can be classified into the following five groups as per the size of the product they produce.

1. **Primary Crushers :** Jaw crusher, Gyratory crusher
2. **Secondary Crushers :** Reduction gyratory, Cone crusher, Rolls crusher
3. **Tertiary Crushers :** Short-head cone crusher
4. **Fine Crushers :** Impact crushers
5. **Special Crushers :** Bradford breaker, Toothed roll crusher, Gravity stamps

Lumps of run-of-mine ore usually of 1-m size is reduced to 100–200 mm in heavy-duty primary crushers. They are also called breakers as they are break the ore to a relatively large size. The usual size of feed to secondary crushers is 600 mm and the product is usually 10–100 mm. In tertiary crushers, 250-mm size particles are reduced to 3–25 mm size.

Fine crushers reduce coarse particles to fine, even to 200 mesh in some cases. Special crushers are designed for specific ores, e.g. rotary breaker and toothed roll crusher for coal and gravity stamps for gold ore milling.

8.1 JAW CRUSHER

A jaw crusher consists of two jaws set at an acute angle—called **angle of nip**—to each other which forms crushing chamber. One jaw is fixed and kept vertical and the other jaw—called **swing jaw**—is moved to approach and recede alternately from the fixed jaw. Motion to the swing jaw is transmitted by pitman working on an eccentric and toggles. The material is fed between the jaws and is alternately nipped and crushed.

Jaw crushers are classified by the method of pivoting the swing jaw (Fig. 43). If the swing jaw is pivoted at the top, it has a fixed receiving area and variable discharge area; it is known as a **Blake crusher**. It gives greatest movement on the smallest lump. If the swing jaw is pivoted at the bottom and has a fixed discharge area and variable receiving area; it is known as a **Dodge crusher**. It provides the greatest movement on the largest lump. If the swing jaw is pivoted at an intermediate position, it has both variable receiving and discharge areas and is known as a **Universal crusher**. It gives equal movement on all sizes. Crushers of universal type are not currently in use.

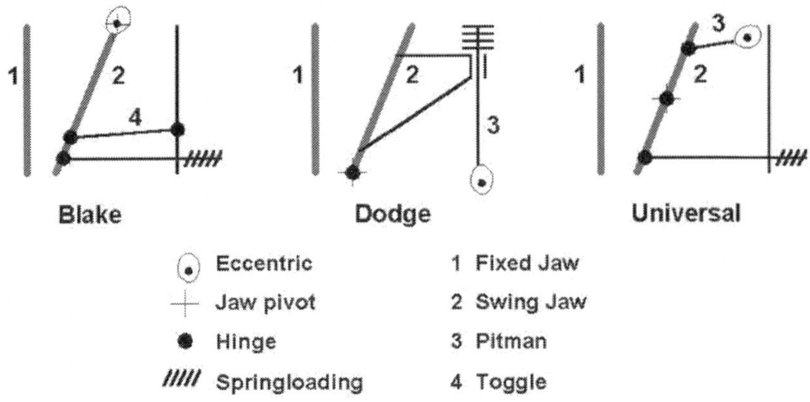

Fig. 43 Types of jaw crushers

In a **double-toggle Blake jaw crusher** (Fig. 44), the swing jaw is actuated through toggles, a pitman and an eccentric shaft. Heavy flywheels are attached to both ends of the eccentric shaft and driven by a pulley attached to one of these. Both fixed and swing jaws are lined with plates to take up wear. The jaw plates are usually corrugated to assist in gripping the ore. Using springs, the tension rod holds the swing jaw against the toggle plates when the pitman is on the downward stroke.

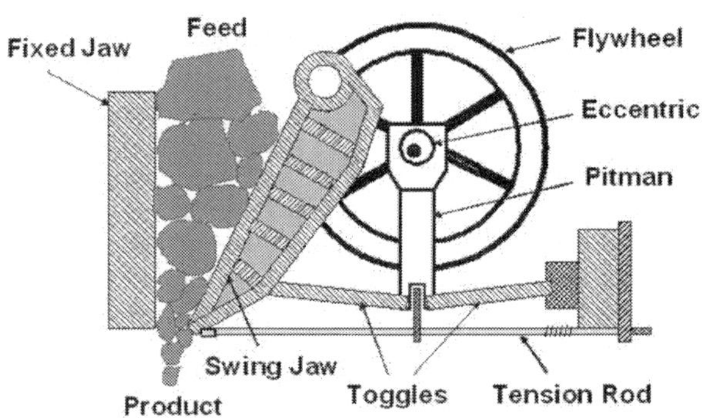

Fig. 44 Double-toggle Blake jaw crusher

In a double-toggle Blake jaw crusher, the movement of the swing jaw is affected by the vertical movement of the pitman. This moves upwards and downwards by the eccentric. The back-toggle plate causes the pitman to move sideways as it is pushed upward. This motion is transferred to front toggle plate; this, in turn, causes the swing jaw to come nearer to the fixed jaw. Similarly, downward movement of the pitman allows the swing jaw to move away from the fixed jaw.

Analysis of the forces will be useful in understanding the working of a jaw crusher. Let F_1 be the force on the pitman, F_2 be the force on toggles and θ be the angle between pitman and toggle (both sides), as shown in Fig. 45.

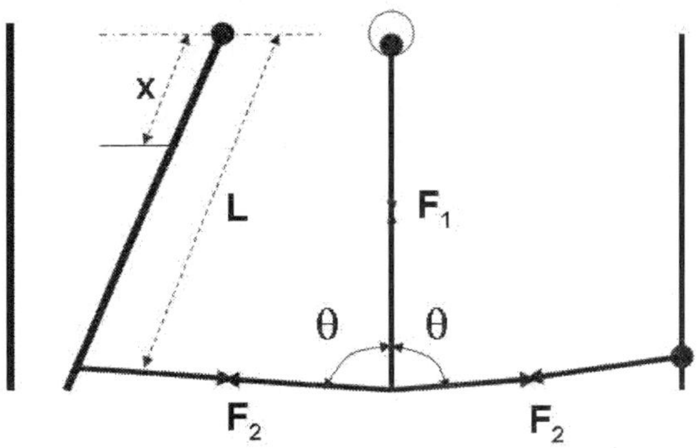

Fig. 45 Force balancing in double-toggle Blake jaw crusher

Sum of the vertical components of F_2 must be equal to F_1.

$$\therefore\ F_1 = F_2 \cos\theta + F_2 \cos\theta$$

$$\Rightarrow\ F_2 = \frac{F_1}{2\cos\theta}$$

The horizontal thrust F_L on the particle by swing jaw at toggle block is equal to the horizontal component of F_2, i.e. $F_2 \sin\theta$. If F_x is the horizontal thrust at any distance x from the pivot, then

$$F_x\, x = F_L\, L$$

$$F_x = \frac{F_L L}{x} = \frac{F_2 \sin\theta\ L}{x} = \frac{F_1 L \tan\theta}{2x}$$

Since θ is nearly equal to 90°, tan θ is very large and hence F_x is large. It means that a small force exerted on the pitman is communicated to the swing jaw as a very large force by the toggle. As x is decreasing, F_x is increasing. Hence it is evident that, in a Blake jaw crusher, maximum force is applied on the larger feed particles. As the maximum movement of the jaw is at the bottom, there is no tendency for the crusher to clog. Of course, some uncrushed material will pass through. As the crushing is accomplished only in one stroke when the swing jaw moves towards the fixed jaw, the particles get compressed against the fixed jaw and get crushed. When the swing jaw recedes from the fixed jaw, no crushing takes place and the crushed material travels down by gravity till it is caught when the swing jaw re-approaches the fixed jaw. Thereby, the material is intermittently caught and crushed between jaw plates several times depending on the length of the jaws and speed of the stroke till the particles having the size less than the set pass through and discharges from the crushing chamber. Therefore, the crushing process is intermittent, i.e. their crushing action is slightly less than half the time which leads to intermittent power requirement and results in uneven loading on the machine. Due to its alternate loading and release of stress, it must be very rugged and needs strong foundations to accommodate the vibrations. The heavy flywheel which is attached to the drive stores energy during idling stroke and delivers energy during crushing stroke.

In operation, the throw of the swing jaw is only a few centimetres and the frequency is between 60 and 360 strokes per minute. The angle between the jaws is approximately 20°.

In a **single-toggle Blake jaw crusher** (Fig. 46), swing jaw is suspended on the eccentric shaft and made to move directly by the flywheel shaft. The flywheel shaft is used both as a pivot for the receiving end of the swing jaw and as an eccentric for to-and-fro movement of the swing jaw. The swing jaw movement is imparted through the inclined toggle plate. The swing jaw not only moves towards the fixed jaw but also vertically as the eccentric rotates. This elliptical jaw motion assists in pushing the ore through the crushing chamber. Hence, it has a higher capacity than the double-toggle jaw crusher. This crusher is known as **overhead eccentric Blake jaw crusher**.

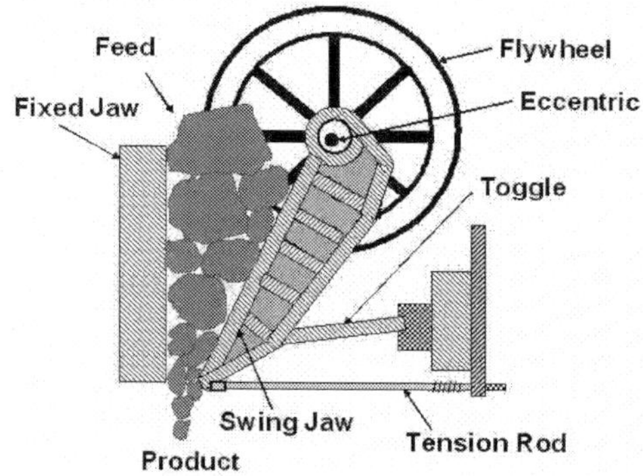

Fig. 46 Single-toggle Blake jaw crusher

In a **dodge jaw crusher** (Fig. 47), jaw movement is reverse to that of a Blake jaw crusher. This crusher does not require toggles and the jaw is operated through the pitman through an eccentric. The maximum movement is applied to the largest piece and the minimum to the smallest. Thus, the choke point coincides with the point of least motion. Consequently, these crushers are of relatively low capacity and the rapid action in the feed area gives the machine the advantage of large reduction ratio and a closely-sized product. Thus, it fits well into low-capacity operations such as sample reduction. Besides law capacity, its principal disadvantages are a tendency to pack and make fines. Therefore, this dodge crusher is restricted to laboratory use.

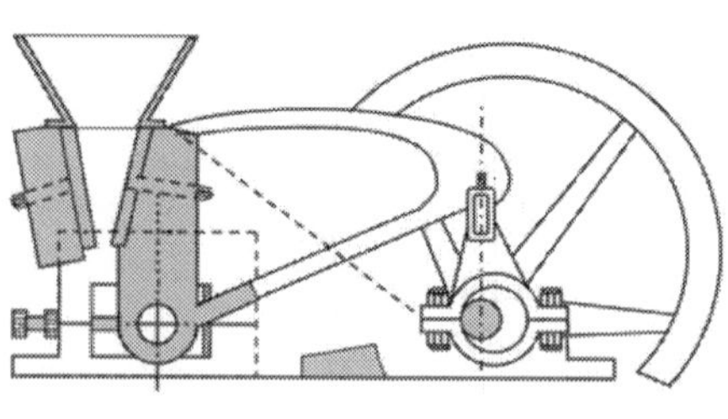

Fig. 47 Dodge jaw crusher

Jaw crushers are intended for use as primary crushers. In mining industry, jaw crushers are used to crush the run-of-mine ore to a size suitable for transportation. They are heavy-duty machines as they must reduce the lumps as big as 1 m. Hence, jaw crushers have a relatively large gape. Size of the crusher product is controlled by the set, which is the maximum opening of the jaws at the discharge end. This can be adjusted by using toggle plates of the required length. The reduction ratio of jaw crushers varies from 4 to 7.

Since the jaw crusher works intermittently in a half cycle, it is limited in capacity. Taggart (1945) gave an empirical formula for the capacity of the crushers as

$$T = 0.6\,L\,S \qquad 8.1$$

where T is the capacity in tons/h, L is the length of the receiving opening and S is the width of the discharge opening (set) both in inches. This formula is fairly accurate except for smallest and largest crushers. For small crushers, the answer is too high and for largest crushers, the answer is too small.

The capacity varies with speed up to a point where the speed is so great the ore cannot drop in between jaws owing to more strokes. As a rule, the crushers are run far below their capacity and for few hours only out of 24 hours. The advantage of a large crusher is that it will do its day's work in a short time and leave the attendant free for other work, thus saving labour. The disadvantage is that it costs more at the start and requires more power but does not consume more power per ton. The cost of breaking and repairs is also less with a large crusher than compared to several smaller ones.

The energy consumption by jaw crushers depends upon the size of feed, the size of the product, capacity of the machine, properties of the ore and percentage of idling time. Broadly speaking, the energy consumption decreases with increasing size of feed at constant reduction ratio, increasing size of product at constant size of feed and increasing capacity of the machine.

The amount of fines produced will largely depend upon the condition of the jaw plates. Smooth jaw plates tend to increase the amount of fines. The production of fines can be minimised using corrugated or curved jaw plates. Except in very small units, vertically corrugated jaw plates are universally used.

8.2 GYRATORY CRUSHER

Gyratory crusher (Fig. 48) consists of two crushing surfaces which are vertical truncated conical shells. The outer shell is a hallow conical shell known as concave. It is stationary and its apex is pointing down. The inner shell is a solid conical shell—known as crushing head or mantle—and its apex is pointing up. The crushing head or mantle is mounted on a spindle and gyrated by the main shaft. The main shaft is driven by a bevelled gear. The gyrating motion is imparted by the eccentric. At the top of the machine, a heavy frame supports the spider which in turn supports the spindle. The action of the machine is as follows: when the bevel gear revolves, the spindle is free to gyrate or rotate in the eccentric. Practically, it rotates until ore is fed between the crushing surfaces, then it gyrates. The gyrating motion causes the crushing head to approach towards and recede from the concave. The maximum movement of the head occurs near the discharge. This tends to relieve the choking. Crushing always takes place at some point in these crushers, making them continuous crushers. They are considered as an infinite number of jaw crushers of infinitely small width.

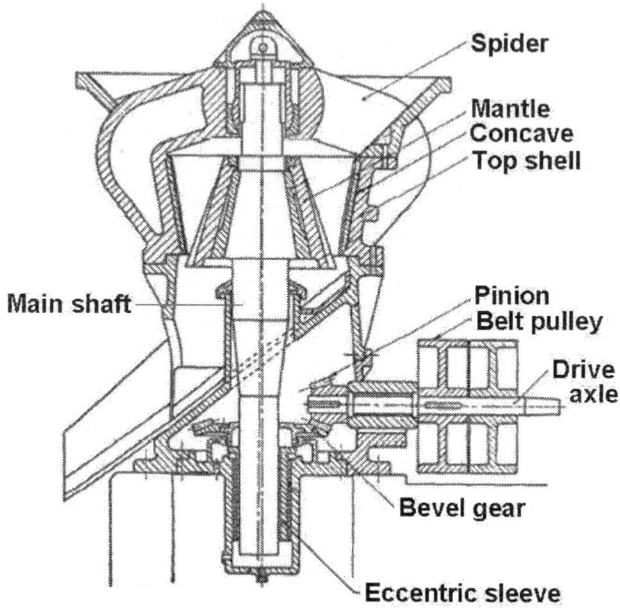

Fig. 48 Gyratory crusher (Courtesy of ISI Specification document)

Many forms of gyratory crushers have been developed. Hydraulic support and suspended spindle types are most important. A hydraulic support gyratory crusher with a hydraulic piston as the supporting member is the most used. In these crushers, when overloading occurs, a valve is tripped which releases the fluid, thus dropping the spindle and allowing the tramp material to pass out between the head and the concave. In suspended-spindle gyratory crushers, the spindle is suspended from a spider and as it rotates, it sweeps out a conical path within the fixed crushing shell due to the gyrating action of the eccentric.

The movement of the crushing head which performs the crushing action in a gyratory crusher can be visualised as in Fig. 49 wherein the feed drops down; the crushing head squeezes it against the concave and crushes the ore. When the crushing head moves away during its cycle of gyration, the crushed ore slips down to be caught again between the crushing head and the concave on the next cycle, resulting in further size reduction. The process is repeated until the sizes of the crushed ore are less than the open set at the bottom of the chamber. The reduction ratio of gyratory crushers varies from 3 to 10.

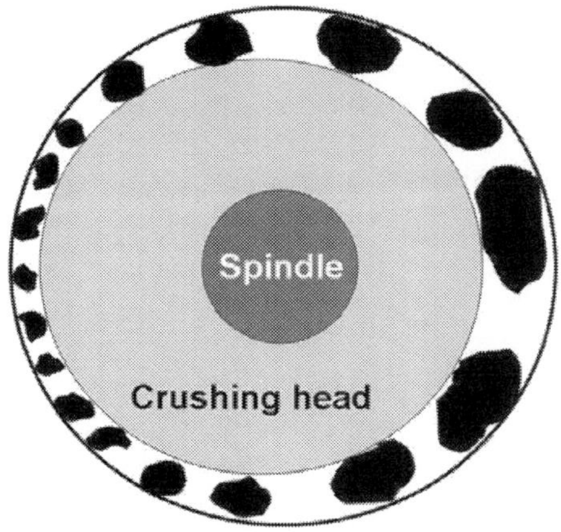

Fig. 49 Section of gyratory crusher

Since the gyratory crusher crushes on full cycle, it has a much higher capacity than a jaw crusher of the same gape handling the same size of feed. Gyratory crushers tolerate various shapes of feed particles, including slabby ore, which are not readily accepted in jaw crushers owing to the shape of the feed opening.

The angle of nip for large gyratory crushers varies between 21° and 24° but it is 27° to 30° for curved surfaces. The distances of opening between the concave and the crushing head at the top and bottom ends are usually used to describe the size of the gyratory crusher.

Most gyratory crushers are used as primary crushers, i.e. they receive the initial feed from the mine or quarry. They can be operated with the head buried in the ore. The same type of crusher with chamber modifications is used as a secondary crusher, particularly when following a large primary crusher. A **reduction gyratory crusher** is the modification of gyratory which has straight or curved heads and concaves and used for secondary crushing. The fine reduction gyratory crusher can also be used for tertiary crushing.

The gyratory crusher has an advantage over the jaw crusher in that it has a more regular power draft. This is because the gyratory crusher is continuously crushing, whereas the jaw crusher is intermittently working. Since jaw and gyratory crushers are used as primary crushers, maximum size of the ore and capacity required are the two factors to be considered while selecting the crusher to be employed. Jaw crushers are used where the crusher gap is more important than the capacity. Gyratory crushers are used where high capacity is required. They are more efficient than jaw crushers as they crush the ore on full cycle.

The capital and maintenance costs of a jaw crusher are slightly less than those of gyratory. Installation costs are less for gyratory. As the gyratory crushers have self-feeding capability, expensive feeding devices can be eliminated in case of gyratory crushers thus saves the capital investment. In case of transportation of crushers to remote locations and for underground use, jaw crushers are preferred as they can be easily sectionalised for convenience in transportation.

The type of the material crushed is also important in selecting the crusher. Gyratory crushers are suitable for hard, abrasive and slabby materials, whereas jaw crushers perform better on sticky, clayey and plastic materials.

8.3 CONE CRUSHER

Cone crushers were originally designed and developed by Symons and are therefore often described as Symon cone crusher. The principle of operation of cone crusher is similar to that of a gyratory crusher. However, the outer crushing surface in a cone crusher flares out unlike a gyratory crusher. The cone angles are much flatter and the slope of the mantle and the concave are parallel to each other. The flatter cone angles help retain the particles longer between the crushing surfaces and thus produce much finer particles. The outer crushing surface is held to the main frame by the series of bolts and springs so that it will be lifted away from the crushing head when an uncrushable particle enters the crushing chamber thus avoids crushing of the tramp iron.

Symons cone crushers utilise certain principles that not taken advantage of in gyrating crushers. In the first place, the crushing head moves considerably faster and has several times the amplitude of movement of standard gyratories. The time of one gyration is less than the time required for pieces of ore to fall clear through the parallel zone by gravity; hence, all the ore particles are crushed at least once in the minimum width of the parallel zone. Efficient and rapid discharge and movement of crushed material through the machine are favoured by the shape of the crushing chamber whose length considerably increases from top to bottom. The effective area of the opening available for the crushing and the fall of finer sizes at the bottom are of the same magnitude as that for the coarser sizes at the top. Furthermore, the relatively high maximum width of opening attained in each gyration allows free fall of material during the time between crushing impacts.

The two types of Symons cone crushers used are **standard** and **short-head**. The outside crushing surface of the standard cone crusher has stepped liners to allow coarser feeds for reduction. The short-head cone crusher has smoother crushing surfaces and steeper cone angles, and it is used for the reduction of finer feeds. The sectional view of the Symons standard cone crusher is shown in Fig. 50. The crushing is done between the gyrating conical crushing head and the curved concave called a bowl. The conical crushing head and the shaft are entirely supported on the spherical socket bearing and are gyrated by the sleeve-type eccentric which is driven by the gear and countershaft assembly. The bowl is threaded on its outer circumference to fit the similar set of threads of the adjustment ring and the bowl may thus be raised or lowered in relation to the cone to take care of wear of the crushing surfaces and regulate the product size.

Ore is fed to the cone crusher through the adjustable feed spout onto the distributor disk which gyrates with the crushing head and distributes the feed around the annular crushing space to be nipped and crushed. The crushed product is usually in the range of 6–60 mm and the reduction ratio varies from 6 to 8.

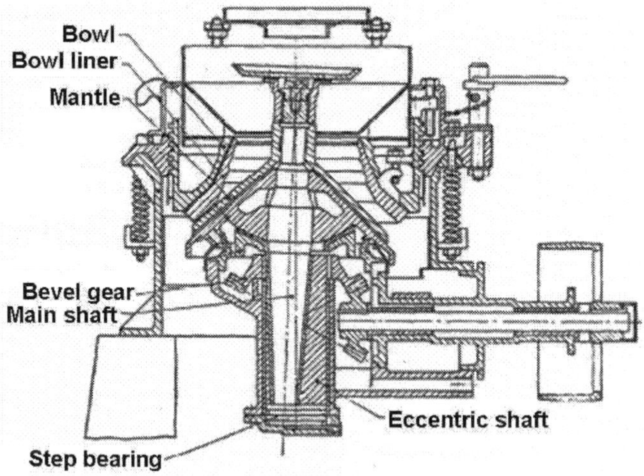

Fig. 50 Sectional view of Symons standard cone crusher (Courtesy of ISI Specification document)

The Symons short head cone crusher is similar in operation to that of the standard type. The head is shortened and the angle of the crushing cavity is steepened. The steeper head angle prevents choking from the much finer material being handled. It also has a narrower feed opening and a longer parallel section at the discharge. The crushed product ranges from 3 to 20 mm and the reduction ratio is 4-6.

In Symons cone crushers, each gyrating cycle is timed so that the descending material and the upwards thrust of the crushing head meet at the moment of maximum impact. The optimum speed of gyration and the large eccentric throw combine to produce two important results:

1. The rapidly closing crusher head catches falling material, rearranges it and then delivers an extremely high crushing force
2. In the part of the crushing chamber where crushing is not taking place, the rapidly receding head allows material to freely fall to the next point of impact.

This combination of superior crushing force and free flow of material provides higher crushing efficiency, resulting in lower power consumption, reduced product cost per ton, less maintenance and high profits.

The Symons cone crusher, with its large eccentric throw and complimentary head angle has a large displacement volume. Displacement volume is the difference between the volume displaced by the crushing head when it is fully closed and fully open. Larger displacement volume results in greater capacity for the following reasons:

1. More material is crushed in each crushing cycle
2. More space is left open as the crushing head recedes, allowing more material to be fed to the crusher at each cycle to fill the larger void

3. Greater eccentric throw coupled with ideal gyrating cycles allows material to drop further into the cavity before the next blow, resulting in greater flow rate through the crushing chamber

The benefits of high efficiency, greater crushing force and high capacity are the principle reasons the Symons cone crusher has been accepted worldwide as the best way to increase productivity.

Standard cone crushers are designed to make a product ranging from half inch to more than 2.5 inches. Symons standard crushers may be applied in large or small operations for single reduction, or as a primary or secondary crusher ahead of short head cone crusher. Short head cone crushers are designed to take a feed of smaller size. Product range is from less than three-sixteenth of an inch to about 1 inch in both open and closed circuits.

8.4 ROLL CRUSHER

The standard roll crusher consists of two horizontal rolls which revolve towards each other. Ore particles introduced between the rolls are nipped and then crushed as they pass between the rolls. Two types of roll crushers are generally designed. In one type both rolls are rigidly fixed to a frame with provision for adjusting the lateral position of one of the rolls to control the gap between them. In another type one roll is spring mounted, forming the driving roll; the other roll rotates by friction (Fig. 51). The distinguished feature of roll crusher is that the material is crushed one time only while it is passing through the crushing chamber. Due to this fact, the reduction ratio of roll crusher varies from 2 to 4—the lowest among all the crushers. Production of fines is minimum. They can well handle friable, dry, wet, sticky, frozen and less-abrasive feeds.

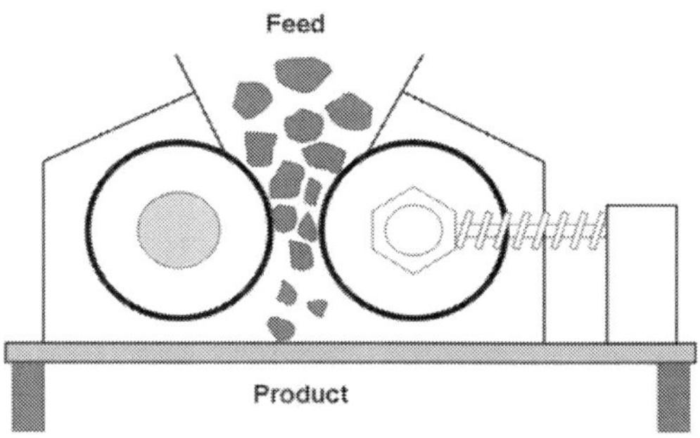

Fig. 51 Roll crusher

The angle of nip is affected by the distance between the rolls. The angle of nip is defined as the angle between the tangents to the roll surface at the points of contact between the rolls and the particle (Fig. 52). Usually, the angle of nip is between 20° and 30°; however, it is up to 40° in some large roll crushers. For the selection of size of roll crusher for the reduction of different sizes of feed, an expression for the angle of nip can be obtained as follows considering a spherical particle. Let D be the diameter of the rolls, d be the diameter of the spherical feed particle, n be the angle of nip and s be the distance between two rolls.

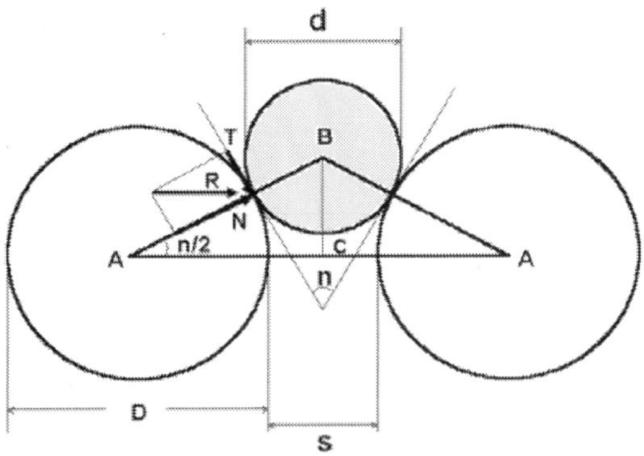

Fig. 52 Angle of nip of Roll Crusher

Neglecting gravity force, the two forces acting on the particle when it is fed between two rolls are tangential force T and normal force N. If the resultant of these two forces R is directed downward, the particle will be nipped and get crushed. If the resultant force R is directed upwards, the particle will be thrown out from the rolls and there will be no crushing. The limiting condition is, therefore, when resultant force R is just horizontal. Under this limiting condition, the angle between the tangents is called angle of nip n because this is the maximum angle beyond which the particle will not be nipped. Under these limiting condition, the vertical components of N and T will balance with each other. Thus,

$$\text{N} \sin \frac{n}{2} = \text{T} \cos \frac{n}{2} \qquad \Rightarrow \tan \frac{n}{2} = \frac{T}{N}$$

since $\frac{T}{N}$ is the coefficient of friction μ then $\tan \frac{n}{2} = \mu$ 8.2

Hence, in a roll crusher, the necessary and sufficient condition for the particle to be nipped is evidently

$$\tan \frac{n}{2} \leq \mu \tag{8.3}$$

Considering the triangle ABC in Fig. 54,

$$\cos \frac{n}{2} = \frac{\frac{D}{2} + \frac{s}{2}}{\frac{D}{2} + \frac{d}{2}}$$

$$\Rightarrow \quad \cos \frac{n}{2} = \frac{D+s}{D+d} \tag{8.4}$$

The theoretical capacity of a roll crusher in kg/h is given by

$$C = 60\pi DNWs\rho \tag{8.5}$$

where N is the number of revolutions of the rolls per minute, D is the diameter of the rolls, W is the width of the roll face, s is the set all in meters and ρ is the density of the particle in kg/m^3. In practice, when allowing for voids between the particles, loss of speed in gripping the feed, etc., the actual capacity is usually about 25% of the theoretical capacity. Smooth-surfaced roll crushers are generally used for fine crushing, whereas corrugated or toothed roll crushers are used for coarse crushing.

Calculations by using above formulae are illustrated in Examples 8.4.1–8.4.4

Example 8.4.1: *A material of 1.5" diameter spherical particles is intended to be crushed in a roll crusher. Assuming coefficient of friction between the rock and the steel to be 0.29 and reduction ratio to be 3:1, calculate the minimum diameter of rolls required.*

Solution:

Given

Size of the feed particle = 1.5″

Reduction ratio = 3

Coefficient of friction between rock and steel = 0.29

Coefficient of friction tan n/2 = 0.29 ⇒ n/2 = 16°10′

$$\therefore \cos n/2 = \cos 16°10' = 0.9611$$

$$\text{Size of the product particle} = \frac{1.5}{3} = 0.5''$$

$$\text{However, } \cos\frac{n}{2} = \frac{D+s}{D+d} \qquad \Rightarrow 0.9611 = \frac{D+0.5}{D+1.5}$$

$$\Rightarrow D = 24.2''$$

Diameter of the roll to reduce 1.5″ pieces of rock to 0.5″ = 24.2″

Example 8.4.2: *A crushing roll of 24" diameter is set at 0.5" apart. If the angle of nip is 31°, calculate the maximum size of the feed particle to the rolls.*

Solution:

Given

Diameter of the roll = D = 24″

Set of the rolls = s = 0.5″

Angle of nip = n = 31°

⇒ n/2 = 31/2 = 15.5°

$\cos n/2 = \cos 15.5° = 0.9636$

However, $\cos\dfrac{n}{2} = \dfrac{D+s}{D+d} \Rightarrow 0.9636 = \dfrac{24+0.5}{24+d} \Rightarrow d = 1.4''$

$\therefore$ Maximum size of the feed particle = 1.4″

Example 8.4.3: *Calculate the angle of nip in case of a roll crusher of the following specifications*

Diameter of the roll	*= 7.5 cm*
Feed size	*= 0.6 cm*
Product size	*= 0.05 cm*

Solution:

Given

Diameter of the roll	= D = 7.5 cm
Diameter of the feed particle	= d = 0.6 cm
Set of the rolls	= s = 0.05 cm

$$\cos\frac{n}{2} = \frac{D+s}{D+d} = \frac{7.5+0.05}{7.5+0.6} = 0.9321$$

$$\Rightarrow \quad \frac{n}{2} = 21°14' \Rightarrow n = 42°28'$$

Angle of nip in roll crusher = 42°28′

Example 8.4.4: *A crushing roll crushes a rock of specific gravity 2.5 from 3.81 cm with a reduction ratio of 3. The angle of nip is 39° and the rolls operate at 60 rpm to give a capacity of 40 tons/h. Calculate the coefficient of friction between steel and the rock, product particles size, the diameter and width of the rolls.*

Solution:

Given

Size of the feed particle	= d = 3.81 cm
Reduction ratio	= RR = 3
Angle of nip	= n = 39°
Speed of the rolls	= N = 60 rpm
Mill capacity	= C = 40 tons/h = 40000 kg/h
Density of the rock	$= \rho = 2.5 \times 1000 = 2500$ kg/m^3
Coefficient of friction	= f = tan n/2 = tan 39/2 = 0.354
Size of the product particle	$= \dfrac{d}{RR} = \dfrac{3.81}{3} = 1.27$ cm

$$\frac{n}{2} = \frac{39}{2} = 19.5 \Rightarrow \cos\frac{n}{2} = 0.9426$$

However, $\cos\frac{n}{2} = \frac{D+s}{D+d} \Rightarrow 0.9426 = \frac{D+1.27}{D+3.81} \Rightarrow \mathrm{D} = 40.44$ cm

∴ Diameter of the roll crusher = 40.44 cm

Theoretical capacity of the crushing rolls = C = 60πDNWsρ

$$\mathrm{W} = \frac{\mathrm{C}}{60\,\pi\,\mathrm{D\,N\,s}\,\rho}$$

$$\mathrm{W} = \frac{40\,000}{60\,\pi.\times 40.44/100 \times 60 \times 1.27/100 \times 2500} = 0.2754 \text{ m} = 27.54 \text{ cm}$$

∴ Width of the roll face = 27.54 cm

8.5 CRUSHING OPERATION

Slow compression is the characteristic action of the nipping crushers. Crushers are usually operated dry. When the material fed to the crusher is at a slow rate, the individual particles are freely crushed. The crushed product is quickly removed from the crushing zone. This type of crushing—known as **free crushing**—avoids the production of excessive fines by limiting the number of contacts.

When the material fed to the crusher is at a high rate, the crusher is choked and prevents the complete discharge of crushed product. This results the crushing between ore particle and crushing surface as well as between ore particles. This type of operation is called **choked crushing** and it increases the amount of fines produced. Choked crushing is preferred in some cases as it reduces the reduction stages. In example 8.5.1, 80% passing size of the feed and the product and reduction ratio are calculated.

Example 8.5.1: *Data obtained from a crushing test conducted on a sample of Iron ore in laboratory is given in Table 22.*

Table 22: Crushing Test Data for example 8.5.1

FEED		PRODUCT	
Screen size	Weight of the material retained (g)	Screen size	Weight of the material retained (g)
5″	1112		
4″	1668	1½″	696
3″	3614	1″	1624
2″	2641	½″	4060
1″	2363	¼″	3132
½″	1807	−¼″	2088
−½″	695		

Considering 80% passing size of the feed and the product, calculate reduction ratio.

Solution:

Weight percentages and Cumulative weight percentages passing are calculated for feed and product and tabulated in the Table 23:

Table 23: Calculated values of Table 22

FEED					PRODUCT				
Screen size	**Wt**	**Wt%**	**Screen size**	**Cum. wt% passing**	**Screen size**	**Wt**	**Wt%**	**Screen size**	**Cum. wt% passing**
5″	1112	8	+5″	100					
4″	1668	12	5″	92	1½″	696	6	+1½″	100
3″	3614	26	4″	80	1″	1624	14	1½″	94
2″	2641	19	3″	54	½″	4060	35	1″	80
1″	2363	17	2″	35	¼″	3132	27	½″	45
½″	1807	13	1″	18	−¼″	2088	18	¼″	18
− ½″	695	5	½″	5					
	13900	100				11600	100		

From the table 80% passing size of feed and product are 4″ and 1″.

$$\text{Reduction ratio} = \frac{\text{80\% passing size of feed}}{\text{80\% passing size of product}} = \frac{4}{1} = 4$$

8.6 OPEN- AND CLOSED-CIRCUIT CRUSHING OPERATIONS

Usually, each stage of size reduction is followed by a screen which forms a circuit. Crushing may be conducted in open or closed circuit as shown in Fig. 53.

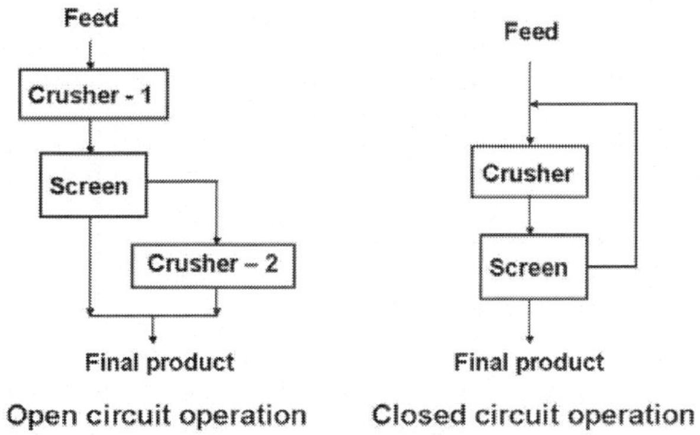

Fig. 53 Open-and closed-circuit crushing operations

In an open-circuit crushing operation, the feed material is reduced by one crusher. The product of this crusher is screened and only oversize material is crushed by another crusher of small size as the **throughput** (the quantity of material crushed in a given period of time) is less. The crushed product from the second crusher and undersize material from the screen together form the final product. In case of a closed-circuit crushing operation, the oversize material from the screen is re-fed to the same crusher. The undersize material from the screen is the required final product.

The primary crushers, i.e. jaw and gyratory crushers, are normally operated in open circuit as their purpose is to reduce the size of ROM ore. The crushed product of the primary crusher is screened and oversize product is crushed in next stage of crushing, i.e. secondary crushing, to the desired size. Secondary crushers accept suitable feed size material either prepared by a primary crusher or occurring naturally and reduce it to a size suitable for marketing or make feed for subsequent crushing or grinding stages. Secondary crushers are also operated in open circuits.

8.6.1 Open-circuit crushing operation

Let us consider a crusher set at 4″ is crushing an ore at the rate of 100 tons/h to a product size consists of 70% of −2″. The product of the crusher is sent to a screen of 2″ square opening. The oversize from the screen is fed to another crusher to give a product of all pass 2″. The circuit details assuming 100% and 90% screen efficiencies are shown in the Figs. 54 and 55, respectively.

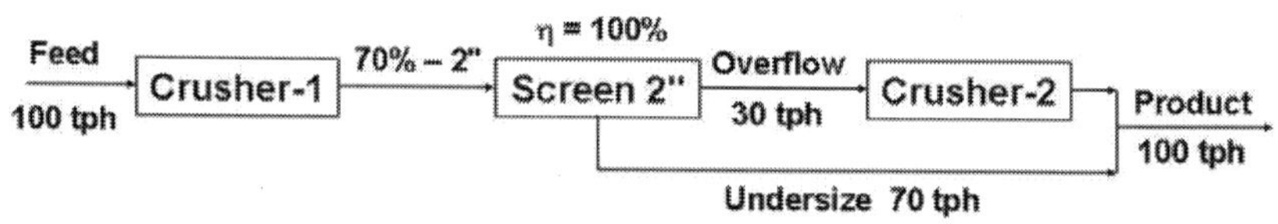

Fig. 54 Open-circuit crushing details when screen efficiency is 100%

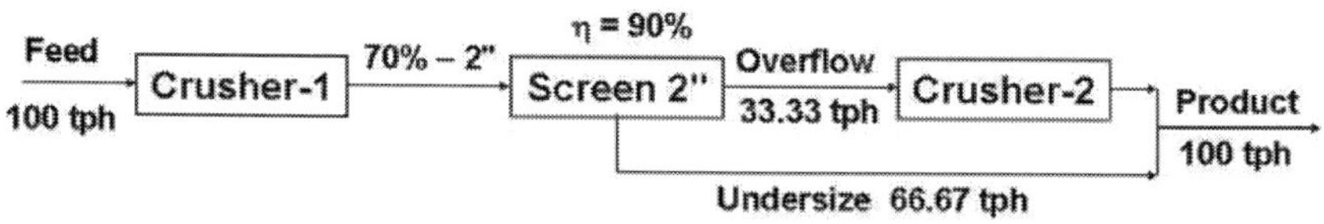

Fig. 55 Open-circuit crushing details when screen efficiency is 90%

It is noteworthy that the overflow of the screen when screen efficiency is 90% also consists of undersize material. When screen efficiency is 90%, 100 tph of material is crushed in primary crusher and 33.33 tph of material is crushed in secondary crusher. This can be arrived by considering screen efficiency as follows:

Efficiency of the screen based on oversize

$$\eta = \frac{\text{Weight of } + 2''\text{ material in crusher product}}{\text{Weight of } + 2''\text{ material obtained from the screen (overflow)}}$$

Therefore, weight of +2″ material obtained from the screen (overflow) and crushed by secondary crusher

$$= \frac{\text{Weight of } + 2''\text{ material in crusher product}}{\eta} = \frac{30}{0.9} = 33.33 \text{ tph}$$

Similar calculations are illustrated in example 8.6.1.

Example 8.6.1: *Crushed product of a primary crusher having 57% of +1″ is sent to 1″ screen. The overflow of the screen is again crushed in a secondary crusher to –1″ size and sent along with underflow of the screen as final product. Draw the crushing circuit with all details. Calculate the rate of the material crushed in secondary crusher if the effectiveness of the screen is 80% based on oversize material and the feed rate to a primary crusher is 100 tons/h.*

Solution: Crushing circuit is drawn as shown in Fig. 56

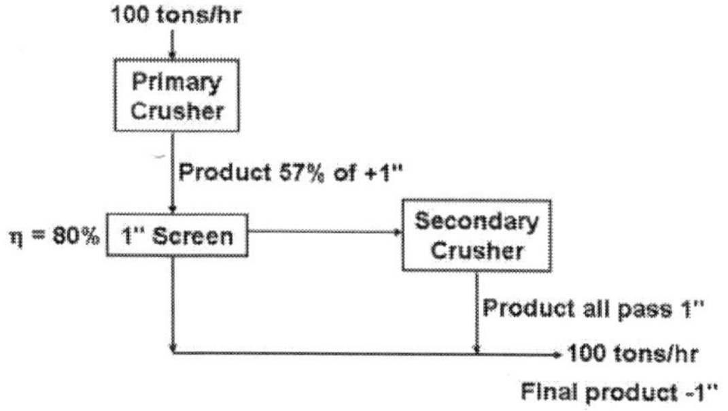

Fig. 56 Crushing circuit for example 26

Weight of +1″ material in feed to screen = 57% of 100 tons/h = 57 tons/h

$$\text{Screen efficiency} = \eta = 0.80 = \frac{\text{Weight of oversize material present in feed}}{\text{Weight of overflow obtained from the screen}}$$

$$\therefore \text{Weight of overflow obtained from the screen} = \frac{57}{0.80} = 71.25 \text{ tons/h}$$

Material crushed in secondary crusher = 71.25 tons/h

Underflow from the screen = 100 – 71.25 = 28.75 tons/h

8.6.2 Closed-circuit crushing operation

In a closed-circuit crushing operation, the oversize material from the screen is fed back to the same crusher. Secondary crushers may be operated in closed circuit with a vibrating screen depending upon the product requirements. Screening ahead of secondary crushers is generally recommended, especially where the feed contains more than 25% material smaller than the desired closed-side setting. Tertiary crushers normally take the secondary crusher product and reduce it to a marketable product or make it suitable for subsequent comminution steps. Tertiary crushers are normally operated in closed circuits with a vibrating screen and make a product smaller than all passing half an inch or five-eighths of an inch.

The two types of closed-circuit crushing operations are **regular** and **reverse**. In the regular type, the new feed is fed to the crusher, the crusher product is fed to the screen and the overflow from the screen is fed back to the same crusher. The quantity of the overflow material fed back to the crusher is called as **circulating load**.

The underflow from the screen is the required final product and is equal to the new feed. This regular type is usually employed when the feed contains low percent undersize. Initially the quantity of final product produced is less than the quantity of the feed material. As the operation proceeds further the quantity of the final product gradually increases and will be equal to the quantity of the feed material after some time. After attaining this equilibrium condition, the quantity of the final product is always equal to the quantity of the feed material and the circulating load is constant. The circulating load, expressed as a percentage of the quantity of feed material is called **percent circulating load**.

This is explained as follows by taking an example. Consider a crusher reduces 75% of feed material to −2″ size and assume 100% screen efficiency. For an initial feed of 100 tons, crusher product contains 75 tons of undersize material and 25 tons of oversize material. The oversize of 25 tons is returned to the same crusher. Furthermore, assume that the crusher reduces 75% of this 25 tons to −2″ on the second time. Hence, 25% of 25 tons (6.25 tons) is returned to the crusher as oversize. On the next pass, 25% of 6.25 tons (1.56 tons) is returned to the crusher as oversize and so on. It reaches a stage where oversize fraction returned to the crusher is constant afterwards. This is called circulating load. The crushing operation when the circulating load is constant can be considered as equilibrium stage. Details of calculations are shown in the Table 24.

Table 24: Closed-circuit crushing calculations

Stage No.	tph new feed	tph Oversize from previous stage	tph Total feed to this stage	Products from this stage	
				tph Oversize	tph Undersize
1	100	---	100.0	25.0	75.0
2	100	25.0	125.0	31.25	93.75
3	100	31.25	131.25	32.81	98.44
4	100	32.81	132.81	33.20	99.61
5	100	33.20	133.20	33.30	99.90
6	100	33.30	133.30	33.325	99.975
7	100	33.325	133.325	33.331	99.994
8	100	33.331	133.331	33.333	99.998
9	100	33.333	133.333	33.3332	99.9998
∞	100	33.3333	133.3333	33.3333	100.0

The objectives of employing closed-circuit operation are to minimise the production of fines and to reduce the energy consumption by avoiding size reduction of already reduced particles to the required size.

Closed-circuit crushing operation when screen efficiency is 100% can be represented as shown in Fig. 57.

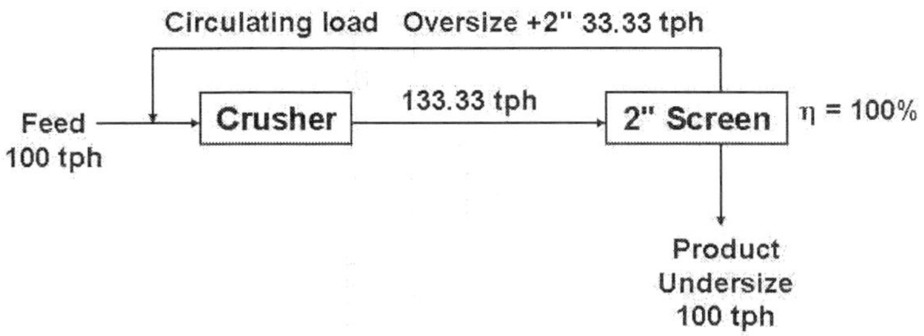

Fig. 57 Closed-circuit crushing details when screen efficiency is 100%

Let C be the circulating load. The amount of material crushed in a crusher at equilibrium stage is (100 + C). As already assumed, the crusher product contains 25% oversize. This oversize is retained on the screen as 100% screen efficiency is assumed.

$$\therefore C = (100 + C) \times 0.25$$

$$\Rightarrow C = 33.33 \text{ tph}$$

Thus, to crush 100 tph of ore by closed-circuit crushing operation, it is necessary to use a crusher of capacity 133.33 tph. The circulating load is 33.33 tph.

When the screen efficiency is assumed as 90%, then

$$C = \frac{(100 + C) \times 0.25}{0.90} \Rightarrow C = 38.46 \text{ tph}$$

In this case, a crusher of 138.46 tph capacity is necessary to crush 100 tph of ore. The circulating load is 38.46 tph. Closed-circuit crushing operation when screen efficiency is 90% can be represented as shown in Fig. 58.

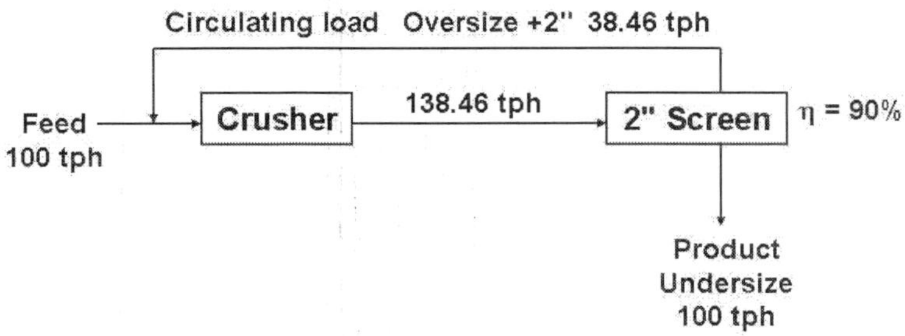

Fig. 58 Closed-circuit crushing details when screen efficiency is 90%

The most common closed-circuit crushing operation is considered for the calculations in example 8.6.2.

Example 8.6.2: *A 3" square screen is in closed circuit with a gyratory crusher which is fed at a rate of 200 tons/h. If the crusher product consists 70% of -3" particles and the screen efficiency is 90% based on undersize material, draw the circuit diagram in detail and calculate the circulating load. Also calculate the circulating load if the screen efficiency is 85% based on oversize.*

Solution:

Circuit diagram is drawn as in Fig. 59

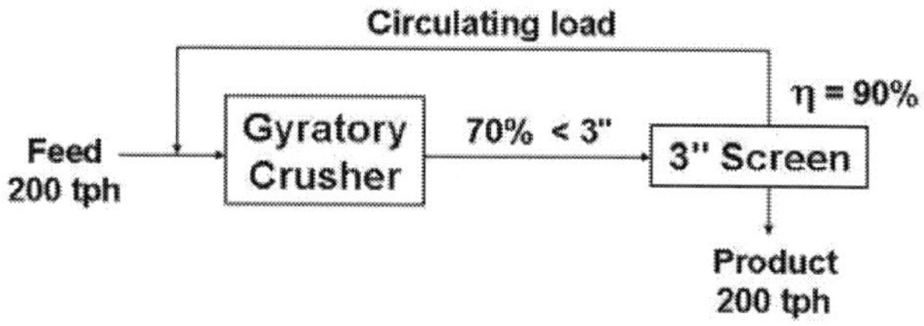

Fig. 59 Circuit diagram for example 27

Let C be the circulating load.

Load on the gyratory crusher = C + 200 tons/h

–3″ material in crusher product = 0.70(C + 200) tons/h

$$\text{Screen efficiency} = \eta = 0.90 = \frac{-3'' \text{ fraction obtained from the screen}}{-3'' \text{ present in the feed to the screen}}$$

$$= \frac{200}{0.70(C + 200)}$$

$$\Rightarrow \qquad C = 117.46 \text{ tons/h}$$

Circulating load if screen efficiency is based on undersize = 117.46 tons/h

+3″ material in crusher product = 0.30 (C+200) tons/h

$$\text{Screen efficiency} = \eta = 0.85 = \frac{+3'' \text{ present in feed to the screen}}{+3'' \text{ fraction obtained from the screen}}$$

$$= \frac{0.30(C + 200)}{C}$$

$$\Rightarrow \qquad C = 109 \text{ tons/h}$$

Circulating load if screen efficiency is based on oversize = 109 tons/h

In the reverse-type closed-circuit crushing operation, the new feed is fed to the screen, overflow from the screen is fed to crusher and crusher product is re-fed to the screen. The screen overflow which is fed to the crusher and fed back to the screen is the circulating load. The screen underflow is the final product and is equal to the new feed. The reverse-type circuit is usually employed in the case where the feed contains a relatively high percent undersize which needs to be removed prior to its introduction into the crusher.

Fig. 60 shows the typical example of reverse-type closed-circuit crushing operation when screen efficiency is assumed as 100%.

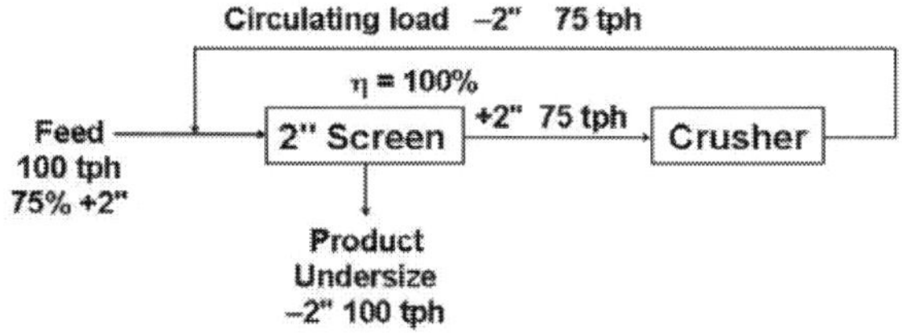

Fig. 60 Closed circuit crushing operation

It is to be noted that in this type of a crushing circuit, not only the crushing of undersize material of the feed material (25% in this case) is avoided but also the lower capacity crusher (75 tph) is required to install.

Example 8.6.3 illustrates the calculations when the screen efficiency is 85% and the crusher in closed circuit crushes the material not to all pass required size (–¾").

Example 8.6.3: *Manganese ore at the rate of 300 tons/h is crushed in a gyratory crusher set at 4" to a size 58% of +1½" and fed to a 1½" screen of 75% efficiency based on oversize material. The overflow of the screen is crushed in another gyratory crusher set at 2" to a size 30% of +1½" and re-fed to the screen along with the product of first gyratory crusher. If the finished product is 300 tons/h of –1½" size, draw the flow diagram and calculate the total load on the screen.*

Solution: Flow diagram is drawn as in Fig. 61

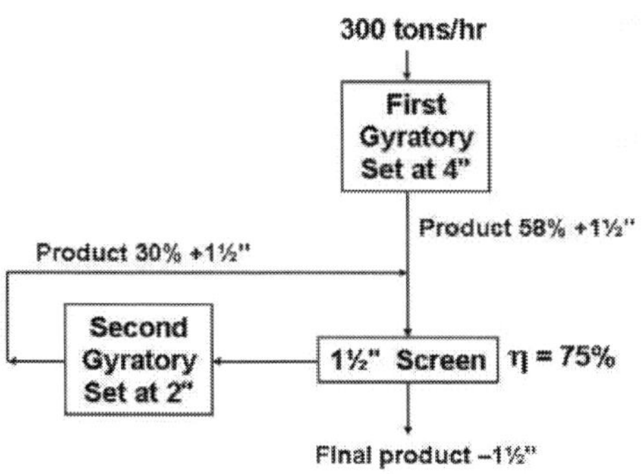

Fig. 61 Flow diagram for example 28

Let C be the overflow of screen (or) circulating load

Weight of +1½″ material in feed to screen = 0.30C + 0.58 × 300 = 0.30C + 174

$$\text{Screen efficiency} = \eta = 0.75 = \frac{\text{Weight of } +1.5'' \text{ material in feed to the screen}}{\text{Weight of } +1.5'' \text{ fraction obtained from the screen}}$$

$$= \frac{0.3C + 174}{C} \Rightarrow C = 386.67 \text{ tons/h}$$

∴ Total load on the screen = 300 + 386.67 = 686.67 tons/h

8.7 HAMMER MILL

Hammer mill, variously known as swing hammer crusher, pulverator or impactor, is a type of impact crushers. Impact crusher reduce the particles by impact forces applied through sharp blows of fixed or free swinging hammers revolving about central rotor at high speed to the free-falling particles against stationary surfaces.

When the material broken by compression pressure there are internal stresses in the material broken which can later cause cracking. Impact causes immediate fracture with no residual stresses. This stress-free condition is particularly valuable in stone used for brick making, building and road making. Impact crushers, therefore, are used widely in the quarrying industry than the metal mining industry. A great advantage in quarrying is that they produce a very good cubic product.

The hammer mill (Fig. 62) consists essentially of a high-speed rotor turning inside a cylindrical casing. The rotor is mounted on a shaft which is usually horizontal. The swing hammers are pivoted to a rotor disk. Several rotor disks each carrying four to eight swing hammers are often mounted on a single shaft. As the hammers are hinged, the presence of any hard material does not cause damage to the equipment. The hammers are readily replaced when they wear out.

The hammer mill operates on the principle that most materials will be crushed or pulverised upon impact. Material fed into the crushing chamber by gravity through a top feed opening falls in to the path of rotation of the hammers. Initial breakage is accomplished in mid-air by collision of the dropping feed material with high-speed hammers. The second stage of breakage occurs when the pieces hit the breaker plates fixed around the periphery of the cylindrical casing as well as particle-on-particle impacts. The crushed product cannot leave the perforated metal screens or bar grates fixed across the outlet until it is smaller than the screen aperture; this results in shearing and attrition action between free-swinging hammers and grates to fracture the particles by abrasion. The particles which are sufficiently broken will pass through the grate as finished product.

Some hammer mills are reversible in the sense that both sides of the mill are identical, permitting the hammer mill to be run in either direction to even out wear. The reversible hammer mill (Fig. 63) is fed at the centre of the top. The material drops to rotating hammers which hit it and drive it to breaker plates at the side of the mill corresponding to the direction of rotation. Breaker plates are often designed to cause the material to be repeatedly struck before it leaves this zone. A product not crushed to size and uncrushable will pass through an opening in the grate at the bottom of the hammer circle.

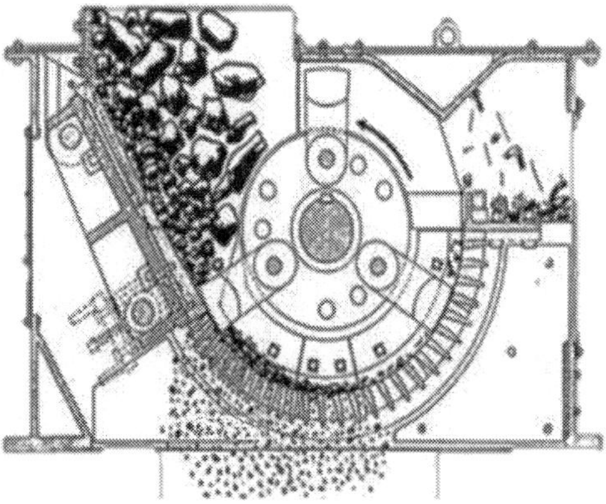

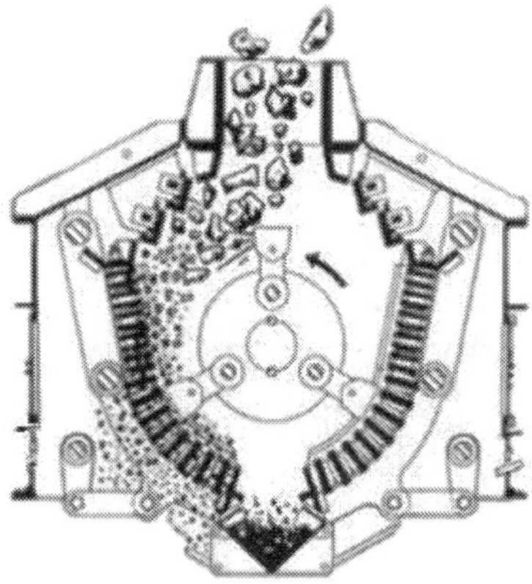

Fig. 62 Hammer mill **Fig. 63** Reversible hammer mill
(Courtesy of Pennsylvania Crusher Corporation)

Varying the screen size, shaft speed or hammer configuration can dramatically alter the finished size of the product being crushed. Faster speed, a smaller screen and more hammers result in a finer end product. Each component can be changed individually or in any combination to produce the precise size of material required.

For fine crushing, non-reversible hammer mills are arranged so that hammers strike the feed on the upswing. As a result, due to arrangement of breaker plates, particles have a greater chance for more than one blow before entering the grate circle. Pivoted hammers exert less force than they would if rigidly attached. Therefore, they are used on smaller impact crushers or for crushing soft material. Fixed hammer impact mill is often used for much coarser crushing.

The hammer mill is used either as a one-step primary crusher for reducing run-of-quarry material to a size less than 1 inch or as a secondary crusher taking a primary crusher product of 4–8 inches down to less than three-fourths of an inch. Its use as a rock crusher is almost wholly confined to the softer, easily crushable materials such as phosphates, gypsum, barite, asbestos rock and the like. As a secondary crusher, it is used for more abrasive material. The mill is particularly useful for clayey material that would clog reciprocating type primary crushers. They are extensively used in limestone quarrying and coal crushing.

8.8 SPECIAL CRUSHERS

Rotary breaker, Toothed roll crusher and Stamp mill are three important crushers specially designed for crushing the materials like coal and gold ores.

8.8.1 Rotary breaker

The rotary breaker was named as the Bradford breaker, after its inventor Hezekiah Bradford, in 1893. It is a combined breaker and cleaner, though it is used primarily as a breaker in some instances. When large tonnages of coal are treated, the rotary coal breaker is often used.

A rotary breaker (Fig. 64) is very similar in operation to the cylindrical trommel screen, rotates around its longitudinal axis in a horizontal position at a speed of 12–18 revolutions per minute. The cylinder is constructed with perforated walls, the size of the perforations being the size to which the coal is to be broken. The breaker utilises differential breakage, while the coal, being much more friable than the associated stones and shales, will be broken to small sizes. The run-of-mine coal, which is fed at one end of the cylinder after being broken to a size less than the perforations, quickly falls through the holes. The larger lumps are retained and, as the breaker rotates, are lifted by side plates—known as longitudinal lifters—within the cylinder. Near the top of the cylinder, the coal slides off the plates and drops back to the bottom. Repeated drops shatter the pieces until they pass through the perforations in the shell. Large pieces of shale and stone do not break easily and are usually discharged from the other end of the breaker, cleaning the coal to a certain degree. Movement of this refuse is facilitated by fixing the side plates at a slight angle on the shell or setting the breaker such that the shaft is not exactly horizontal. In the latter case, the discharge end is little lower than the feed end.

Fig. 64 Rotary Breaker (Courtesy of Pennsylvania Crusher Corporation)

As the broken coal is quickly removed from the breaker and, owing to the relatively slight blows, exerted on the coal by drops, the production of fines in a Bradford breaker is at minimum. For hard coal, the length to diameter ratio is increased.

8.8.2 Toothed roll crusher

There are two varieties of Toothed Roll Crusher used for crushing the coal (a) Single-toothed roll crusher and (b) Double-toothed roll crusher.

(a) Single-toothed roll crusher

A single-toothed roll crusher (Fig. 65) is a simplest type of crusher and consists of a heavy cast-iron frame on which the crushing roll and stationary crushing plate are mounted. The roll shaft is gear driven by the counter shaft, which in turn is driven by the driving pulley. The crushing roll consists of a drum to which toothed segments are bolted. These segments usually have a series of long teeth spaced at intervals and various short teeth covering the entire crushing surface. The coal is squeezed between the revolving roll and the crushing plate. The long teeth act as feeders and penetrate the lumps of coal, splitting them into smaller pieces, whereas the smaller teeth make the appropriate size reduction. The design, spacing and general arrangement of these teeth considerably influence the type of product made. The crushing plate is provided with renewable wear plates bolted to the crushing plate. The crushing plate is held in position by a tension rod with a spring relief mechanism for protection against tramp iron.

Fig. 65 Sectional view of a single-toothed roll crusher (Courtesy of aimehq.org)

A single roll crusher can reduce run-of-mine coal to 1.5 inches in one stage. However, production of fines at the time of crushing is more in a single-toothed roll crusher than that of a double-toothed roll crusher. Single-toothed roll crushers have the advantages of simplicity, compactness and the ability to operate satisfactorily at high reduction ratios. The crushing action involves a combination of impact, squeezing and abrasion.

The capacity of a roll crusher varies with the setting and nature of the coal. Power consumption is very high owing to frictional forces set up by the abrasive action.

(b) Double-toothed roll crusher

A double-toothed roll crusher (Fig. 66) is similar to smooth roll crusher, but with teeth over the rolls. It consists of pair of iron or steel cylinders; one roll usually is belt driven and second roll is gear driven from the first. Floating gearmatic drive permits crushing roll adjustment over as much as 5 inches and the roll setting can be changed without stopping the crusher when a change in product size is desirable.

Tooth design and spacing are important features in the design of the toothed roll crushers. Dull and worn teeth may reduce crushing capacity, increase power consumption and decrease the yield of the desired coarser sizes. The roll face is usually made up of segments that can be replaced when necessary.

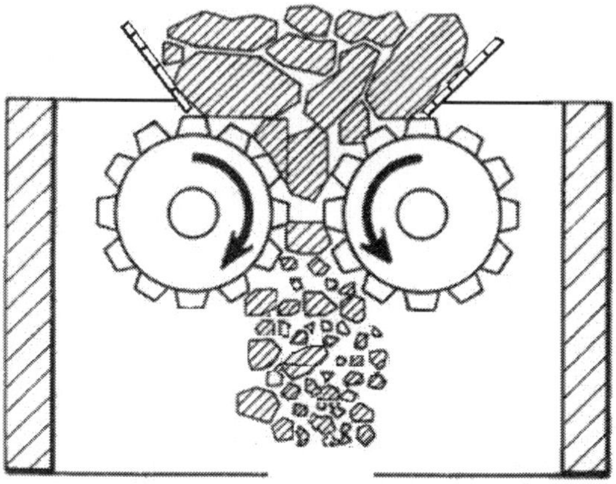

Fig. 66 Double-toothed roll crusher

Double-toothed roll crushers can handle run-of-mine coal and make a product to about 1.5-inch size and larger by suitable adjustment of the gap. To take care of stone and iron pieces damaging the crusher, one of the roll bearing blocks are provided with suitable relief springs.

Double-toothed roll crushers have certain advantages over single-toothed roll crushers:

1. Materially less power is required for its operation
2. Less fines are produced during crushing
3. By way of suitable adjustment of the opening between the rolls, a crushed coal will be obtained which will contain very little oversize coal

8.8.3 Stamp mill

A stamp mill, in principle, is a large-size mortar and pestle. It is the oldest device for fine crushing preparatory to concentration. In a stamp mill, crushing is done by the blow struck by a pestle or stamp upon the ore. The stamp comes down with accelerated motion, reaching its maximum velocity when it strikes the ore placed on the hard metal block in the mortar. The momentum of the stamp is then spent in crushing the ore. The crushed ore is splashed against an upright screen set in the mortar wall. The finer material pass out through the screen and the coarser portion is retained in the mortar for further crushing. The ore is regularly fed in the mortar and falls on the metal block. The operation is conducted in an aqueous pulp.

The two types of stamp mills are gravity stamps and steam stamps. In gravity stamps, the stamp is raised by a cam working under collar. After the stamp has been raised to the highest point, it is allowed to drop on the crushing block. Gravity stamps are not generally used except in small gold milling units where amalgamation is also performed. In a steam stamp, the raising and lowering of the stamp is actuated by the steam piston. Steam stamps are used in the treatment of the native copper ore. In both the stamp mills, the crushing is accomplished by multiple stamps acting alternatively to obtain the best possible agitation of the pulp.

8.9 HIGH-PRESSURE GRINDING ROLLS (HPGR)

High-pressure grinding rolls (Fig. 67) consist of a pair of counter-rotating rollers, i.e. a fixed and movable roller. The hydraulic pressure system acts on a piston of the movable roller and makes it move towards the fixed roller, applying high pressure for a short time on the material being crushed. The high pressure causes the formation of micro-cracks in the feed particles and generates a substantial amount of fine material. Tiny micro-cracks weaken the particles so that even if the ore is not shattered, subsequent fine grinding requires less energy to achieve final product size. The feed introduced to the gap in between the rollers is crushed by the high-pressure compressive forces as well as by the mechanism of inter-particle breakage as the feed travels between the two rollers. The force applied to the crushing zone is controlled by the hydraulic pressure system. Speeds of the rolls are also adjustable to obtain optimum grinding conditions.

The product from the HPGR can be directly fed to a ball mill. As a significant amount of work has been affected before the ball mill, it facilitates increase in throughput of the ball mill and, in most cases, significantly reduces the specific power consumption of the total milling system. Applying the HPGR process as a replacement for the conventional SAG milling circuit results in significant savings in energy costs, reduced grinding media consumption and operating costs, smaller footprint, faster equipment delivery schedules and finer product.

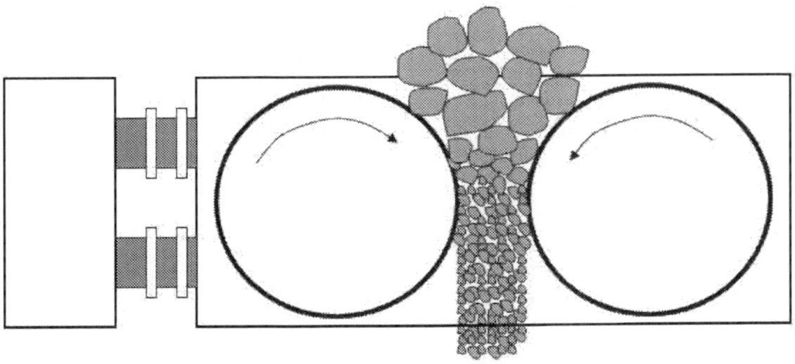

Fig. 67 High-pressure grinding rolls (HPGR)

HPGR is ideal for crushing hard rock with relatively low moisture content and can provide superior results in the size reduction of variety minerals. Currently,

HPGR technology is successfully operating in numerous applications including diamonds, iron ore, copper, gold, platinum, molybdenum and industrial minerals.

8.10 PROBLEMS FOR PRACTICE

8.10.1: *Determine the minimum diameter of the rolls in a roll crusher to reduce 3.81cm pieces of rock to 1.27 cm if the coefficient of friction between the rock and steel is 0.4.* *[31.72 cm]*

8.10.2: *In a smoothed rolls crusher, crushing rolls of 100 cm diameter each are set 12.5 cm apart. If the angle of nip is 31°, calculate the maximum size of the feed particle to the roll crusher.* *[16.75 cm]*

8.10.3: *Determine the maximum size of the limestone particles fed to crushing rolls run at 100 rpm, having rolls of 150 cm diameter and 50 cm width of the roll set at 1 cm. The angle of nip is 30° and the specific gravity of limestone is 2.66. Also calculate the theoretical capacity of the crushing rolls.*

[6.33 cm, 376.2 tph]

8.10.4: *Rolls of 100 cm diameter and 38 cm face width of a crushing rolls are set so that crushing surfaces are 1.25 cm apart. They are to crush a rock of 2.35 specific gravity and the angle of nip of 30°. As per manufacturer recommendation, they may be run at 50–100 rpm. What is the permissible size of the feed to the crusher? If the actual capacity of the crusher is 12% of the theoretical capacity, calculate the maximum actual capacity of the crusher in tons/h.* *[4.82 cm, 25.26 tph]*

8.10.5: *Data from a crushing test conducted on a sample of copper ore when it is crushed in a jaw crusher is given in Table 25.*

Table 25: Crushing test data for problem 8.10.5

FEED		PRODUCT	
Screen size	**Weight of material (g)**	**Screen size**	**Weight of material (g)**
+2.0″	3556		
−2.0″ + 1.5″	4214	+1/2″	563
−1.5″ + 1.0″	14355	−1/2″ + 3/8″	3322
−1.0″ + 3/8″	9592	−3/8″ + 4 mesh	9456
−3/8″ + 4 mesh	3370	−4 + 10 mesh	2272
−4 + 10 mesh	1857	−10	3812
−10	1906		

The aperture sizes of 4 and 10 mesh are 4760 and 1680 microns, respectively. Considering 80% passing size of the feed and the product, calculate the reduction ratio of the Jaw crusher. [4]

8.10.6: *The details of crushing plant employing gyratory crusher are as follows:*

2″ square screen is in closed circuit with crusher

New feed to crusher = 100 T.P.H

Crusher product contains 54% < 2" and 46% > 2" fed to screen

Screen efficiency = 85% (based on oversize)

Draw the flow diagram and find circulating load.

Find circulating load if the screen efficiency is based on undersize.

[117.95 tph, 117.86 tph]

8.10.7: *The circuit shown in Fig. 68 is adopted to crush an ore.*

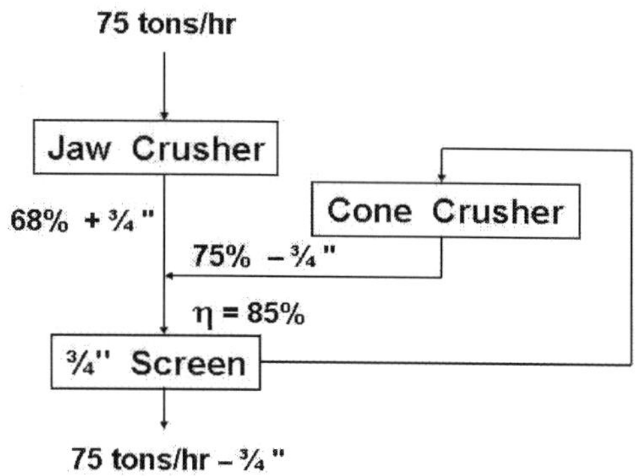

Fig. 68 Crushing circuit for problem 8.10.7

a) *At steady state condition, what is the total load goes through the cone crusher?*

b) *What is the percent circulating load?*

c) *Of the original 75 tph of jaw crusher product, how many tons bypass the cone crusher?*

d) *At steady state conditions, what is the total load which goes over the screen?*

Efficiency of the screen is based on oversize material.

[85 tons/h, 113.3%, 15 tons/h, 160 tons/h]

■■■■

9 Grinding

Grinding is the last stage of the comminution process. The particles are reduced from a maximum upper feed size of 3/8 inch, to an upper-limiting product size ranging between 35 and 200 mesh (420 and 74 microns, respectively).

Grinding is performed in rotating steel vessels known as **tumbling** or **grinding mills**. A grinding mill reduces particle size by applying impact and attrition stresses to the materials to be ground. It is designed to strike the particles with sharp blows of short duration and produce a rubbing action under as high a unit pressure as possible. Accordingly, grinding mill consists of a horizontal rotating steel shell supported by end bearings on which hallow trunnions revolve. Bearings are fastened directly to concrete piers. Loose crushing bodies—known as **grinding medium**—are placed inside the shell. Either steel balls, rods or pebbles are used as the grinding medium. They are free to move inside the rotating shell, making the particles to break by repetitive blows and by rolling and sliding one over the other. Attrition or shearing forces which result from the application of forces by rolling and sliding bodies tend to produce more fine particles than impact forces applied on particles by repetitive blows.

Grinding mills used in practice are continuous machines. Crushed ore is fed at one end of the mill and discharged at the other end or through the periphery. They can be operated wet or dry. The feed is introduced by a feeder through a hallow trunnion at the centre of the feed end. In a dry-grinding mill, the feed is by a vibrating feeder. In a wet-grinding mill, the function of a feeder is to transport pulp from a receiving point outside the mill to the mill barrel, and to do this smoothly, certainly and with sufficient driving force to overcome any tendency for the pulp to move in the opposite direction. The feeder may be required to elevate from receiving to delivery point, excavate at the receiving point and deliver grinding medium as well as ore. Four types of feeders are in use to feed the material to the wet grinding mills:

1. Spout feeder (Fig. 70 cylindrical ball mill with spout feeder)
2. Drum feeder (Fig. 69 cylindro-conical ball mill with drum feeder)
3. Scoop feeder (Fig. 78 cylindrical rod mill with scoop feeder)
4. Combination scoop-drum feeder

In a **spout feeder,** the material is fed by gravity through the spout which consists of a cylindrical chute projecting directly in to the trunnion liner. In a **drum feeder**, the entire mill feed enters the drum and an internal spiral carries it and fed to the mill. Grinding balls are conveniently added through this feeder during operation. In case of a **scoop feeder**, material is fed to the feed box and the scoop picks it up and feeds the mill. A **combination feeder** adds a scoop to a drum feeder. The scoop may be mounted on the periphery of the drum or on the spool between the drum and the mill trunnion.

Drum feeders are preferable for coarse dry feeds as they keep large particles out of the feed box, avoiding digging shocks to both box and feeder. Scoop feeders have the great advantage of serving both as feeders and elevators. The combination feeder has the advantages of both drum and scoop feeders.

The grinding mill is robust and the loose grinding medium can usually be added without stopping the mill. They are relatively high in power consumption and require expensive foundations. Grinding mills are classified as **Ball**, **Rod**, **Tube** or **Pebble**, and **Autogenous mills** on the basis of grinding medium and shell proportions.

9.1 BALL MILL

A ball mill uses spherical steel or iron balls as the grinding medium. Since balls have a greater surface area per unit weight, they are better suited for fine grinding. Ball mills usually have a length to diameter ratio of 1.0–1.5. According to the shape of the mill, ball mills are classified as cylindro-conical and cylindrical mills.

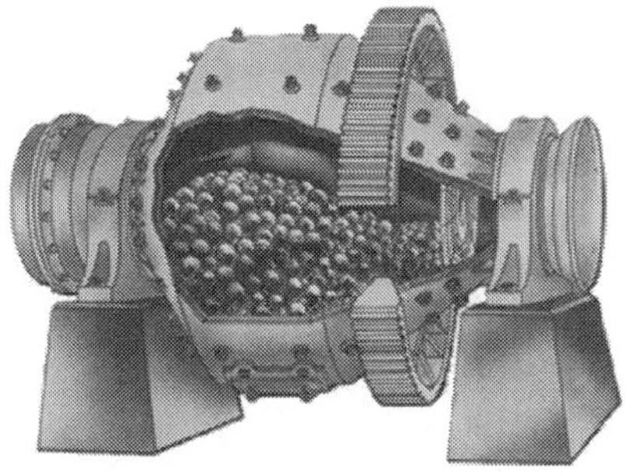

Fig. 69 Cylindro-conical ball mill (Courtesy of Metso Minerals)

Fig. 70 Cylindrical ball mill (Courtesy of aimehq.org)

A **cylindro-conical ball mill** (Fig. 69) consists of two conical sections connected by a short cylindrical section which is supported by end bearings on which hollow trunnions revolve. The conical section towards the feed end is obtuse, and that towards the discharge end is acute. The conical shape compels coarse particles and large balls occupy the cylindrical section of the mill and fine particles and small balls occupy the conical sections, especially near the discharge end of the mill. That is how the preferential grinding of coarse particles by large balls and fine particles by small balls take place. In the metallurgical field, these mills are adapted for wet grinding. They are also used for dry grinding in conjunction with pneumatic classifier for pulverizing coal, clay, limestone and cement clinker.

As shown in Fig. 70, a **cylindrical ball mill** is characterised by the cylindrical shape of the shell having uniform diameter throughout. When the grinding mill is rotated, the mixture of grinding medium, raw material, and water (in case of wet grinding) is lifted, the magnitude depends on the rotational speed of the mill.

Due to the rotation and friction of the mill shell, the balls are lifted along the rising side of the mill until a position of dynamic equilibrium is reached and then drop down to the toe of the mill charge. Grinding of ore particles takes place due to simple rolling of one ball over the other (cascading) and by the free fall of balls (cataracting). Cascading leads to fine grinding whereas cataracting leads to coarse grinding. Fig. 71 illustrates the motion of the charge in ball mill.

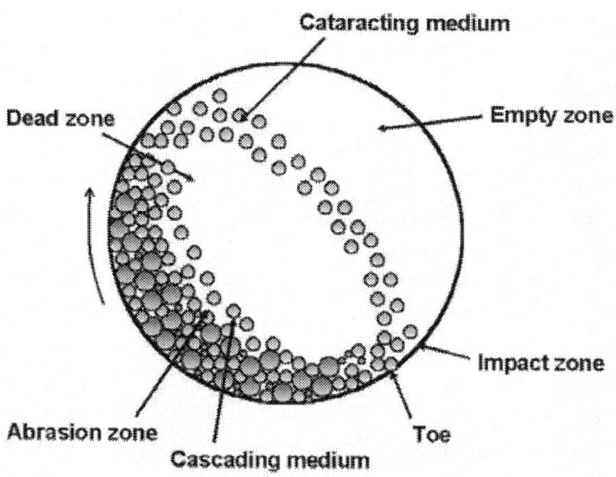

Fig. 71 Motion of the charge in ball mill

As the speed of the mill increases, the balls are lifted to a greater height and a stage is reached where the balls are carried around the shell and never allowed to fall. That means centrifuging occurs. The balls will rotate as if they are part of the shell. There will be no relative movement of the balls. The speed at which centrifuging occurs is known as **critical speed**.

Let m = mass of the ball in kg

R = Radius of the mill shell in metres

N = speed of the mill in rpm

v = linear velocity of the ball, m/s = $\frac{2\pi R N}{60}$

Let us consider a ball which is lifted in the shell. When the ball moves around inside the mill, it follows a circular path till it reaches certain height and then changes its circular path and follows a parabolic path while dropping on to the toe (Fig. 72).

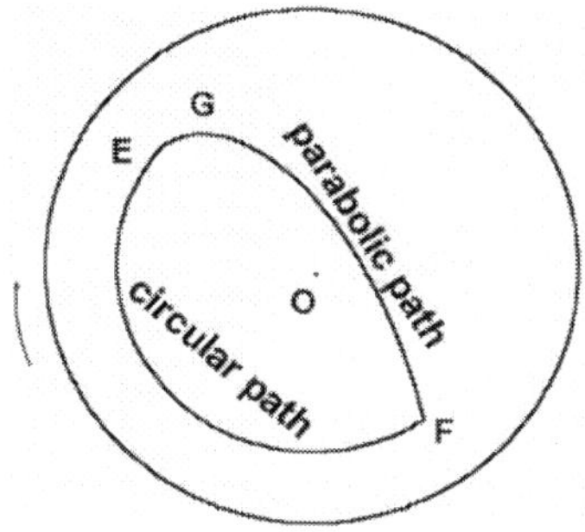

Fig. 72 Path of a ball

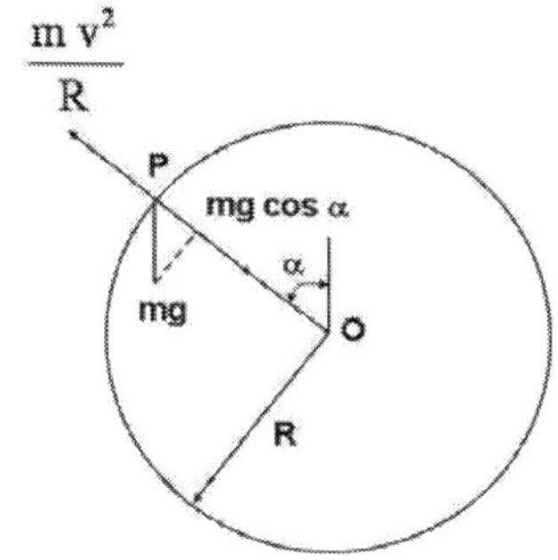

Fig. 73 Forces on a ball

When the ball is at the point of changing its path, centrifugal and centripetal forces acting on it will balance. The centrifugal force acting on the ball is $\frac{\mathrm{m\ v^2}}{\mathrm{R}}$. The component of weight of the body opposite to the direction of centrifugal force is the centripetal force, i.e. mgcosα, where α is the angle made by the line of action of centripetal force with the vertical. P is a point where the ball changes its circular path to parabolic path. At this point, both the forces are equal (Fig. 73).

Therefore, $$\frac{\mathrm{m\ v^2}}{\mathrm{R}} = mg\cos\alpha$$

$$\Rightarrow \frac{\left(\frac{2\,\pi\,\mathrm{R\,N}}{60}\right)^2}{\mathrm{R}} = g\cos\alpha \Rightarrow N^2 = \frac{(60)^2\ \mathrm{g}\cos\alpha}{4\,\pi^2 R}$$

If D is the diameter of the mill and d is the diameter of the ball, the radius of the circular path of outer most ball is $\frac{(\mathrm{D-d})}{2}$. On substitution, we get

$$N^2 = \frac{(60)^2\ \mathrm{g}\cos\alpha}{2\,\pi^2(D-d)} \quad 9.1$$

Critical speed of the mill occurs when $\alpha = 0$, i.e. the ball will rotate along with the mill. At this point, $\cos\alpha = 1$. Then, $\mathrm{N = N_c}$.

$$\therefore\ N_c{}^2 = \frac{(60)^2\ \mathrm{g}}{2\,\pi^2(D-d)} \quad 9.2$$

If D and d are taken in metres, it is reduced to

$$\text{Critical speed} = N_c = \frac{42.3}{\sqrt{(D-d)}}\ \text{revolutions/min} \quad 9.3$$

If D and d are taken in feet, the same expression is reduced to

$$\text{Critical speed} = N_c = \frac{76.65}{\sqrt{(D-d)}} \text{ revolutions/min} \qquad 9.4$$

Davis has provided an exhaustive mathematical analysis of the action in a ball mill and explained various zones in a ball mill (Fig. 74).

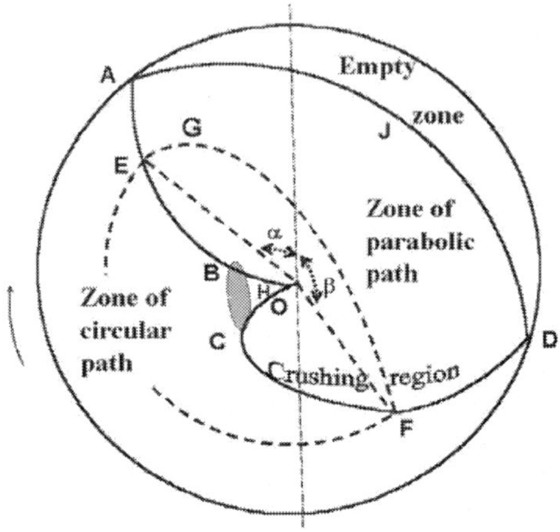

Fig. 74 Zones in a ball mill

As shown in Fig. 74, FE is the circular path and EGF is the parabolic path of a ball placed at point E. The locus of the point E where the ball changes from a circular to parabolic path for all positions of the balls from the centre of the mill to the periphery is OBEA. The locus of the point F, i.e. the end of the parabolic path for all positions of the balls, is DFCO. Davis has shown that the arcs CO and BO correspond to unstable equilibrium. The zone BCH is a dead zone where there is no effective motion, hence no grinding takes place. Thus, the inside of the ball mill consists of four zones:

(a) an empty zone where no balls occupy this zone during operation

(b) a dead zone where no grinding takes place

(c) a zone of circular path where balls roll on each other and grinding takes place by slippage between ball layers before they are lifted

(d) a zone of parabolic path where the balls spread out and down

When mill speed exceeds critical speed, all these zones disappear, the balls will centrifuge and no grinding takes place anywhere inside the mill. Therefore, the mill should be operated below critical speed. Speed of ball mills should be as high as possible without centrifuging the charge (grinding medium, ore and water). As the speed of the mill is increased, energy input increases at first in proportion to the speed. Then, as slippage increases, the energy input increases much slower than the speed. The energy input increases until a critical speed is reached. Beyond critical speed, the energy input rapidly decreases to vanishing point, as shown in the graph of

Fig. 75. When this condition is reached, the solids are centrifuged on the mill shell and there is no work done by the mill. The usual range of the ball mill speed is 60–80% of the critical speed.

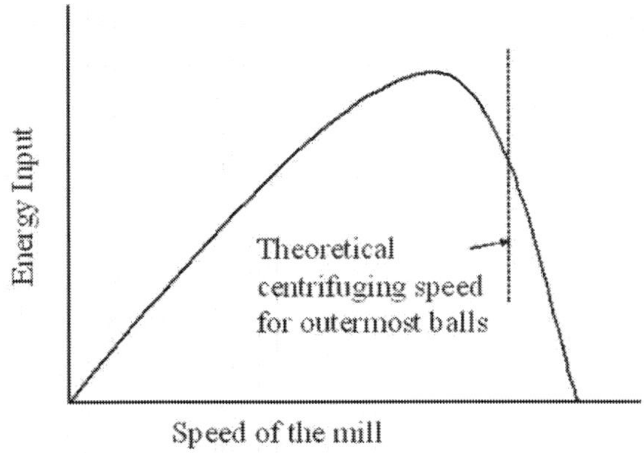

Fig. 75 Relationship of energy input to the ball mill to speed of the mill

In examples 9.1.1 and 9.1.2, calculations concerned to critical speed are illustrated.

Example 9.1.1: *Calculate the critical speed of a ball mill having 600 mm diameter with 25-mm-size balls.*

Solution:

Diameter of the ball mill = 600 mm = 0.6 m

Diameter of the balls = 25 mm = 0.02 m

Critical speed of the ball mill $= N_c = \dfrac{42.3}{\sqrt{(D-d)}}$

$$= \frac{42.3}{\sqrt{(0.6-0.02)}} = 55.51 \text{ rpm}$$

Example 9.1.2: *Determine the operating speed of a ball mill of 1800 mm diameter containing steel balls of 90 mm diameter. The mill operates at 70% critical speed.*

Solution:

Diameter of the ball mill = 1800 mm = 1.8 m

Diameter of the balls = 90 mm = 0.09 m

Critical speed of the ball mill $= N_c = \dfrac{42.3}{\sqrt{(D-d)}}$

$$= \frac{42.3}{\sqrt{(1.8 - 0.09)}} = 32.29 \text{ rpm}$$

Operating speed $= 70\%$ of critical speed

$$= 0.7 \times 32.29 = 22.6 \text{ rpm}$$

The discharge of the ground product from the mill is primarily of two types: overflow discharge (Fig. 78) and grate discharge (Fig. 70). In an overflow discharge mill, the ground product pass through a hallow trunnion at the discharge end of the mill. In a grate discharge mill, the ground product passes through the slots of a retaining grid—or grate —fixed at the discharge end of the mill. Another method for discharging ground material especially in rod mills is by peripheral discharge—either end- or centre-peripheral discharge. This type of product discharge is used in both wet and dry grinding operations. However, it is more common in dry grinding applications.

The ball load is the working part of the mill. It draws substantially the same amount of power whether it does any useful work or not. The amount of useful work that it does differs with the shape of the grinding bodies, their size relative to the grains being ground, their quantity, and the kind of material they are made of. Loss by wear is one of the principal items of grinding expense. Wear of balls depends upon the material of the ball and the method of manufacture, the material being ground, mill diameter, speed of the mill, feed rate, pulp density, liner surface and ball diameter. The ball wear is usually between 1 and 3 pounds of metal per ton of ore ground.

Balls vary in size from 1 to 6 inches. The largest balls are used for coarser grinding. Initially, at the starting of a ball mill, balls of various sizes, known as **seasoned charge**, are introduced. As the balls are wear out gradually, only largest balls are added as make up medium.

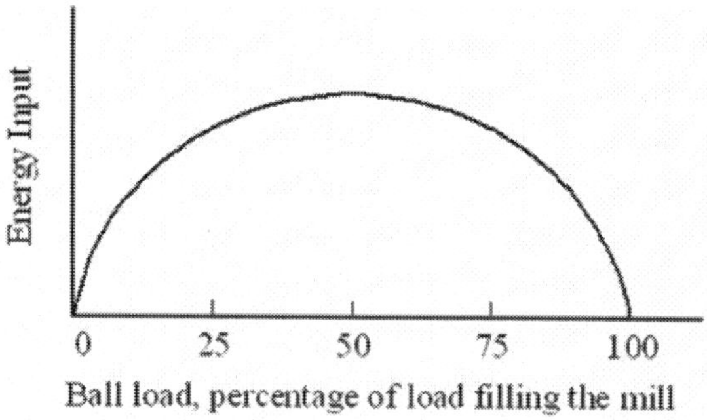

Fig. 76 Relationship of energy input to the ball mill to ball load

The volume of the grinding medium usually occupies 50% of the mill volume including voids. Normal operating loads range from 50% down to 40% of the mill volume. In case of ball mill, the mill should be slightly more than half full with material and balls. However, the volume contributed by water and material will be comparatively very small since they occupy the interstices between the balls. If the ball load increases, the power consumed by the mill increases at first until maximum is reached and after this stage the power consumption decreases gradually to zero as the centre of gravity of the load approaches the axis of rotation (Fig. 76).

Replaceable lining is used on all surfaces with which pulp comes in contact to protect the mill body from wear and damage. Shell lining acts as a final link in the transmission of energy to the grinding load. When the load slips, energy input and consequently grinding are decreased. Slip increases with decrease in grinding load, speed, pulp density and hardness of the ore. Roughness of liner is more necessary for maximum capacity of the mill. Shell liners differ in type according to whether the mill feed is coarse or fine. In general, ribbed liners are used for coarse feeds, non-ribbed for fine products and smooth for ultra-fine products. Some types of liners are smooth, ribbed, wave, shiplap, wedge bar, stepped, etc. (Fig. 77).

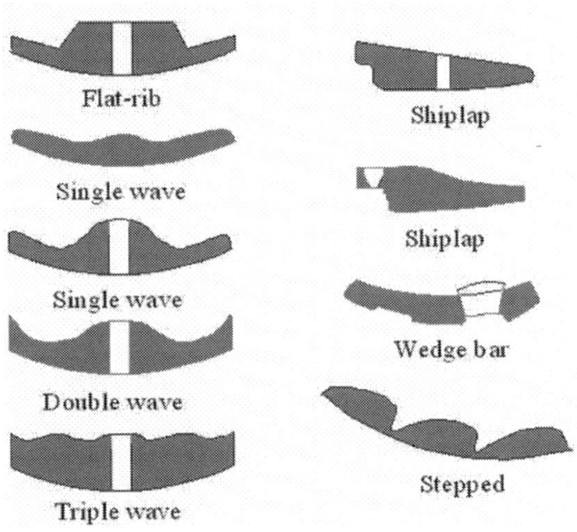

Fig. 77 Few types of shell liners

Smooth liners favour abrasion resulting fine grinding but of high metal wear. Liners other than smooth are designed to help in lifting the ball load as the mill is revolved and to minimize the slip between the layers of balls. They are usually made in blocks or other shapes of such size that they can be introduced into the interior of the mill through a manhole or the discharge end opening. End liners are sectioned radially or in rings or may be cast in one piece. They are used for protective purposes and are usually smooth. Liners are usually made of alloy steel but sometimes of rubber.

Liner wear ranges from 0.1 to 0.5 pounds per ton of ore ground. Liners are more expensive per pound than balls because of their shape. In addition, their installation involves a shutdown expense. Liner wear is more variable than ball wear perhaps because it is affected to a greater extent by mill size and design. Overflow mills and mills of largest volume, other things being equal, have the least liner wear.

The grinding may be wet or dry depending on the subsequent process and the nature of the product. Wet grinding is generally used in mineral processing plants as subsequent operations for most of the ores are carried out wet. Wet grinding is usually carried out with 60%–75% solids by weight. Following are the advantages of wet grinding when compared to dry grinding:

1. Power consumption per ton is less. Less power consumption is due to the penetration of water into the cracks of the particles which reduces the bond strength at the crack tip. Bond index is 75% of dry index.
2. Capacity of the mill is increased by as much as 15%.
3. As the grinding takes place in water, dust is not produced which is harmful to human health.
4. Higher speeds of rotations are possible. Mill can run up to 75%–80% of the critical speed in the case of wet grinding whereas in dry grinding maximum limit is only 65%–70%.
5. As the material can be introduced as a slurry, feed control is easily automated and pumps, pipelines and launders can be used for distribution.
6. Strict control of particle size can be exercised by using wet screening or classification.
7. Surface active agents can be more easily introduced and dispersed to assist rate of grinding.
8. Certain materials may be resistant for wetting in a finely divided state. In such cases wetting can be easily obtained as the grinding itself is wet.
9. It requires less space than for dry grinding.
10. Installation costs are generally less for a closed-circuit operation.

Disadvantages compared with dry grinding are:

1. The wear on the grinding medium and the liners is usually more.
2. Less material in the finest size range is produced.
3. Sometimes more flocculation of particles takes place.
4. Drying a ground material is expensive.
5. Wet process cannot be instantaneously shut down and therefore is less convenient for shift operations.
6. Requires large quantities of water and pump maintenance is also high.

Solids–liquid ratio in wet grinding influences the grinding efficiency and power consumption. The energy input is maximum for a certain critical range of solids-liquid ratio of the pulp, which is usually obtained in a pulp consisting of 60%–75% solids. However, this range depends on the chemical composition, particle size, and specific gravity of ore. To make the most efficient use of a grinding mill to grind ore, the ore particles must form a coating on both the media and liners. At the same time, the pulp must be fluid enough to flow through the mill. At the appropriate solids to liquid ratio there is enough interference with settling to prevent even quite large particles from settling out. If the pulp becomes too dilute, that is it contains too much water, settling and centrifuging of coarse ore particles takes place and the particle coating on the media is greatly reduced. In such circumstances, void spaces in the outer strings of balls become filled with solids, ball movement relative to each other and to the mill shell becomes restricted and the grinding rate is reduced. Finely ground dry ore will flow like a fluid, but in the presence of 2%–15% moisture the mass loses its fluidity and becomes quite sticky and viscous. Under these conditions the pulp cannot flow and a blockage follows. When the moisture reaches 20% by weight of the pulp, the flow through the mill is good and at about 25% moisture by weight the grinding capacity nears its maximum all other conditions being constant. When moisture goes above 40%, the pulp becomes watery, chances of coating the grinding media reduces and results in high steel consumption and lowered capacity. Generally speaking and within reasonable limits, the thicker the pulp the less wear and tear on liners and the higher the capacity.

Dry grinding is used whenever physical or chemical changes in the material occur if water is added. It causes less wear on the liners and grinding medium. Dry grinding mills are often employed to produce an extremely fine product. This arises from the high settling rate of solids suspended in air as compared with solids suspended in water.

9.2 ROD MILL

Rod mill (Fig. 78) is similar to ball mill in appearance and in general principle. Rod mill uses rods as grinding medium. Length to diameter ratio is between 1.5 and 2.5. Rods are few centimetres shorter than the length of the mill to avoid jamming of rods in the mill.

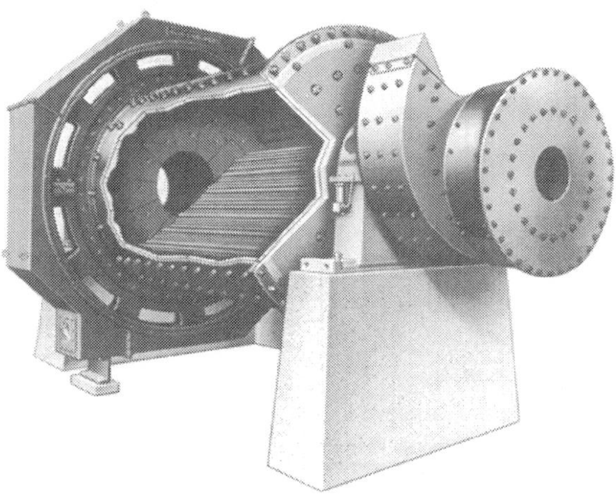

Fig. 78 Rod mill (Courtesy of aimehq.org)

The rods are kept apart by the coarsest particles (Fig. 79). The grinding action results from line contact of the rods on the ore particles and is exerted preferentially on the coarsest particles. Smaller and fine particles do not ground till the coarsest particle is reduced in size. Thus, the rod mill produces a more closely sized product with little oversize or slimes. Hence the rod mills may be considered as coarse grinding machines. They are used to produce feed material for the concentration processes that fail on fine particles.

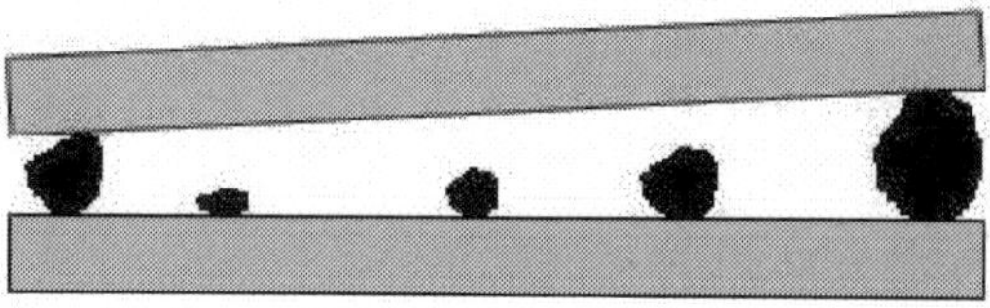

Fig. 79 Grinding action of rods

If the material inside the mill is made to discharge through the periphery of the mill shell, it passes quickly and less over-grinding takes place. Rod mills are classed according to the method of the discharge of ground product as **Overflow, Grate** and **Peripheral discharge mills**. Overflow and Grate discharge mills are described in article 9.1. There are two types in peripheral discharge mills: (a) end peripheral discharge and (b) centre peripheral discharge mills as shown in Fig. 80.

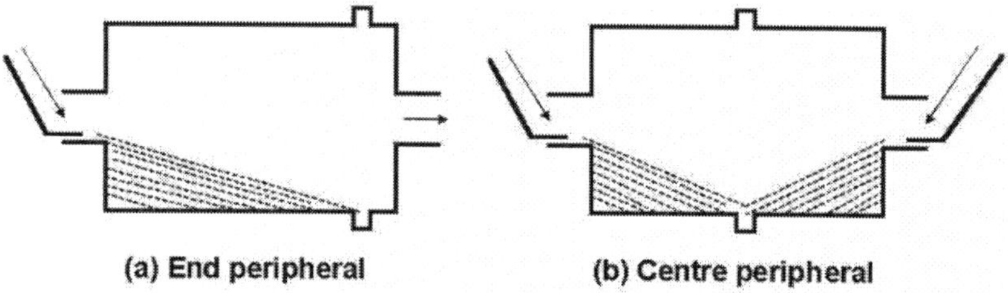

Fig. 80 Peripheral discharge methods

In end-peripheral discharge mills, the material is fed at one end of the mill and the ground product is discharged from the other end by means of several peripheral apertures into a close-fitting circumferential chute. In centre-peripheral discharge mills, the material is fed at both ends of the mill and the ground product is discharged through circumferential ports at the centre of the shell.

Rod mills are normally run at 50%–65% of the critical speed so that the rods cascade rather than cataract. In wet grinding rod mills, pulp density ranges from about 75% solids for coarse feeds to 60% in fine-intermediate grinding. The rod wear is usually between 0.2 and 2 pounds of metal per ton of ore ground.

9.3 TUBE OR PEBBLE MILL

Tube mill is the grinding mill where length to diameter ratio is 3 to 5. It is also known as pebble mill as it uses ceramic pebbles made of flint or porcelain as the grinding medium. Flint pebbles are fine grained quartzite. As pebbles are fragile, pebble mills are of smaller in diameter. Because of low specific gravity of pebbles, large quantity of pebbles must be used to attain required grinding for equivalent duty as compared with ball mills. Pebble mills are used when iron contamination in the product is highly objectionable such as in the manufacture of paints, pigments, cosmetics etc.

Tube mills are sometimes divided into several longitudinal compartments, each having a different charge composition. The charge can be steel balls, rods or pebbles and they are often used dry to grind cement clinker, gypsum and phosphate.

9.4 AUTOGENOUS MILL

Autogenous mills use coarse ore particles as grinding medium. Autogenous grinding is defined as the action of a material grinding upon itself, as occurs when pieces of ore of different sizes are rotated together in a tumbling mill. It is also called ore-on-ore grinding. However, the same action takes place in pebble mill grinding when sized pieces of the ore are used as grinding pebbles. Grinding mills that receive sized ore as grinding medium, including pieces of ore of selected sizes separated by screening, are called pebble mills. Autogenous mills are of very large diameter range up to 36 feet, with lengths as little as one-third, or even one-fourth of the mill diameter. The larger pieces in the feed must be sufficient in size and number to break the smaller pieces as fast as they themselves are broken down in the mill. The only metal wear is that of the mill linings, including the mill grate.

Autogenous grinding greatly reduces metal wear and also reduces the number of crushing and grinding stages as compared with conventional size-reduction operations, thus it offers the possibilities of large savings in capital and operating costs. However, it is practical for only a limited number of ore types. The large pieces of ore must be abundant in the broken mill feed, and they must be relatively resistant to breakage in the tumbling mill. The finer sizes of the ore should not be harder to break than in the coarse, preferably the fines should break more easily. Ores of high specific gravity, such as iron ores, grind well autogenously, especially so if the large pieces do not have appreciably lower specific gravities than the small particles.

Autogenous grinding differs from grinding with metal balls or rods in that the breakage is much more confined to zones of weakness in the rock, such as crystal surfaces and fine cracks. An ore with a pronounced natural grain size, or crystal size, may be ground autogenously to that size with comparative ease, but is ground finer only with difficulty.

It is one practice to do only the coarse grinding autogenously, and finish-grind the autogenous mill product in ball mills. This reduces the cost saving by increasing the amount of metal wear loss and the number of size reduction stages. Another practice is to obtain a coarse discharge from the autogenous mill by partial or overall discharge through peripheral ports. The peripheral discharge material is screened at the appropriate ball mill feed size, and the screen oversize is returned to the autogenous mill in closed circuit.

Autogenous mills designed for coarse-ore grinding have been modified by adding large diameter steel balls to the charge, making them partially autogenous, called **Semi-autogenous grinding**. Such mills are called semi-autogenous mills, in short, SAG mills. Those autogenous mills use only coarse ore are called fully autogenous mills, in short, FAG mills.

9.5 GRINDING CIRCUITS

Mesh of grind (m.o.g) is the term used to designate the size of the grounded product in terms of the percentage of the material passing a given mesh. **Optimum mesh of grind** defines the mesh of grind at which a maximum profit is made on sales, when both the working costs and the effect of grinding on the recovery of values have been considered. In grinding, there are always some particles which may repeatedly be reduced to fine size whereas some other particles may not be reduced. The primary objective of a grinding mill is to reduce all particles to the stated size i.e. mesh of grind. When a grinding mill is fed with a material, it should be at a rate calculated to produce the correct product in one pass in which case it is known as **Open circuit grinding** (Fig. 81).

Fig. 81 Open circuit grinding

There is no control on product size distribution in the open circuit grinding. If the residence time of the material in the mill is reduced, it results underground product with low degree of liberation. If the residence time is increased, it results over-ground product. Much energy is wasted in the process by grinding the material to a size below the size required. If the material is removed from the mill as soon as it is grounded to the required size, over-grinding can be eliminated or minimized. This is done by employing a classifier. This operation is known as **Closed-circuit grinding** operation (Fig. 82).

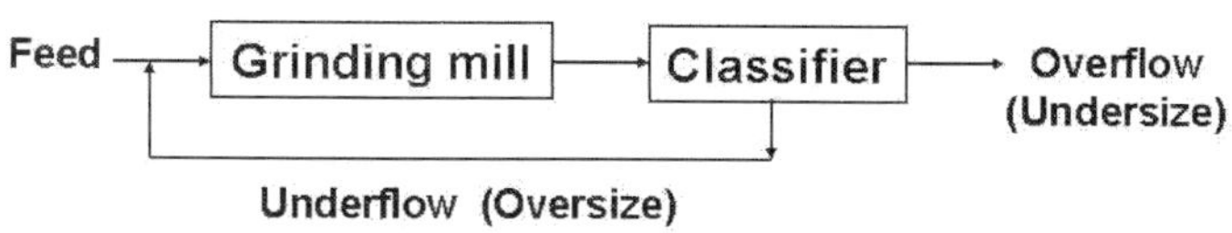

Fig. 82 Closed-circuit grinding

Separation takes place in the classifier based on the principle of settling of solids in fluids. Classifier separates the feed material into two fractions—fine or undersize material and coarse or oversize material—wherein oversize material is returned to the same grinding mill and undersize material is the finished product. The tonnage or weight flow rate of the overflow or undersize solids of a classifier is always equal to the tonnage or weight flow rate of the solids fed to the grinding mill in a closed-circuit operation under steady state conditions.

A material balance equation can be written over the classifier as follows:

Slurry balance $\quad F = P + U \quad$ 9.5

Solids balance $\quad F\,f = P\,p + U\,u \quad$ 9.6

where F, P and U are the pulp or slurry flow rates of the feed, overflow (or undersize) and underflow (or oversize) of the classifier respectively and f, p and u are the fraction or percent solids in the feed, overflow and underflow of the classifier.

$$\text{Solving 9.5 and 9.6} \quad \Rightarrow \quad \frac{U}{P} = \frac{f-p}{u-f} \qquad 9.7$$

The tonnage or weight flow rate of the solids fed back or returned to the grinding mill is called **circulating load** and its weight expressed as a percentage of the tonnage or weight flow rate of new feed solids fed to the grinding mill is called **%circulating load**.

% circulating load

$$= \frac{\text{Tonnage or weight flow rate of underflow solids from classifier}}{\text{Tonnage or weight flow rate of new feed solids fed to grinding mill}} \times 100$$

$$= \frac{Uu}{Pp} \times 100 = \frac{u(f-p)}{p(u-f)} \times 100 \qquad 9.8$$

If F, P and U are the weight flow rates of solids in feed, overflow and underflow, and DR_F, DR_P and DR_U are the dilution ratios of feed, overflow and underflow, then

Solids balance $\quad F = P + U \quad$ 9.9

Water balance $\quad F \times DR_F = P \times DR_P + U \times DR_U \quad$ 9.10

$$\text{\% circulating load} = \frac{U}{NF} \times 100 = \frac{U}{P} \times 100 \qquad 9.11$$

where NF = Weight flow rate of new feed solids fed to the mill

= Weight flow rate of overflow solids discharged from the classifier

Percent circulating load is called as **circulating ratio** if expressed as ratio.

Solving equations 9.9 and 9.10,

$$\text{Circulation ratio} = \frac{U}{P} = \frac{DR_P - DR_F}{DR_F - DR_U} \qquad 9.12$$

The grinding mills are generally operated at circulating load of 200 – 500% to have a grinding correctly to the required size.

As the particle size decreases, the energy required to fracture each particle decreases but the energy per unit mass rises more rapidly. Hence, in closed-circuit grinding, considerable energy is saved by avoiding over grinding. Grinding in the mining industry is almost always in closed circuit.

Grinding circuits in which all of the plant feed is ground are called **primary grinding circuits**. The principal circuits in which ball mills are used are single stage ball mill, rod mill-ball mill, and autogenous or semi-autogenous ball mill circuits.

The single stage ball mill circuit is the simplest primary grinding circuit in which crushed ore is ground in closed circuit in a single stage ball mill to the size required for mineral liberation.

As discussed in closed circuit crushing, the closed-circuit grinding is of **regular** type when new feed is fed to grinding mill, grounded product is classified in a classifier and classifier underflow is re-fed to grinding mill. Regular type closed circuit grinding is considered in examples 9.5.1 and 9.5.2 and illustrated performing calculations based on percent solids and size analysis of the streams.

Example 9.5.1: *If the percent solids by weight in the feed to classifier, operated in closed circuit with a ball mill, in the classifier overflow and underflow are 40%, 20% and 80%, calculate the % circulating load. The rate of feed to the ball mill is 100 dry tons per hour*

Solution:

Given

Solids feed rate to the ball mill = NF = 100 tons/h

Solids in feed to the classifier = f = 40% = 0.4

Solids in overflow from classifier = p = 20% = 0.2

Solids in underflow from classifier = u = 80% = 0.8

Closed circuit grinding diagram is shown in Fig. 83

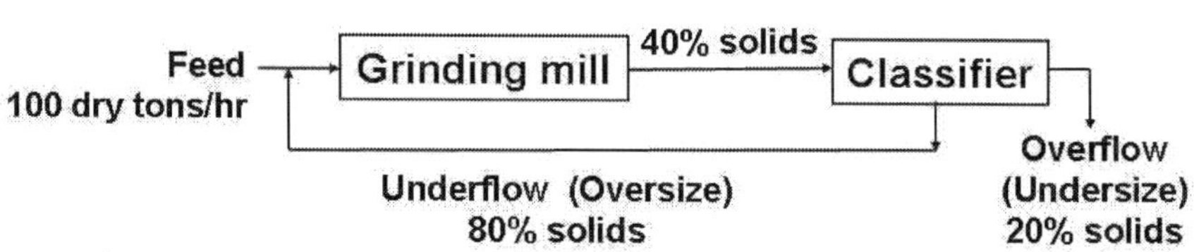

Fig. 83 Closed-circuit grinding for example 31

In closed-circuit grinding operation,

New feed solids (Feed solids to the grinding mill)

= Product solids (Overflow solids from the classifier) = 100 dry tons/h

Total flow rate of overflow (water + solids) product = $P = \dfrac{100}{0.20} = 500$ tons/h

Let F = Total weight flow rate of feed (water + solids) to the classifier

U = Total weight flow rate of underflow (water + solids) from the classifier

Total material (solids + water) balance $F = P + U$

$\Rightarrow F = 500 + U$

Solids balance $F f = P p + U u$

$\Rightarrow F (0.4) = 500 (0.2) + U (0.8)$

Solving above two equations U = 250 tons/h

Solids in underflow = 0.8 × 250 = 200 dry tons/h

$$\% \text{ circulating load} = \frac{\text{Cirulating load}}{\text{New feed}} \times 100 = \frac{200}{100} \times 100 = 200\%$$

Example 9.5.2: *Calculate the circulating load of a mill which is in closed circuit with a classifier and grinds 500 tons of dry ore per day. The screen analysis of the mill discharge, classifier overflow and underflow gives 6.1, 12.4 and 4.1 percent −150+200 mesh material by weight.*

Solution:

Let F, P and U be the feed, overflow and underflow solids, respectively.

$$\therefore \; F = P + U$$

New feed to the mill = overflow from classifier = P = 500 dry tons/day

$$\therefore \; F = P + U = 500 + U$$

Let f, p and u be % solids of any size in feed, overflow and underflow, respectively

$$\therefore \; Ff = Pp + Uu$$

Balance of −150+200 mesh material gives

$$F\,(6.1) = P\,(12.4) + U\,(4.1)$$

$$\Rightarrow (500+U)\,(6.1) = 500\,(12.4) + U\,(4.1)$$

$$\Rightarrow U = 1575 \text{ dry tons/day}$$

Circulating load = 1575 dry tons/day

In **reverse**-type closed-circuit grinding operation, the new feed enters the classifier, underflow of the classifier is fed to the grinding mill, the mill discharges to the classifier and the classifier overflow is the finished product of the grinding circuit. Calculation of reverse type of circuit is illustrated in example 9.5.3.

Example 9.5.3: *An ore at the rate of 80 tons/h is grounded in rod mill, and its product is further grounded in a classifier ball mill circuit as shown in Fig. 84. If the percent of –150 mesh material in the rod mill discharge, ball mill discharge, classifier overflow and underflow are 20, 50, 75 and 25, respectively, calculate the percent circulating load.*

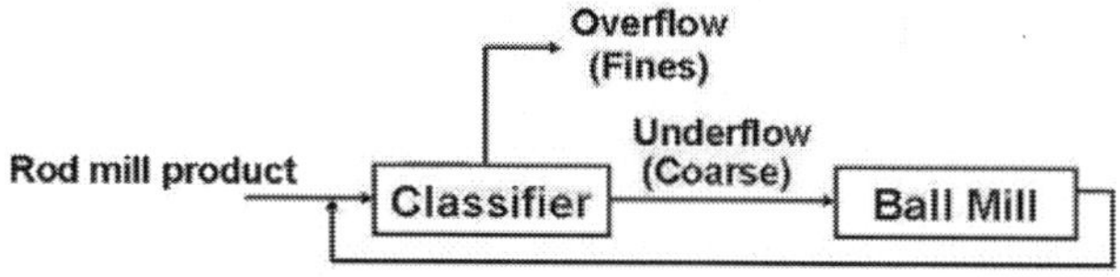

Fig. 84 Circuit diagram for example 33

Solution:

In closed-circuit grinding operation feed solids to the grinding circuit will be equal to the overflow solids from the classifier at steady state.

Therefore, tonnage of classifier overflow solids = 80 tons/h

Let the tonnage of classifier underflow solids be S tons/h.

Balance of –150 mesh material over the classifier

$$80 \times 20 + S \times 50 = 80 \times 75 + S \times 25$$

$$\Rightarrow \quad S = \text{Classifier underflow solids} = 176 \text{ tons/h}$$

$$\text{Percent circulating load in the ball mill circuit} = \frac{176}{80} \times 100 = 220\%$$

Different primary grinding circuits with rod mill-ball mill are used depending on characteristics of the ore and the product size requirements. For these circuits (Fig. 85) the raw ore needs crushing to feed to the rod mill.

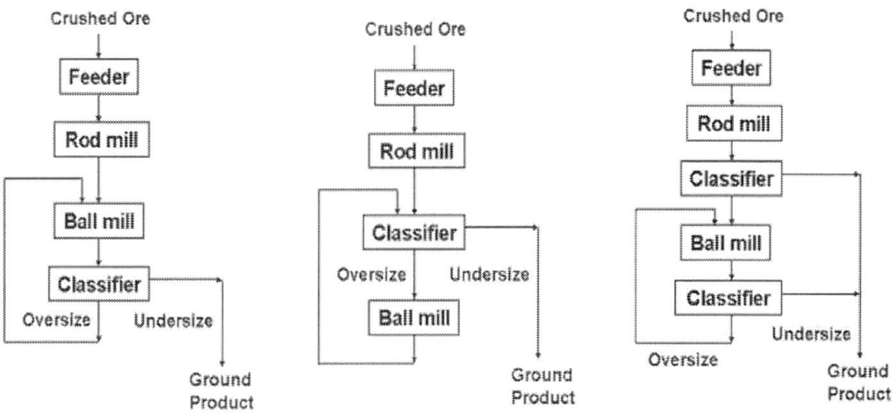

Fig. 85 Rod mill–Ball mill primary grinding circuits

As seen in all the three-rod mill-ball mill circuits, rod mills are operated in open circuit and ball mills are operated in closed circuit in both regular and reverse types. In autogenous or a semi-autogenous ball mill circuit, autogenous mill is operated in closed circuit with a screen and ball mill is operated in closed circuit with a classifier.

In some processing plants, the products from the beneficiation operation are required to grind for further liberation of the mineral to be recovered, to obtain higher concentrate grade, to reduce tailing losses and/or to obtain liberation of other minerals such as pyrite in Lead–Zinc circuit and magnetite in a Copper circuit. In these cases, circuits are called **regrinding circuits**. Another example of regrinding circuit is grinding the iron ore concentrate to the required particle size specifications as practiced in pelletisation.

9.6 PROBLEMS FOR PRACTICE

9.6.1: *What is the critical speed of a ball mill of 48" I.D. charged with 3" balls?* *[39.58 rpm]*

9.6.2: *A lead beneficiation plant installed a ball mill of 2-m diameter. If it wishes to use 10 cm balls as a grinding media and operate at 70% of the critical speed, what would be the operating speed of the ball mill.* *[21.48 rpm]*

9.6.3: *What rotational speed, in revolutions per minute, would you recommend for a ball mill of 1200 mm in diameter charged with 75-mm balls?* *[24 to 32 rpm]*

9.6.4: *Hundred dry tons per hour of new crude ore are fed to a ball mill which is in closed circuit with a classifier. On sampling and analyses, the percent solids by weight in the feed to the classifier, in the classifier overflow (fines) and underflow (coarse) are found as 50, 25 and 84, respectively. Calculate the circulating load tonnage.* *[247.1%]*

9.6.5: *A mill in closed circuit with a classifier receives 300 dry tons of crude ore per day. The sieve analyses of three samples from mill discharge, classifier overflow and underflow are shown in Table 26.*

Table 26: Screen analyses of three samples for problem 9.6.5

mesh	Mill discharge wt%	Classifier overflow wt%	Classifier underflow wt%
+48	42.30	1.20	55.70
48/150	30.50	26.20	31.90
150/200	6.10	12.40	4.05
−200	21.10	60.20	8.35

Calculate the tonnage of the circulating load. *[920 tons/day]*

■■■■

10 Settling of Solid Particles in Fluids

The settling of solid particles in a fluid is a phenomenon that plays an important role in classification wherein ore particles are separated by their sizes, gravity concentration wherein ore particles are separated by their densities and thickening wherein all the solid particles are separated from the pulp. A theoretical analysis on the settling conditions and knowledge of the parameters that influence the motion of particles in a fluid is essential to analyse these processes, to design the suitable equipment and to apply these processes in mineral processing plants.

The settling of solid particles in a fluid depends on many parameters and the inter-relation among these parameters is pretty complex. A simplified analysis of the movement of the particle can be made by the following assumptions:

1. The shape and size of the particle are defined. For the simplest case, it is assumed that the particle is sphere of diameter ‘d’
2. The particle is non-porous and incompressible (i.e. its density remains constant). The particle is insoluble in the fluid and does not react chemically with it
3. The density and viscosity of the fluid are constant
4. The effect of surface characteristics or interfacial conditions between the solid and the fluid on the dynamics of the particle is neglected
5. The particle is freely settling under gravity. Other particles are either absent or even if presents, do not interfere with the motion of the particle under consideration
6. The fluid forms an infinite medium. In other words, the particle under consideration is at an infinite distance (very long distance) from the fluid boundaries and therefore the boundary effect or commonly called the wall effects on the dynamics of the particle can be neglected
7. The fluid is a continuous medium, and the particle size is much larger than the mean free path of the fluid molecules. Thus, the effect of slip between the particles and the fluid molecules can be neglected

Consider a single homogeneous spherical particle of diameter 'd' and density 'ρ_p' falling under gravity in a viscous fluid of density 'ρ_f' and viscosity 'μ_f'. There are three forces acts on a particle:

1. **Gravity force** ($m_p g$) product of the mass of the particle (m_p) and acceleration due to gravity (g), acts downwards
2. **Buoyancy force** ($m_f g$) (by Archimedes' principle) product of the mass of the fluid displaced by the particle (m_f) and the acceleration due to gravity (g), which acts parallel and opposite to the gravity force
3. **Drag force** (F_R) (resistance to the motion), which acts on the surface of the particle and is parallel and opposite to the gravity force.

According to the Newton's second law of motion, the equation of motion of the particle is

$$m_p g - m_f g - F_R = m_p \frac{dV}{dt} \qquad 10.1$$

where V is the velocity of the particle and $\frac{dV}{dt}$ is the net or effective acceleration of the particle.

If the total downward force (gravity force) acting on the particle becomes just equal in magnitude and opposite in direction to the total upward force (the sum of the buoyancy force and drag force or resistance force) acting on the particle in a fluid, the net force acting on the particle is zero, the acceleration of the particle is also zero and the particle moves with a constant velocity.

This velocity is the maximum velocity attained by the particle. It is known as **maximum velocity** or **terminal velocity (V_m)**. When once the particle attains this velocity, it will fall with the same velocity thereafter.

When the acceleration is zero, the particle attains the terminal velocity.

Hence, when $\frac{dV}{dt} = 0$ equation 10.1 becomes

$$F_R = g(m_p - m_f) \qquad 10.2$$

$$\Rightarrow \quad F_R = g\left[\frac{\pi}{6}d^3\rho_p - \frac{\pi}{6}d^3\rho_f\right] \text{ as the particle is assumed as sphere}$$

$$\Rightarrow \quad F_R = \frac{\pi}{6}gd^3(\rho_p - \rho_f) \qquad 10.3$$

10.1 LAMINAR AND TURBULENT FLOWS

When a particle is falling in a fluid, the fluid flows around the particle. This flow may be either laminar (streamline flow round a particle) or turbulent. In laminar flow the fluid flows around the particle without forming any eddies or swirls. In turbulent flow the fluid breaks into eddies and swirls, around the particle. With increased rate of flow, eddies become larger and more complex and the flow becomes more turbulent. If the particle is small, the flow is usually viscous or laminar and the resistance to the fall of the particle mainly depends on the viscosity of the fluid. If the particle is large, the flow is usually turbulent and accompanied by the formation of eddies and vortices in the fluid. These eddies give large resistance to the fall of particle and the viscosity of the fluid becomes less important and can be neglected in determining the resistance under fully turbulent conditions.

Reynolds number was derived by Osborne Reynolds (1883) and is given by

$$N_{Re} = \frac{dV\rho_f}{\mu_f} \qquad 10.4$$

The type of flow depends on the Reynolds number.

If $N_{Re} < 1.0$, the flow is laminar

If $N_{Re} > 1000$, the flow is turbulent

If $1.0 > N_{Re} < 1000$, the flow is neither laminar nor turbulent

The laminar flow around the particle is possible when the particles are fine, whereas turbulent flow is possible with comparatively coarser particles. However, in most of the mineral processing operations, intermediate range of particle sizes exists.

10.2 FLUID RESISTANCE

The nature of the resistance (or drag) depends on the velocity of descent. At low velocities, motion is smooth because the layer of fluid in contact with the body, moves with it, while the fluid, which is a short distance away, is motionless. Between these two positions is a zone of intense shear in the fluid all around the descending particle. Hence, the resistance to the motion is due to the shear forces or viscosity of the fluid and is called **viscous resistance**. As the size of the particle increases, settling velocity increases. At high velocities, the main resistance is due to the displacement of fluid by the particle and is known as **turbulent resistance**. In this case, the viscous resistance is relatively small.

Drag force increases with increase in the velocity of the particle and is often called the kinematic force. At the initial stages of settling, when the particle velocity is small, F_R is negligibly small. As a result, the magnitude of the effective acceleration is quite large. Thus, the particle has a high acceleration and its velocity starts increasing rapidly. As the velocity of the particle increases, the kinematic force also increases and the rate of change of velocity diminishes. Ultimately, a stage is reached when the total downward force acting on the particle (gravity force) becomes just equal to the total upward force acting on it (the sum of the buoyancy force and kinematic force). Then the net force acting on the particle is zero, the particle moves with a constant velocity or zero acceleration.

The kinematic force F_R can be expressed in general as

$$F_R = AKC_D \qquad 10.5$$

where A = characteristic area of the system

K = characteristic kinetic energy per unit volume

C_D = a dimensionless parameter

The terms A and K are to be defined depending on the situation, the system and the process under consideration. For the free fall of a spherical particle in a fluid, Newton proposed that A is defined as projected area of the particle, measured in a plane perpendicular to the direction of motion of the particle.

For the spherical particle $A = \frac{\pi d^2}{4}$ 10.6

$$K = \frac{1}{2}\rho V^2 \quad 10.7$$

where V is the relative velocity between the particle and the fluid. As it is assumed that the fluid is stationary, V is the free settling velocity of the particle. At the point of maximum velocity of the particle, it is V_m. C_D is the frictional drag or drag coefficient or coefficient of resistance.

Therefore $$F_R = \frac{\pi d^2}{4}\frac{1}{2}\rho_f (V_m)^2 C_D \quad 10.8$$

10.3 TERMINAL VELOCITY

Equation 10.3 can be written as

$$\frac{\pi d^2}{4}\frac{1}{2}\rho_f V_m^2 C_D = \frac{\pi}{6} g d^3 (\rho_p - \rho_f)$$

$$\Rightarrow \quad C_D = \frac{4}{3}\frac{gd}{V_m^2}\frac{(\rho_p - \rho_f)}{\rho_f} \quad 10.9$$

or $$V_m = \sqrt{\frac{4}{3}\frac{gd}{C_D}\frac{(\rho_p - \rho_f)}{\rho_f}} \quad 10.10$$

This is a Newton's equation. This equation is applicable for the $N_{Re} > 1000$, and particle sizes of more than 2000 microns in general.

If the particle velocity is quite low, the resistance to the motion is due to the shear forces or viscosity of the fluid and is called **viscous resistance**. Stokes, an eminent English Physicist, deduced an expression for the drag force as $3\pi d\mu_f V$ (Stokes 1891). Then equation 10.3 becomes, after replacing V by V_m, terminal velocity of the particle,

$$3\pi d\mu_f V_m = \frac{\pi}{6} g d^3 (\rho_p - \rho_f) \quad 10.11$$

$$\text{On computation, } V_m = \frac{d^2 g(\rho_p - \rho_f)}{18\mu_f} \qquad 10.12$$

This is a Stokes' equation. This equation describes settling of particles under laminar flow conditions where fluid viscosity provides the resistance to flow. This equation holds good up to 100 microns size of quartz particles and up to 74 microns size of galena particles.

Since the Stokes' equation has been developed for viscous resistance or laminar free settling of particles, its range of validity can be analysed by substituting this equation in equation 10.9

$$C_D = \frac{4}{3}\frac{gd}{V_m}\frac{(\rho_s - \rho_f)}{\rho_f}\frac{18\mu_f}{d^2 g(\rho_p - \rho_f)}$$

$$= \frac{24\mu_f}{dV_m\rho_f} = \frac{24}{dV_m\rho_f / \mu_f} = \frac{24}{N_{\text{Re}}}$$

$$C_D = \frac{24}{N_{\text{Re}}} \qquad 10.13$$

$N_{\text{Re}} = \dfrac{dV_m\rho_f}{\mu_f}$ is called particle Reynolds number

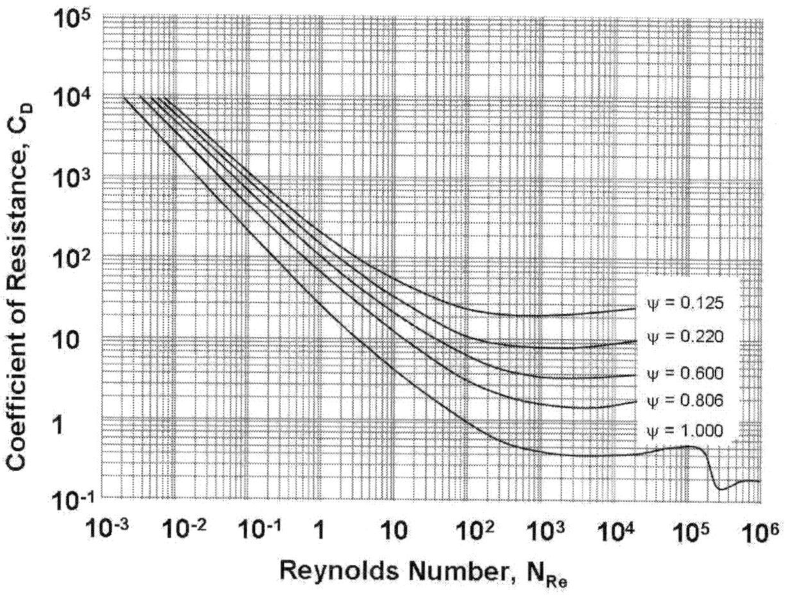

Fig. 86 Relationship of coefficient of resistance to Reynolds number

A chart of log C_D vs. log N_{Re} is a diagonal line of slope –1 for this Stokes relationship for spherical particles. Fig. 86 shows a standard plot reproduced from 1937 report of the Committee on Sedimentation of the National Research Council (Rouse 1937).

From this plot, it can be observed that the plot, for sphericity of 1, is linear for N_{Re} less than 1.0 which can be marked as laminar settling region within which the Stokes' law is applicable. For large values of particle Reynolds number ($N_{Re} > 1000$), the plot is almost a horizontal line parallel to log N_{Re} axis which indicates that coefficient of resistance is constant and independent of the Reynolds number. This is called turbulent settling region.

For this condition, Sir Isaac Newton (1729) assumed that the resistance is entirely due to turbulent resistance and deduced as $F_R = 0.055\pi d^2(V_m)^2\rho_f$. After substitution in equation 10.3

$$0.055\pi d^2(V_m)^2\rho_f = \frac{\pi}{6} g d^3(\rho_p - \rho_f)$$

On computation $$V_m = \sqrt{\frac{3gd(\rho_p - \rho_f)}{\rho_f}} \qquad 10.14$$

This is a modified form of Newton's equation. This is applicable for the particle Reynolds number from 1000 to 200,000 and particle sizes of more than 2000 microns in general. Beyond 200,000, the coefficient of resistance decreases sharply. This is due to the formation of an eddy in the fluid behind the particle which travels with particle, resulting in sharp decrease in the drag on the particle.

For the most common sizes of particles encountered in mineral processing, the Reynolds number for settling under gravity in water falls into the transitional region, where the flow is neither fully turbulent nor fully laminar and the resistance to flow changes non-linearly with velocity. For the transition region ($1.0 < N_{Re} < 1000$) it is difficult to propose a hard and fast correlation between C_D and N_{Re}. A large number of empirical equations based on experimental data have been proposed in the literature for estimating the drag coefficient in the transition region. Newton's equation can be applied for this transition region to arrive at approximate value of settling velocity.

Terminal settling velocity can be calculated through an equation formed between C_D and N_{Re}. Equation can be formed as follows:

Equation 10.9 is $$C_D = \frac{4}{3}\frac{gd}{V_m^2}\frac{(\rho_p - \rho_f)}{\rho_f}$$

Taking logarithm on both sides

$$\log C_D = \log\left[\frac{4gd(\rho_p - \rho_f)}{3\rho_f}\right] - 2\log V_m \qquad 10.15$$

Reynolds number at the terminal velocity is $N_{Re} = \dfrac{dV_m\rho_f}{\mu_f}$

Taking logarithm on both sides

$$\log N_{Re} = \log \frac{d\rho_f}{\mu_f} + \log V_m \qquad 10.16$$

Elimination of V_m from equations 10.15 and 10.16 gives

$$\log C_D = -2 \log N_{Re} + \log \left[\frac{4gd^3\rho_f(\rho_p - \rho_f)}{3\mu_f^2} \right] \qquad 10.17$$

At $N_{Re} = 1$, equation 10.17 becomes

$$C_D = \frac{4gd^3\rho_f(\rho_p - \rho_f)}{3\mu_f^2} \qquad 10.18$$

For the above example, C_D is to be calculated first

$$C_D = \frac{4gd^3\rho_f(\rho_p - \rho_f)}{3\mu_f^2} = \frac{4 \times 981 \times (0.1)^3 \times 1.0(2.65 - 1.0)}{3(0.01)^2}$$

$$= 21{,}852$$

At $N_{Re} = 1$ and $C_D = 21{,}852$, a straight line is to be drawn with slope of –2 on Fig. 87. This straight line intersects the C_D vs N_{Re} curve. Corresponding to this intersection point $N_{Re} = 170$ (Fig. 87)

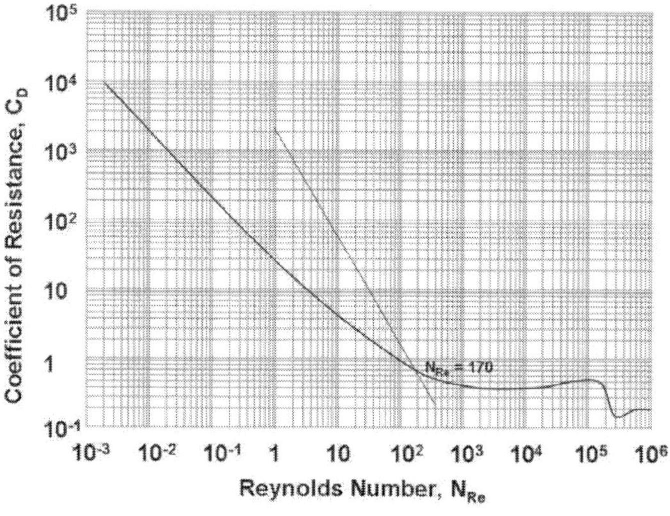

Fig. 87 Determination of Reynolds number

Therefore

$$\text{Terminal settling velocity} = V_m = \frac{N_{Re}\mu_f}{d\rho_f} = \frac{170\times 0.01}{0.1\times 1.0} = 17 \text{ cm/s}$$

A chart of log C_D vs. log N_{Re} is given in Fig. 86 for different values of sphericity and can be used for estimation of terminal settling velocity values for non-spherical particles.

Examples 10.3.1 and 10.3.2 illustrates the calculations of settling velocities.

Example 10.3.1: *Compute the maximum velocity at which particles of silica of 0.01 cm in diameter and 2.65 in specific gravity will fall from rest through quiet water.*

Solution:

Given

Density of the silica particle	$= \rho_p = 2.65$ g/cc
Diameter of the silica particle	$= d = 0.01$ cm
Density of water	$= \rho_w = 1.0$ g/cc
Viscosity of water	$= \mu_w = 0.01$ poise

Stokes' law is applicable here as the size of particle is 0.01 cm

$$\text{Terminal settling velocity} = v_m = \frac{d^2 g(\rho_p - \rho_w)}{18\mu_w}$$

$$= \frac{(0.01)^2(980)(2.65-1.0)}{18(0.01)} = 0.9 \text{ cm/s}$$

Maximum settling velocity of 0.01 cm silica particle = 0.9 cm/s

Example 10.3.2: *Calculate the particle size of a spherical galena particle of specific gravity 7.5 settling in water from rest with a terminal velocity of 0.57 cm/s.*

Solution:

Given

Density of the galena particle	$= \rho_p = 7.5$ g/cc
Density of water	$= \rho_w = 1.0$ g/cc
Viscosity of water	$= \mu_w = 0.01$ poise
Terminal settling velocity	$= v_m = 0.57$ cm/s

$$\text{As per Stokes' law} \quad v_m = \frac{d^2 g(\rho_p - \rho_w)}{18\mu_w}$$

$$d = \sqrt{\frac{v_m 18\mu_w}{(\rho_p - \rho_w)g}} = \sqrt{\frac{0.57 \times 18 \times 0.01}{(7.5 - 1.0)980}} = 0.004 \text{ cm}$$

$$\text{Reynolds Number} = N_{Re} = \frac{dV_m \rho_f}{\mu_f} = \frac{(0.004)(0.5)(1.0)}{0.01} = 0.2 < 1.0$$

Hence Stokes' law is applicable

Diameter of galena particle of 0.57 cm/s terminal velocity = 40 microns

The terminal settling velocity of a spherical particle is a function of size and specific gravity (density) of the particle. If two particles have the same specific gravity, then the larger diameter particle has higher terminal velocity and if two particles have the same diameter, then the heavier particle has higher terminal velocity. The velocity of an irregularly shaped particle with which it is settling in a fluid medium also depends on its shape. As almost all natural particles are irregular in shape, principle can be stated as

> *"The coarser, heavier and rounder particles settle faster than the finer, lighter and more angular particles"*

10.4 FREE SETTLING

In all the foregoing discussions, the particle is settling in a large volume of fluid by its own specific gravity, size and shape and uninfluenced by the surrounding particles as particles are not crowded. Such settling process is called **Free settling**. Free settling predominates in well-dispersed pulps where the percent solids by weight is less than 10.

10.5 HINDERED SETTLING

When the particles settle in relatively small volume of fluid, they are crowded in the pulp. Particles collide each other during their settling and this collision affects their settling velocities. Such settling process is called **Hindered settling**. Hindered settling predominates when the percent solids by weight is more than 15. Under hindered settling conditions, the settling velocity or rate of settling of each individual particle will be considerably less than that for the free settling conditions. While the motion of large particles gets hindered, the small particles tend to get dragged downwards by the large ones and thus get accelerated. It must be noted that each particle is in fact settling through a suspension of other particles in the liquid rather than through the simple liquid itself.

The effective density and viscosity of a concentrated suspension or slurry are much larger than those of clear liquid. The settling medium therefore offers high resistance and this resistance to fall is mainly due to turbulence created. Hindered settling velocity can be approximately estimated from Stokes' equation or Newton's equation after replacing ρ_f and μ_f by ρ_{sl} and μ_{sl}.

Density of the slurry ρ_{sl} can be determined by the following formula knowing the volume fraction of the solids C_v in the slurry

$$\rho_{sl} = C_v \rho_p + (1 - C_v) \rho_f \qquad 10.19$$

To be precise, slurries are non-Newtonian fluids. The apparent viscosity of the slurry is a function of the shear rate and depends on its rheological characteristics. The viscosity (μ_{sl}) of the slurry containing spherical particles is related to the volume fraction of the solids in the slurry by the formula (Brown et al 1950)

$$\frac{\mu_{sl}}{\mu_f} = \frac{10^{1.82C_v}}{1-C_v} \qquad 10.20$$

By substituting ρ_{sl} and μ_{sl} in Stokes' equation (10.12), an expression for hindered settling velocity V_H can be obtained as follows for the fine particles:

$$V_H = \frac{d^2 g(\rho_p - \rho_{sl})}{18\mu_{sl}} \qquad 10.21$$

$$= \frac{d^2 g(\rho_p - \rho_f)}{18\mu_f}\left[\frac{(1-C_v)^2}{10^{1.82C_v}}\right] = V_m F_s$$

$$\Rightarrow \frac{V_H}{V_m} = F_s \qquad 10.22$$

where $F_s = (1 - C_V)^2 / 10^{1.82C_V}$

F_s is therefore the ratio of hindered settling velocity (V_H) of the particle to its free settling velocity (V_m) and its magnitude will evidently be less than 1.0. Standard plots of F_s versus C_v and the viscosity ratio (μ_{sl}/μ_f) versus C_v are available in the literature (Brown et al 1950).

Similarly, an expression for hindered settling velocity V_H for coarse particles can be obtained by replacing ρ_f by ρ_{sl} in Newton's equation (10.10):

$$V_H = \sqrt{\frac{4}{3}\frac{gd}{C_D}\frac{(\rho_p - \rho_{sl})}{\rho_{sl}}} \qquad 10.23$$

The lower the density of the particle, the more marked is the effect of reduction of the effective density, ($\rho_p - \rho_{sl}$), and the greater is the reduction in falling velocity. Similarly, the larger the particle, the greater is the reduction in falling rate as the pulp density increases.

Fig. 88 shows how the particles of different sizes and two specific gravities settle under free and hindered settling conditions. By comparing Figs. 88(A) and (B), it is evident that heavier (or lighter) particles can be separated when they settle by hindered settling. This is possible because hindered settling reduces the effect of size and increases the effect of specific gravity.

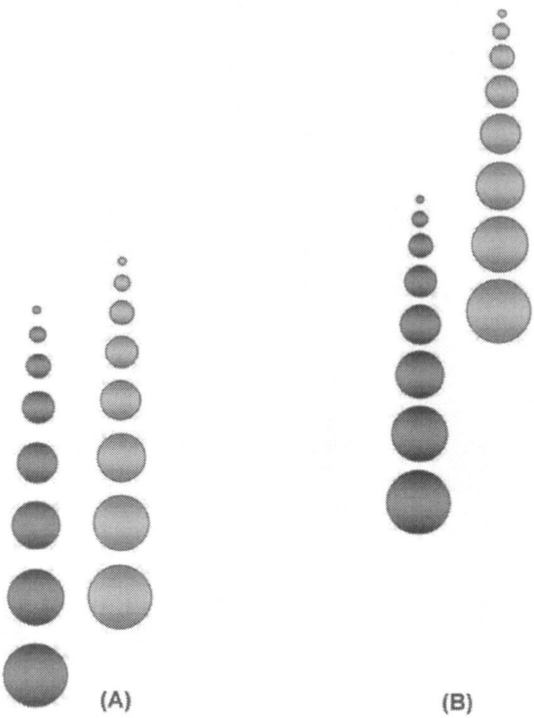

Fig. 88 Settling of particles under (A) Free settling (B) Hindered settling

10.6 EQUAL SETTLING PARTICLES

Particles are said to be equal settling if they have the same terminal velocities in the same fluid and in the same field of force.

10.7 SETTLING RATIO

Settling ratio is the ratio of the sizes of two particles of different specific gravities fall at equal rates.

Under free settling conditions, settling ratio is known as free settling ratio and can be obtained by equating the terminal velocities of lighter and heavier particles of different sizes. Let d_1 and d_2 are the diameters of lighter and heavier particles and ρ_{p1} and ρ_{p2} are the densities of lighter and heavier particles. When the terminal velocities of these two particles are same and the particles are fine obeying Stokes' law of settling, the equation for the terminal settling velocity (equation 10.12) can be written as

$$V_m = \frac{d_1^2 g(\rho_{p1} - \rho_f)}{18\mu_f} = \frac{d_2^2 g(\rho_{p2} - \rho_f)}{18\mu_f} \qquad 10.24$$

$$\Rightarrow \text{Free settling ratio} = \frac{d_1}{d_2} = \left(\frac{\rho_{p_2} - \rho_f}{\rho_{p_1} - \rho_f}\right)^{1/2} \qquad \ldots 10.25$$

When the terminal velocities of the two particles are same and the particles are coarse obeying Newton's law of settling, equation 10.10 can be written as

$$V_m = \sqrt{\frac{4}{3}\frac{gd_1}{C_D}\frac{(\rho_{p_1} - \rho_f)}{\rho_f}} = \sqrt{\frac{4}{3}\frac{gd_2}{C_D}\frac{(\rho_{p_2} - \rho_f)}{\rho_f}} \qquad 10.26$$

$$\Rightarrow \text{Free-settling ratio} = \frac{d_1}{d_2} = \frac{\rho_{p_2} - \rho_f}{\rho_{p_1} - \rho_f} \qquad 10.27$$

The general expression for free-settling ratio can be deduced from equations 10.25 and 10.27 as

$$\text{Free-settling ratio} = \frac{d_1}{d_2} = \left(\frac{\rho_{p_2} - \rho_f}{\rho_{p_1} - \rho_f}\right)^{n} \qquad 10.28$$

where n = 0.5 for fine particles obeying Stokes' law and n = 1 for coarse particles obeying Newton's law. The value of n lies in the range 0.5–1 for particles in the intermediate size range where neither Stokes' law nor Newton's law is applicable.

Consider a mixture of galena (density 7.5 g/cc) and quartz (density 2.65 g/cc) particles settling in water. For fine particles, obeying Stokes' law, the free settling ratio is

$$\sqrt{\frac{7.5 - 1}{2.65 - 1}} = 1.98$$

i.e. a fine galena particle will settle at the same rate as fine quartz particle of diameter 1.98 times larger than galena particle. Fig. 89(A) shows the settling of fine particles.

For coarse particles, obeying Newton's law, the free settling ratio is

$$\frac{7.5 - 1}{2.65 - 1} = 3.94$$

i.e. a coarse galena particle will settle at the same rate as coarse quartz particle of diameter 3.94 times larger than galena particle. Fig. 89(B) shows the settling of coarser particles.

Therefore, the free-settling ratio for coarse particles is larger than for fine particles. This means that density difference between the particles has more effect at coarser size ranges when they settle.

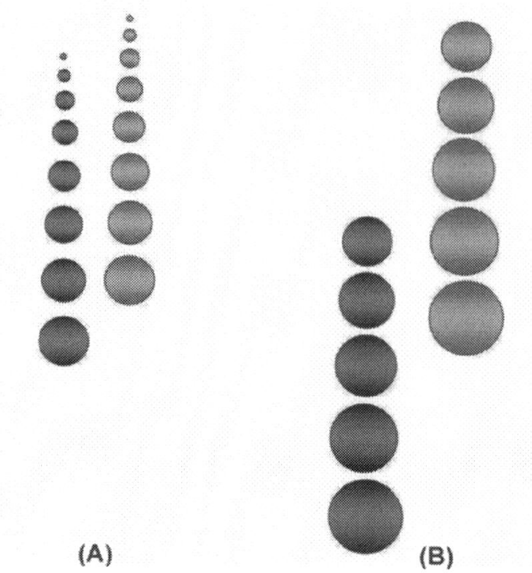

Fig. 89 Free settling of (A) Fine particles (B) Coarse particles

Hindered settling ratio can be obtained by equating the terminal velocities of lighter and heavier particles of different sizes in Stokes' equation (equation 10.21)

$$V_m = \frac{d_1^2 g(\rho_{p_1} - \rho_{sl})}{18\mu_{sl}} = \frac{d_2^2 g(\rho_{p2} - \rho_{sl})}{18\mu_{sl}} \qquad 10.29$$

$$\Rightarrow \text{Hindered settling ratio} = \frac{d_1}{d_2} = \left(\frac{\rho_{p_2} - \rho_{sl}}{\rho_{p_1} - \rho_{sl}}\right)^{1/2} \qquad 10.30$$

Similarly hindered settling ratio is obtained by equating the terminal velocities of lighter and heavier particles of different sizes in Newton's equation (equation 10.23)

$$V_m = \sqrt{\frac{4}{3}\frac{gd_1}{C_D}\frac{(\rho_{p_1} - \rho_{sl})}{\rho_{sl}}} = \sqrt{\frac{4}{3}\frac{gd_2}{C_D}\frac{(\rho_{p_2} - \rho_{sl})}{\rho_{sl}}} \qquad 10.31$$

$$\Rightarrow \text{Hindered settling ratio} = \frac{d_1}{d_2} = \frac{\rho_{p_2} - \rho_{sl}}{\rho_{p_1} - \rho_{sl}} \qquad 10.32$$

For mixture of galena and quartz particles settling in a pulp of density 1.5, the hindered settling ratio is

$$\text{for fine particles } \left(\frac{7.5 - 1.5}{2.65 - 1.5}\right)^{1/2} = 2.28$$

for coarse particles $\dfrac{7.5 - 1.5}{2.65 - 1.5} = 5.22$

i.e. a galena particle will settle at the same rate as quartz particle of diameter 2.28 times larger than galena particle if the particles are fine obeying Stokes' law and it is 5.22 times larger than galena particle if the particles are coarse obeying Newton's law.

When the hindered settling ratio of 5.22 is compared with the free settling ratio of 3.94, it is evident that hindered settling reduces the effect of size, while increasing the effect of density, which means that more heavy (or light) particles can be separated in hindered settling. Hindered settling ratio is always greater than the free settling ratio. As the pulp density increases, this ratio also increases.

In example 10.7.1, settling ratios are calculated.

Example 10.7.1: *Calculate the free settling ratio of magnetite and quartz particles having specific gravity of 5.6 and 2.65, respectively.*

Solution:

Given

Density of the quartz particle $= \rho_q = 2.65$ g/cc

Density of the magnetite particle $= \rho_m = 5.60$ g/cc

Density of water $= \rho_w = 1.0$ g/cc

Viscosity of water $= \mu_w = 0.01$ poise

Let the diameters of the quartz and magnetite particles be d_q and d_m

$$\text{By Stokes' law, free-settling ratio} = \frac{d_q}{d_m} = \left(\frac{\rho_m - \rho_w}{\rho_q - \rho_w}\right)^{1/2} = \left(\frac{5.6 - 1.0}{2.65 - 1.0}\right)^{1/2} = 1.67$$

$$\text{By Newton's law, free-settling ratio} = \frac{d_q}{d_m} = \left(\frac{\rho_m - \rho_w}{\rho_q - \rho_w}\right) = \left(\frac{5.6 - 1.0}{2.65 - 1.0}\right) = 2.79$$

Free settling conditions are used in classifiers, in which case they are called Free settling classifiers (Mechanical classifiers or Horizontal current classifiers), to increase the effect of size on separation. Hindered settling conditions are used in classifiers, in which case they are called Hindered settling classifiers (Hydraulic classifiers or Vertical current classifiers), to increase the effect of density on separation.

By assigning the numbers in the order of increasing size of the particles for both free and hindered settling particles as shown in the Fig. 88, separation can be explained theoretically as follows:

In case of free settling, if sufficient time for the light particle 5 is not given for settling, all the light particles of 1–5 and heavy particles of 1–4 will be at the top of the classifier, and light particles of 6–8 and heavy particles of 5–8 will be at the bottom of the classifier. This separation is illustrated in Fig. 90A. Each fraction contains both light and heavy particles and almost of the same size or closely sized particles. It means that all the particles are separated into two size fractions. Hence this type of classification is a sizing classification. Classifier used is called **sizing classifier**.

In case of hindered settling, if top and bottom fractions are removed without allowing light particle 8 to settle, the bottom fraction contains solely of heavy particles 6–8. As the settling velocity of all the light particles is less than the settling velocity of light particle 8, they remain at the top along with the heavy particles 1–5 and get discharged as overflow product. If these particles are again classified without allowing light particle 5 to settle, the bottom fraction contains heavy particles 1–5 and light particles 6–8. Only light particles 1–5 remain at the top and get discharged as overflow product. This separation is illustrated in Fig. 90B. Top and bottom fractions contain all light and heavy particles respectively. However, middle fraction containing heavy particles of 1–5 and light particles of 6–8 are together under hindered settling conditions which means that they cannot be separated. Classifier used for this kind of separation is called **sorting classifier**. This separation is obtained by using rising current of water known as hydraulic water in two separating columns. In first column, the velocity of hydraulic water is less than the settling velocity of heavy particle 6 and more than settling velocity of heavy particle 5. In second column, the velocity of hydraulic water is less than the settling velocity of light particle 6 and more than settling velocity of light particle 5.

It is very interesting to note that, under free settling conditions, only heavy particle 8 and light particles 1 and 2 can be separated whereas in hindered settling conditions, heavy particles of 6–8 and light particles of 1–5 can be separated.

It can be reiterated that free settling conditions are to be maintained to separate the mixed density and mixed size particles according to their sizes which increases the effect of size and decreases the effect of density on separation. Hindered settling conditions are to be maintained to separate the mixed density and mixed size particles according to their densities which increases the effect of density and decreases the effect of size on separation.

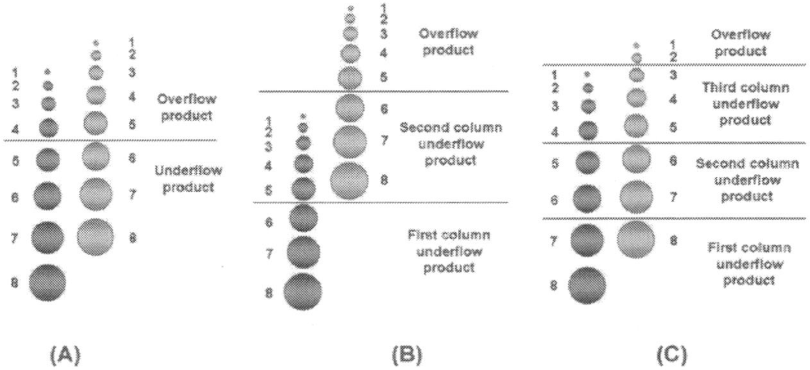

Fig. 90 Separation using free and hindered settling

Even though sorting classifiers are not truly sizing classifiers, they are sometimes used to sort out the particles in close size range by maintaining the free settling conditions as shown in Fig. 90C. If velocity of hydraulic water used in first column is less than the settling velocity of heavy particle 7 and more than the settling velocity of heavy particle 6, the bottom fraction contains heavy particles 7 and 8 and light particle 8. Heavy particles 1–6 and light particles 1–7 remain at the top and get discharged as overflow product and enter the second column. The velocity of hydraulic water used in second column is less than the settling velocity of light particle 6 and more than the settling velocity of light particle 5 so that the bottom fraction contains heavy particles 5 and 6 and light particles 6 and 7. Heavy particles 1–4 and light particles 1–5 remain at the top and get discharged as overflow product and enter the third column. The velocity of hydraulic water used in third column is less than the settling velocity of light particle 3 and more than the settling velocity of light particle 2. The final separation takes place with the bottom fraction contains heavy particles 1 and 4 and light particles 3 and 5. Only light particles 1 and 2 remain at the top and get discharged as overflow product.

The sorting classifier commonly called as hydrosizer is used to sort the feed into different size fractions which are further fed to separate concentrating tables as practiced in chrome ore beneficiation plants.

Almost all gravity concentration operations and many sizing devices make use of the hindered settling phenomenon during the separation of particles.

In example 10.7.2 and 10.7.3, separation calculations are illustrated.

Example 10.7.2: *An ore consists of valuables of 5.8 specific gravity and gangue of 2.6 specific gravity having size range of 10–25 microns is separated in a mechanical classifier. Can all valuables be separated in underflow? If so, will they gangue free? Explain.*

Solution:

Given

Density of gangue	$= \rho_g = 2.6$ g/cc
Density of valuables	$= \rho_v = 5.8$ g/cc
Density of water	$= \rho_w = 1.0$ g/cc
Size range of the particles	= 10 to 25 microns

Settling velocity of coarsest heavy particle by Stokes' law

$$= v_m = \frac{d_v^{\,2} g(\rho_v - \rho_w)}{18\mu_w} = \frac{(0.0025)^2 981(5.8-1.0)}{18 \times 0.01} = 0.164 \text{ cm/s}$$

$$N_{Re} = \frac{d_v v_m \rho_w}{\mu_w} = \frac{0.0025 \times 0.164 \times 1.0}{0.01} = 0.041$$

As Reynolds number (N_{Re}) for coarsest heavy particle is less than 1, all particles settle under laminar conditions. Hence Stokes' law is applicable.

Size of the gangue particle that settles equally with the smallest valuable particle

$$= d_g = \left(\frac{\rho_v - \rho_w}{\rho_g - \rho_w}\right)^{1/2} d_v = \left(\frac{5.8 - 1.0}{2.6 - 1.0}\right)^{1/2} \times 10 = 17.32 \text{ microns}$$

It means that gangue particles of 17.32–25 microns will also settle by the time smallest valuable particle settle. Therefore, underflow of classifier contains all valuable particles and gangue particles of 17.32–25 microns. Hence underflow product is not gangue free. Evidently, overflow product contains only gangue particles of 10–17.32 microns. Hence, overflow product is valuable free.

Example 10.7.3: *It is desired to separate quartz particles of specific gravity of 2.6 from galena of specific gravity 7.5 in a free settling classifier. The size range is from 5.2 to 25 microns. Three fractions are obtained. One of pure quartz, another of pure galena, and the third a mixture. Calculate the size range of the substances in the mixture.*

Solution:

Given

Density of quartz	$= \rho_q = 2.6$ g/cc
Density of galena	$= \rho_g = 7.5$ g/cc
Density of water	$= \rho_w = 1.0$ g/cc
Size range of the particles	= 5.2 to 25 microns

As the maximum size of the particle is 25 microns, the flow zone is laminar.

Let d_q and d_g be the sizes of quartz and galena particles

The size of the quartz particle that settles equally with the smallest galena particle

$$= d_q = \left(\frac{\rho_g - \rho_w}{\rho_q - \rho_w}\right)^{1/2} d_g = \left(\frac{7.5 - 1.0}{2.6 - 1.0}\right)^{1/2} \times 5.2 = 10.4 \text{ microns}$$

10.4 microns size of quartz particles settle with the same velocity as the smallest galena particles. Quartz particles smaller than 10.4 microns size will settle only after the smallest galena particle 5.2 microns has settled. Hence all quartz particles of size less than 10.4 microns size will constitute an overflow fraction of the classifier.

The size of the galena particle that settles equally with the largest quartz particle

$$= d_g = \left(\frac{\rho_q - \rho_w}{\rho_g - \rho_w}\right)^{1/2} d_q = \left(\frac{2.6 - 1.0}{7.5 - 1.0}\right)^{1/2} \times 25 = 12.4 \text{ microns}$$

It means that largest quartz particle will settle only after all the galena particles of size more than 12.4 microns have settled. Hence, all galena particles of size more than 12.4 microns will constitute an underflow fraction.

The middle fraction will, therefore, contain the remaining galena particles having the size from 5.2 to 12.4 microns and the remaining quartz particles having the size from 10.4 to 25 microns.

Therefore,

All galena particles of size more than 12.4 microns are pure galena fraction.

All quartz particles of size less than 10.4 microns are pure quartz fraction

The third fraction, i.e. the mixture, contains

Quartz particles of size 10.4–25 microns

Galena particles of size 5.2–12.4 microns

The motion of a particle, when it starts settling in the fluid and move through the fluid, can be divided into two stages—acceleration period and terminal velocity period. Initially, the velocity of the particle is zero with respect to the fluid and increases to the terminal velocity during a short period, usually of the order of one tenth of a second or less. During this first stage of short period, there are initial-acceleration effects. When once particle reaches its terminal velocity, second stage starts and continued as long as the particle continues to settle. Classification and thickening processes make use of the terminal velocity period, whereas jigging makes use of differences in particle behaviour during the acceleration period for separation.

10.8 SETTLING OF LARGE SPHERES IN A SUSPENSION OF FINE SPHERES

Large spheres settle in a suspension of fine spheres as if the fine spheres were a part of the fluid. When a water-solid suspension is made, the system begins to behave as a heavy liquid whose density is that of the pulp rather than that of carrier liquid. This is the principle used in Heavy Medium Separation operations in cleaning coal with a suspension of sand in water. Thus, a large sphere of quartz falling in a suspension of quartz sand having an average density of ρ_{sl} falls at a speed determined by the equation 10.23.

$$V = \sqrt{\frac{4}{3}\frac{gd}{C_D}\frac{(\rho_p - \rho_{sl})}{\rho_{sl}}} \tag{10.23}$$

10.9 PROBLEMS FOR PRACTICE

10.9.1: *Calculate the terminal settling velocity of a particle having 1.4 specific gravity and 20 microns in size settling in water.* *[0.0087 cm/s]*

10.9.2: *Calculate the maximum velocity at which a spherical particle of galena having specific gravity of 7.5 and 6 mm in diameter will fall in water. Also, find the type of flow. Density and viscosity of water are 1 g/cc and 0.82 cp, respectively.* *[107.1 cm/s]*

10.9.3: *Calculate the size of a spherical silica particle settling in water from rest at 20°C with a terminal settling velocity of 0.5 cm/s. Specific gravity of silica is 2.65.* *[78 microns]*

10.9.4: *Determine the ratio of diameters for spherical particles of sphalerite of specific gravity of 4.0 and quartz of specific gravity of 2.65 that have the same terminal settling velocities in water at 68°F considering free settling conditions.* *[1.35, 1.82]*

10.9.5: *An ore contains valuables of specific gravity 5.1 and gangue of specific gravity 2.0 has size range of 10 microns to 100 microns. When the ore is classified in a free settling hydraulic classifier, find the upward water velocity to remove all gangue particles in overflow. Will this overflow fraction contain any valuables? If so find the size range of valuables collected in overflow.* *[0.545 cm/s, 10 to 49.4 microns]*

10.9.6: *Quartz and pyrite are to be separated in a hydraulic free settling classifier. The feed to the classifier ranges in size between 10 microns and 300 microns. The specific gravity of quartz is 2.65 and that of pyrite is 5.1. If the mixture is best separated into three fractions such as one containing only quartz, another containing only pyrite and the third a mixture of quartz and pyrite, estimate the size ranges of the two materials in these fractions. Assume the flow zone to be essentially laminar. [Overflow quartz–10 to 15.76 μm, Underflow pyrite–190.31 to 300 μm, Middle pyrite–10 to 190.31 μm and quartz– 15.76 to 300 μm]*

■■■■

11 Classification

Classification is a method of separating mixtures of particles of different sizes, shapes and specific gravities into two or more products on the basis of the velocity with which the particles fall through a fluid medium, i.e. settling velocity. Generally, classification is employed for those particles which are considered too fine to be separated efficiently by screening.

The basic principle of classification is

> *"The coarser, heavier and rounder particles settle faster than the finer, lighter and more angular particles"*

In classification, certain particles are only allowed to settle in the fluid medium to separate the particles into two fractions. Classification is thought of as a size separation, but is actually made on the basis of particle settling rate, and so is affected by any variations in particle density and shape. There are two basic types of classification—**pool** classification and **hydraulic** classification. In pool classification, a suspension of feed particles is fed into a pool at such a rate that only those particles—the coarser and faster settling particles—has time to settle out. The reminder overflows as fines. In hydraulic classification, the feed suspension is introduced into or above one or more columns through which water is rising at a controlled velocity. Coarse particles subside through the columns and are removed through spigots. Those which settle slower than the rising velocity in any column are prevented from subsiding and so must overflow. Hydraulic classifiers are capable of making far sharper separations than pool classifiers. These classifiers are known as **sizing classifiers**.

Classification in hydraulic classifiers will also be performed by altering the conditions of settling of particles in such a way that the effect of size is reduced and effect of density predominates so that the particles will be separated according to their densities rather than sizes. Such classifiers are known as **sorting classifiers**.

11.1 POOL CLASSIFIERS

Classification in pool classifiers is sizing classification. Pool classifiers consist of settling tank with parallel sides and a sloping bottom equipped with a mechanical means which continuously agitates the pulp and removes the settled solids. Pool classifiers are mechanical classifiers. Feed slurry is introduced at feed point shown in

Fig. 91. Fines overflow at overflow weir. The rate of feed and distance between feed point and overflow weir is selected in such a way that the rate of travel of required fine particles must be more than the rate of their settling so that all the required fine particles overflow at overflow weir. The coarse particles settle to the bottom. A sand raking mechanism is provided at the bottom of the tank to remove the settled coarse particles.

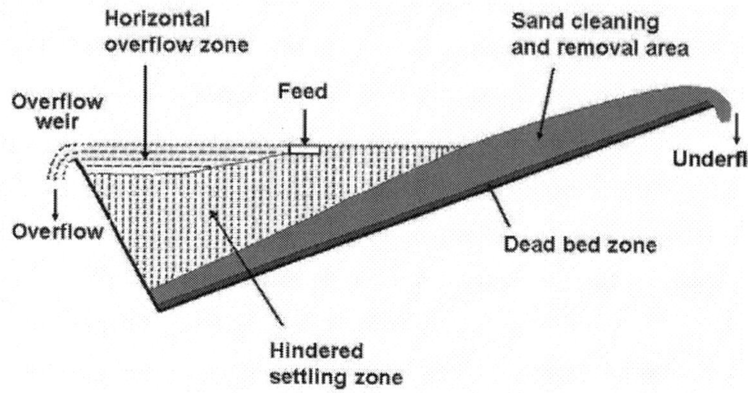

Fig. 91 Separation in sizing classifier

The body of pulp in pool classifier is divided into several zones having different particle-size and pulp-density characteristics as shown in Fig. 91.

1. **Horizontal overflow zone** consists of a stream of pulp flowing horizontally from feed opening to the overflow weir. In this zone, free settling conditions prevail with dilute feed.
2. **Hindered settling zone** exists below the horizontal overflow zone. The size of the particles and the density of the pulp increases from top to bottom of this zone.
3. **Sand cleaning and removal zone** exists at the bottom of the hindered settling zone. The particles settling through the hindered settling zone enter this zone. Such particles are collected in this zone. Agitation generated by the sand raking mechanism makes much of the entrapped water and fine particles come out of the sand, rise and report in the overflow product. Thus, the sand product contains only a small fraction of fines. This is the unique benefit of pool classifiers.
4. **Dead zone** is at the bottom of the settling pool below the reach of the sand raking mechanism. It is the layer of inactive stationary sand, the first to settle after the classifier started operation. The dead zone serves to protect the tank bottom and has no further part in the classification process.

Spiral, rake and drag classifiers are the three important pool classifiers. A bowl classifier is the variant of a rake classifier. Settling cone is a non-mechanical classifier.

A **spiral classifier** (Akins or Hardinge spiral classifier) comprises usual sloping trough and one or two spirals mounted on a trough shaft parallel to the tank bottom, rotate without contacting the sides or bottom of the tank (Fig. 92). Feed is introduced at pool level through one or both side walls. Pool level is maintained as desired by adjusting the height of the overflow weir. A stream of pulp consists of fines having low settling velocities flow horizontally from the feed inlet to the overflow weir and get discharged. Discharged overflow drops into a box and is piped away to the required point usually a feed box of a concentrator.

The settled sands are turned over again and again by the rotation of the spiral till they are finally discharged. This action gives good opportunity for complete de-sliming of the sands. Thus, the spiral also elevates the bottom sand and discharged. The speed of the spiral is usually 3–20 rpm depending on the size of the spiral. For large spiral, the rpm is lower and it is higher for small spiral. Steeper slopes of classifier tank are also possible due to the continuous and gentle action of the spiral. Steeper slopes aid the drainage of sands giving a cleaner and dried product. The spiral classifier may be simplex having one spiral or duplex having two spirals.

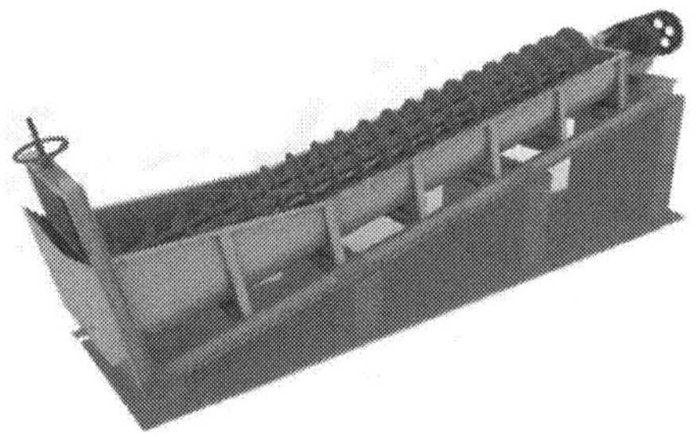

Fig. 92 Spiral classifier
(Courtesy of balling machinery.com)

Fig. 93 Rake classifier
(Courtesy of Dione oil Ltd.)

A **rake classifier** (Dorr rake classifier) (Fig. 93) comprises a sloping trough and one or more rakes. Feed is introduced through either end of the transverse trough near the overflow on to the apron which distributes the feed evenly across the full width of the tank. By means of bell-crank mechanism, the rakes are given a motion so that they dip down into the settled sands and drag it up the incline a short distance. The rakes are then drawn out of the settled sands, returned to the starting point and lowered to repeat the cycle. The settled sands thus dragged slowly up the incline till they are discharged. While the sands are dragging up the inclined trough they are repeatedly turned over and thoroughly drained by the time they are discharged. It is obvious that dragging of the sands is intermittent, not continuous.

The rake classifiers may be simplex having one compartment trough, duplex having two compartments trough or multi rake unit having many compartments. Each compartment is separately raked.

A **bowl classifier** is variant of rake classifier. It consists of, a bowl, a large shallow cylindrical settling tank with a slightly conical bottom sloping towards the centre and a rake classifier at the lower end of the bowl. Feed is introduced at the centre of the bowl through a cylindrical well. Classification takes place in two stages. In the bowl fines overflow into a peripheral launder and sands settle, raked into the centre, fall through an opening into the rake compartment against a rising current of water. Further classification of sands takes place in the rake classifier. The pool in the raking compartment will not overflow except through the bowl.

A **drag** or **Esperanza classifier**, is similar to rake and spiral classifier, and consists of large number of drag flights mounted on an endless belt or link chain. The flights agitate, drain, and remove the underflow from the settling tank.

In adjusting a classifier for a particular application, the size at which the separation is made and the quality of the separation will be determined by several factors, among which the following are important.

(a) Slope of tank bottom: For coarser separations, the classifier slope is usually from 2.5 to 3.5 inches per foot; for finer separations, the slope is adjusted to 1.5 to 2.5 inches per foot. If the slope is too great on rake classifiers, the sand will slip back and accumulate in the tank, eventually resulting in sticking and non-uniform operation. Within the limits given, steeper slope favours cleaner sand products.

(b) Feed rate: The feed rate is determined by its overflow capacity. The overflow capacity depends upon the length of overflow lip. In the straight mechanical classifier it is determined by the width of the classifier tank and in bowl classifier by the circumference of the bowl. The overflow capacity for a given size classifier is greater, the coarser the mesh of separation and the higher the specific gravity of the ore.

(c) Dilution: This is an essential factor to control the size of separation. It is important that dilution water be added uniformly and that the pulp density in the classifier tank be kept constant. The higher the pulp density is, the lower the settling rate of the ore particles and, hence, the coarser the mesh of separation and vice versa. On fine separation, bowl classifiers are capable of giving overflows of higher percentage solids than standard classifiers.

(d) Speed: The speed of the rakes or respective sand moving devices, determines the degree of agitation of the pulp. For coarser mesh separations, a higher degree of agitation is necessary to keep coarse material in suspension; for finer separations, less agitation and lower raking speeds are necessary to avoid oversize in the overflow product. Agitation must be at a minimum near the overflow lip and greatest near the sand end of the tank to ensure washing of sands.

(e) Height of overflow weir: The addition of strips to the overflow weir to increase its height a few inches is sometimes used as an auxiliary means of obtaining finer overflows and reducing the oversize in the overflow.

(f) Primary slimes: The presence of a favourable proportion of primary slimes in the classifier feed is desirable for best results. Absence of primary slimes may lead to surging in the classifier tank and results non-uniform separation. An excess of primary slimes may make it necessary to operate with high dilution to prevent passage of oversize into the overflow product.

Spiral and rake classifiers are extensively used in closed circuit grinding operation with a ball mill where underflow coarse product is directly fed to the feed box of the ball mill.

Settling cone is a simplest form of classifier and it consists of a cone with central cylindrical feed well at the top of the cone. The spigot at the bottom of the cone is initially closed. The feed is distributed in the feed well. When the cone is full, water and slimes will overflow and a bed of settled sand builds up until it reaches the level shown in Fig. 94. If the spigot valve is now opened and the sand discharge is maintained at a rate equal to the feed rate, classification by horizontal current takes place radially from feed well to overflow lip. Settling cone is a non-mechanical classifier as it needs no mechanical means to remove settled sands.

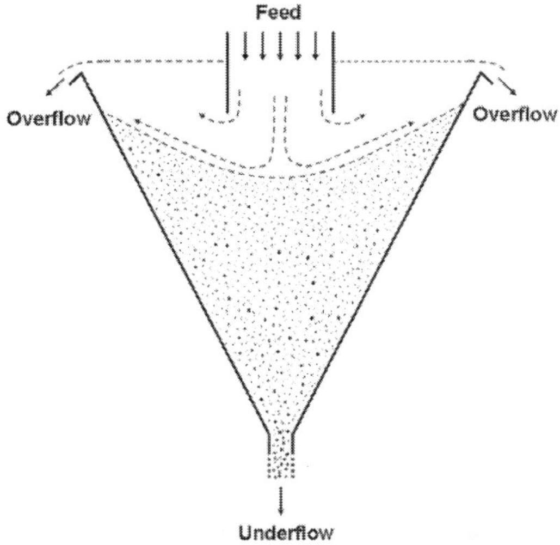

Fig. 94 Settling cone

It is impossible to maintain a regular discharge of sand from the sand bed at the bottom of the cone through an open pipe under the influence of gravity. The spigot is manually opened intermittently with time allowed between discharges. Various forms of automatic discharges have been used in the later developed settling cones such as an Allen cone.

11.2 HYDRAULIC CLASSIFIERS

Hydraulic classifiers are usually hindered settling type. They normally consist of a series of sorting columns. Vertical current of water called hydraulic water is rising through each column and particles are settling out.

As the pulp density increases, a point is reached where each mineral particle is covered only with a thin film of water. This condition is called a **quicksand**, and because of the surface tension, the mixture is a perfect suspension and does not tend to separate. The particles are in a condition of full teeter, which means that each particle is free to move but is unable to do so without colliding with other particles and as a result stays in place. The mass acts as a viscous liquid and can be penetrated by high density particles. A condition of teeter can be produced in a classifier sorting column by putting a constriction in the column either by tapering the column or by inserting a grid at the base of the column as shown in Fig. 95. Such hindered settling sorting columns are known as **teeter chambers**.

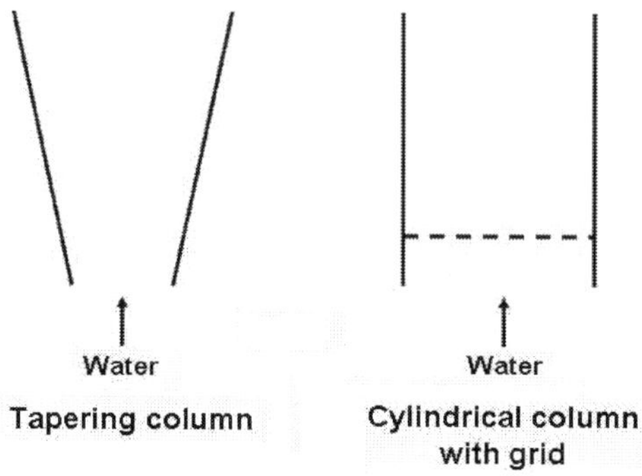

Fig. 95 Constriction in sorting columns

In hindered settling columns of a hydraulic classifier, the velocity of the hydraulic water passing the constriction is higher than that in the part of the column above it. Consequently, falling particles here meet greater resistance. Particles fall until they reach a point where their falling velocity equal to that of rising current. They can now fall no further. As a result, a mass of particles becomes trapped above the constriction and pressure builds up in the mass. The particles thereupon move upward along the path of least resistance, which is usually the centre of the column, until they reach a region of lower pressure at or near the top of the settled mass. Here, under conditions in which they previously fall, they fall again.

As the particles from the bottom rise at the centre, those from the sides fall into the resulting void. A general circulation is thus build up, the particles are said to teeter. The part of the classifier column above the constriction in which the teetering conditions exists is called the **teetering column**. A teeter column acts as though it were a liquid. The lighter particles resented cannot penetrate, but appear to float on it. Heavier particles penetrate and become a part of it but cannot leave it at the bottom. The heaviest particles sink through it and fall through the constriction. Its effective density is higher than the density calculated from its volume and the weights of its solid and liquid. The constant jostling of teetering particles has a scouring effect, removes any entrained or adhering slime particles then leave the teeter chamber and passes out through the classifier overflow. Thus, cleaner separations are made in such classifiers.

In hydraulic classifiers, the velocity of rising hydraulic water is relatively high in the first sorting column, decreased velocities in the subsequent sorting columns and lowest velocity in the last sorting column so that a series of spigot products are obtained as shown in Fig. 96. The size of each successive vessel is increased because the amount of liquid to be handled includes all the water used for classification in the previous vessels and because it is desired to reduce the surface velocity of the liquid flowing from one vessel to the next.

Hydraulic classifiers are of free and hindered settling based on the degree of crowding of particles in the separating zone. If the separating zone is relatively uncrowded so that collision between the particles is relatively infrequent and has no appreciable effects on the results of the operation, it is free settling hydraulic classifier. If the separating zone is crowded and collisions are frequent so that no particle can pass through without hindrance of particles, it is hindered settling hydraulic classifier.

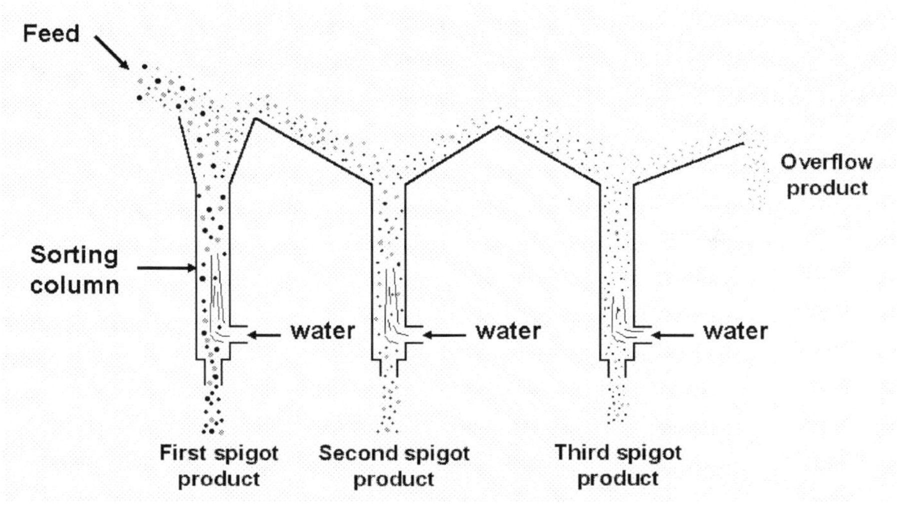

Fig. 96 Separation in hydraulic classifier

Free settling hydraulic classifiers are characterised by the fact that the sorting column is of the same cross sectional area throughout its length. They may be grouped as launder and tank type. Launder classifier is essentially a launder with sorting columns attached to the bottom at convenient intervals. Tank classifier consists of relatively deep V shaped trough or tank with the sorting columns attached to the bottom. The free settling hydraulic classifiers are inefficient in both sizing and sorting. When rough work at relatively high capacity is desired, they are better than the hindered settling hydraulic classifiers.

Hindered settling classifiers differ from free settling classifiers; in that the sorting columns are constricted at the lower end. The hindered settling classifiers use much less water than the free settling classifier and is more selective in its action. Since the ratio of equal settling particles is high, the classifier is capable of performing a concentrating effect and the first spigot product is normally of higher grade than the other products.

Modern hindered settling hydraulic classifiers are designed to eliminate, as far as possible, the necessity for manual control of spigot discharges. This is done by hydrostatic balancing of the quicksands in the teeter columns against some form of automatic control on the spigot discharge ports.

Hydraulic classifiers are used primarily to sort out and group together particles of different densities. This is the first of the two supplementary steps, sorting and sizing, in certain methods of gravity concentration. They are occasionally used to sort sized products containing particles of different densities. They are sometimes used for sizing non-metallic feeds consisting particles of the same densities, because they do closer work than is done by the pool classifiers.

11.3 EFFICIENCY OF A CLASSIFIER

The efficiency of classifiers is difficult to quantify. The usual method consists of screen sizing the classifier overflow and underflow, and of calculating efficiency from those data. The following formula, like formula 4.4 used in screening, can be used to express the efficiency η as a percentage

$$\eta = \frac{p(f-u)}{f(p-u)} \times 100 \tag{11.1}$$

where f, p and u are the percent of minus material in the feed, overflow and underflow respectively, the minus material being any size such that neither f, p, nor u is nil. Since some feed is by-passed into the overflow, the efficiency given by the equation 11.1.1 is increased. To dispose of this objection, it has been proposed to use the following equation (Gaudin 2005)

$$\eta = \frac{10{,}000(p-f)(f-u)}{f(100-f)(p-u)} \tag{11.2}$$

This formula expresses efficiency as the ratio, on percentage basis, of the classified material in the overflow to classifiable material in the feed. The efficiency of classifiers is usually in the range of 50% to 80%, but it would be appreciably greater if gauged by sedimentation analyses rather than by screen analyses. This is because of partly flocculated condition during classification.

Example 11.3.1 illustrates the calculation of classifier efficiency by using the formula 11.2.

Example 11.3.1: *A mill in closed circuit with a classifier is producing –200 mesh material. Screen analysis data of ball mill discharge, classifier overflow and underflow is given in the Table 27. Calculate the efficiency of a classifier.*

Table 27: Screen analysis data for example 11.3.1

Mesh number	Ball mill discharge wt% retained this size	Classifier	
		Overflow wt% retained this size	Underflow wt% retained this size
+18	1.5	0.5	2.7
−18 + 25	4.8	0.8	7.6
−25 + 36	6.8	1.7	16.2
− 36 + 52	15.6	3.4	21.7
− 52 + 72	15.8	8.5	17.4
− 72 + 100	17.5	7.6	15.6
−100 + 150	11.2	9.3	6.8
−150 + 200	5.7	8.0	3.8
−200	21.1	60.2	8.2

Solution:

Given

% of –200 mesh material in feed = f = 21.1%

% of –200 mesh material in overflow = p = 60.2%

% of –200 mesh material in underflow = u = 8.2%

$$\text{Efficiency of a classifier} = \eta = \frac{10{,}000(p-f)(f-u)}{f(100-f)(p-u)}$$

$$= \frac{10{,}000\times(60.2-21.1)(21.1-8.2)}{21.1\times(100-21.1)(60.2-8.2)} = 58.3\%$$

11.4 HYDROCYCLONE

Hydrocyclone, a liquid solid cyclone, was first introduced at the Dutch Stat Mines in Holland in 1939, was taken up by the South Africans in the late 1940s and made its appearance in North America in the early 1950s. A hydrocyclone is a device which by utilizing centrifugal force causes the separation of solids suspended in the liquid (pulp or slurry) fed to it. Due to its tremendous capacity in tons per square foot of space occupied, when compared with the mechanical classifier, as well as some metallurgical advantages accruing from the mechanics of separation, acceptance and use of the hydrocyclone is universal.

Under gravity force, the settling rate of a particle varies as its effective mass. If centrifugal force is applied, the effective mass increases and therefore settling rate increases. As particles are ground smaller they reach a size where the surface drag against the surrounding fluid almost neutralises the gravitational pull, with the result that the particle may need hours, or even days, to fall a few inches through still water. This slowing down of settling rate reduces the tonnage that can be handled and increases the quantity of machinery and plant required. By superimposing centrifugal force, the gravitational pull can be increased from 50 to 500 times depending on the pressure at which the pulp is fed and the size of the vessel. The hydrocyclone is one which utilises centrifugal force to accelerate the settling rate of particles.

Hydrocyclone has no moving parts. It consists of an upright cylindrical section joined to a conical section. The conical section is open at the bottom, variously called as underflow nozzle, discharge orifice, apex or spigot. The top of the cylindrical section is closed with a plate. An open-ended cylinder called vortex finder pass through the plate and concentrically mounted. The vortex finder is extended into the body of the cyclone to prevent short-circuiting of feed directly into the overflow. Cylindrical section is provided with a tangential feed inlet at the upper part. Operating hydrocyclones are usually fitted with variable diameter apex and vortex finder for control of underflow and overflow. For size separation, both outlets are generally open to the atmosphere. Hydrocyclones are generally operated vertically with the spigot at the lower end, hence the coarse product is called the underflow and the fine product, leaving the vortex finder, is called the overflow.

The size of the hydrocyclone is the diameter of its cylindrical section. Fig. 97A shows typical cyclone geometrical configuration. The cylindrical section usually has a length equal to the cyclone diameter. In most cases the length of the vortex finder is the same as the length of the cylindrical section. For the conical section, cone angle varies from 10° to 20°. The hydrocyclone most commonly operates with a continuous flow centrifugal pump.

When slurry is fed tangentially into a cyclone under pressure, a vortex is generated about the longitudinal axis. The accompanying centrifugal acceleration increases the settling rates of the particles, the coarser of which reach the cone's wall. Here they enter a zone of reduced pressure and flow downward to the apex, through which they are discharged with a fraction of the liquid.

At the centre of the cyclone is a zone of low pressure and low centrifugal force which surrounds an air-filled vortex. Part of the slurry, carrying the finer particles with major portion of feed water, moves inward toward this vortex and reaches the gathering zone surrounding the air pocket. Here it is picked up by the vortex finder, and removed through a central overflow orifice. Fig. 97B illustrates generally the action of a hydrocyclone.

The reference point which has been found useful in describing hydrocyclone efficiency is that particle size which has a centrifugal efficiency of 50%. In other words, it is that particle size in the hydrocyclone feed, 50% of whose weight goes to overflow and 50% to underflow. This is called as cut size or separation size. This size is usually referred to as the d_{50} size.

The hydrocyclone is simple to build and operate. A cyclone-pump installation costs 30% to 35% of an equivalent mechanical classifier. The floor area required for a cyclone installation is small compared to a classifier. A cyclone has about 250 times the capacity (per unit area) of a mechanical classifier. It is versatile and can perform a wide variety of separations.

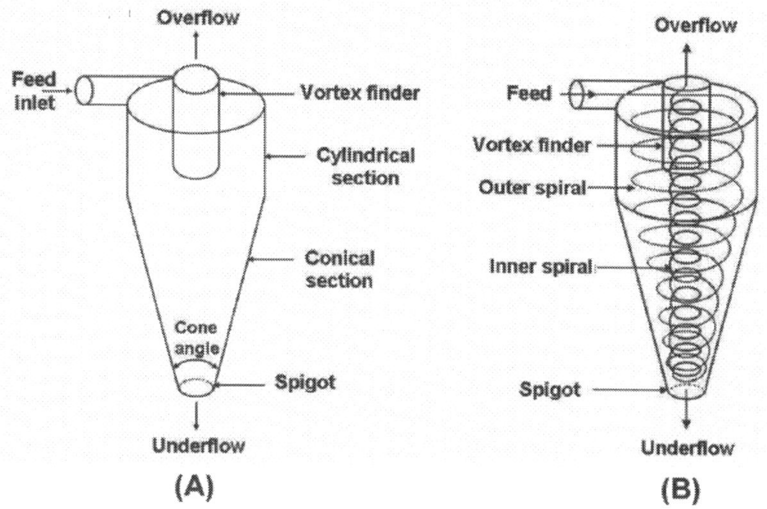

Fig. 97 Hydrocyclone

Following are the design and operating variables of hydrocyclone.

1. **Cyclone diameter**: Increase in cyclone diameter increases the flow rate and the cut size. Since fine separation require small cyclone having small capacity, several cyclones are to be connected in parallel to have high capacity.
2. **Length of the cyclone**: Increase in length of the cyclone increases classification efficiency and residence time. The average time of residence of a particle in a cyclone is the volume of the cyclone divided by the throughput. The residence time in a 10-mm cyclone is of the order of 0.06 seconds and in a low-pressure 300-mm cyclone the residence time might go up to 11 seconds.
3. **Feed inlet diameter**: Increase in feed inlet diameter, increases the flow rate and decreases the cut size.
4. **Vortex finder diameter**: Increase in vortex finder diameter, increases the cut size as more feed material report to the overflow. Hence it is important that the vortex finder diameter is monitored regularly and replaced if excessively worn.
5. **Spigot diameter**: Increase in spigot diameter, decreases the cut size as more fines and excess water report to the underflow. Smaller spigot diameter increases the pressure of the vortex and force more heavy coarse particles to the overflow.
6. **Cone angle**: Increase in cone angle increases the air core diameter which decreases the underflow volume, as a result cut size increases.
7. **Operating pressure**: Increase in operating pressure decreases the cut size. This is because the centrifugal force on the particles is increased, forcing

more particles to the cyclone wall and hence report to underflow. However, too great pressure may exceed the operating capacity of the cyclone causing coarse particles to report to the overflow. If the pressure falls too far, most of the slurry inside the cyclone will report to the underflow. For most operations, the feed pressure ranges from 345 to 700 kPa and depends on cyclone diameter in practice. The minimum pressure for a stable air core is around 30–35 kPa. The feed velocity is 3.7–6.1 m/s and its acceleration in the feed chamber is inversely proportional to that in the hydrocyclone diameter.

8. **Feed size distribution**: The cyclone feed size distribution will affect the cyclone cut size and flow rates. A fine feed size distribution results a fine product size with most of the feed report to the overflow. A coarse feed size distribution results a coarse product size with most of the feed report to the underflow. It is necessary to adjust the pressure, feed rate and feed density to achieve the required separation size to account for changes in feed size distribution.
9. **Feed pulp density (% solids)**: Increased in feed % solids increases cut size as the drag force is increased on each particle causing the particles to remain nearer the centre. The cyclone should ideally be operated at lowest feed density possible while maintaining the required separation.
10. **Feed flow rate**: Feed flow rate increases with pressure. An increase in feed flow rate increases the efficiency of hydrocyclone.

Hydrocyclones are used in many applications owing to the following advantages:

1. Sharper separation
2. Saving of floor space
3. Less power consumption
4. Less maintenance
5. Ability to shut down the mill immediately under load
6. Ability to bring the circuit rapidly into balance
7. Elimination of cyclic surging

Main use of hydrocyclone in mineral processing is as a classifier, which has proved extremely efficient at fine separation sizes (between 150 and 5 microns). It is used increasingly in closed-circuit grinding operations. It is also used for many other purposes such as de-sliming, de-gritting and thickening. It has also found wide acceptance for the washing of fine coal in the form of Heavy Medium Cyclone and Water Only Cyclone.

Material balance equations over hydrocyclone for total material as well as for particular size of the material as percent or fraction of solids can be written as given in equations 4.1 and 4.2 to analyse or to know the performance of the hydrocyclone. Example 11.4.1 illustrates the material balance calculation for hydrocyclone.

Example 11.4.1: *Thirty dry tons of solids per hour is fed to a cyclone. The percent solids in cyclone feed, underflow and overflow are 30%, 55% and 20% by weight, respectively. Calculate the tonnage of solids per hour in the underflow.*

Solution:

Given

Feed rate to the cyclone	= 30 tons/h dry solids
%solids in the feed	= f = 30%
%solids in the underflow	= u = 55%
%solids in the overflow	= p = 20%

Let P and U be the weights of overflow and underflow

$$\text{Total weight of the feed} = \frac{30}{0.3} \text{ tons/h}$$

$$\text{Slurry balance} \qquad P + U = F \Rightarrow P + U = \frac{30}{0.3}$$

$$\text{Solid balance} \qquad 0.20P + 0.55U = 30$$

On solving above two equations U = 28.57 tons/h

Tonnage of solids per hour in underflow = 28.57 × 0.55 = 15.7 tons/h

11.5 CLASSIFICATION AS CONCENTRATION OPERATION

Classification operation can become concentration operation in case of some ores where valuable constituents in the crushed or grounded ore are present in one size range and waste in another size range. Iron ore and clay are the two examples.

In iron ores, siliceous or clayey impurities are generally present in the finer sizes of the crushed ore. Classification of grounded iron ore by using a classifier such as hydrocyclone gives fine and coarse fractions as overflow and underflow. Coarse fraction contains high iron values and fine fraction contains low iron values.

Clays consists of silicate minerals in extremely fine sizes and contaminated by coarse impurities principally quartz, feldspars, micas and pyrite. Deflocculation of raw clay followed by classification yields sedimented coarse residue and suspended washed clay. Thus, the classification results concentration.

11.6 PROBLEMS FOR PRACTICE

11.6.1: *It is intended to produce −200 mesh material using a ball mill classifier closed-circuit operation. Screen analyses data is given in the Table 28. Calculate the efficiency of a classifier.* *[45.7%]*

Table 28: Screen analyses data for problem 11.6.1

Tyler screen mesh	Ball mill discharge wt% retained this size	Classifier	
		Overflow wt% retained this size	Underflow wt% retained this size
20	1.5	0.5	2.7
28	4.8	0.9	7.6
35	6.8	2.3	16.2
48	15.6	4.7	22.7
65	15.8	11.5	18.5
100	17.1	18.6	15.6
150	10.2	15.3	6.8
200	5.7	8.2	3.8
−200	22.5	38.0	6.1

11.6.2: *Table 29 shows the sizing analyses of feed, overflow and underflow solids of a hydrocyclone.*

Table 29: Size analyses for problem 11.6.2

Mesh (Tyler)	+100	100/150	150/270	−270
Feed	*21.0*	*16.9*	*34.8*	*27.3*
Overflow	*6.1*	*14.9*	*42.4*	*36.6*
Underflow	*42.7*	*19.7*	*23.9*	*13.7*

If the separation is aimed at 150 mesh, calculate the efficiency of the hydrocyclone. *[42.5%]*

11.6.3: *Rougher spiral tailings of an Iron Ore Beneficiation plant have been fed to Hydrocyclone at the rate of 100 tons/h for further recovery of Iron from the fines by using HGMS. Hydrocyclone feed, overflow and underflow are sampled. The results of analyses of samples are given in Table 30.*

Table 30: Analyses results of feed, overflow and underflow

	% solids
Feed to hydrocyclone	16
Hydrocyclone overflow	10
Hydrocyclone underflow	40

Calculate the tonnage of solids per hour in the underflow. *[50 tons/h]*

11.6.4: *A classifier was fed with ground silica at the rate of 50 tph. The screen analyses of feed, product and tailings from a classifier are as shown in Table 31:*

Table 31: Screen analyses for problem 11.6.4

Size (microns)	Feed%	Overflow%	Underflow%
850	15.2	0.0	18.3
500	8.5	0.0	10.2
250	33.6	2.9	39.9
125	34.2	65.4	27.8
75	8.5	31.8	3.7
	100.0	100.0	100.0

Estimate the tons of dry ore per day in classifier underflow.

[243.7 tons/day]

■■■■

12 Beneficiation Operations

Beneficiation operations are physical or mechanical unit operations where valuable mineral particles are separated from the mixture of liberated valuable mineral particles and gangue mineral particles. Every beneficiation operation is based on one or more physical properties in which valuable and gangue minerals differ in that property.

Major beneficiation operations and the basis of separation are given in Table 32.

Table 32: Beneficiation operations and Basis of separation

Beneficiation Operations	Basis of Separation
Gravity Separation	Sp.gr, Size and Shape
Flotation	Wettability and Sp.gr
Magnetic Separation	Magnetic susceptibility
Electrical Separation	Electrical conductivity

The beneficiation operations separate the valuable mineral particles from gangue or waste mineral particles into two products. The valuable mineral product obtained from the beneficiation operation is called **concentrate** and the gangue mineral product obtained from the beneficiation operation is called **tailing**. Practically all the valuable mineral particles cannot be separated by a beneficiation operation due to

- Incomplete liberation (degree of liberation < 100%)
- Inefficiency of beneficiation operation (efficiency of operation < 100%)

Because of incomplete liberation, the locked particles may be placed in both the concentrate and the tailing products depending upon their size, shape and specific gravity. Because of inefficiency of beneficiation operation, the particles composed wholly of valuable mineral or of gangue minerals escape the separation owing to their shape or relative size and misplaced in both concentrate and tailing products. Hence, the concentrate contains not only valuable mineral particles but also the locked particles and gangue mineral particles. Similarly, the tailing contains gangue mineral particles, locked particles and valuable mineral particles. The beneficiation operation often yields third product called **middling** and it contains primarily locked particles and small quantities of valuable and gangue mineral particles that have escaped the separation. Therefore, the middling product needs further beneficiation which may consists of size reduction step to detach locked particles present in middling and beneficiation operation to separate as maximum valuable mineral particles as possible as a concentrate product.

The separation operations can generally be divided into preliminary and final. The preliminary separation is unable to do finished work. They simply divide the ore into a set of preliminary products which are well suited for final beneficiation operation. Washing the ore is an example for preliminary separation and usually called as preliminary washing or scrubbing and the equipment used are called as preliminary washers or scrubbers.

12.1 PRELIMINARY WASHERS

Washing the ore with water sprays while it is screened to remove slimes and colloids generated and coated the ore particle surfaces while the ore is comminuted is called de-sliming the ore. The external dirt and dust that might have included in the ore coating the surfaces of the ore particles while the ore is transported over long distances can also be removed by water sprays. Sometimes when they need some kind of rupture force to remove slimes from the surfaces of ore particles, the ore is scrubbed in a trommel screen. Lifter bars, sometimes in the form of bolted in angle iron, are attached to the interior of the trommel. The ore is fed into the elevated end of the trommel. The ore particles are repeatedly turned over while traveling from feed end to the discharge end of the trommel. Water, often under pressure, is provided to the trommel and the combination of water and mechanical action frees the slimes and colloids from the ore. Whenever the greater rupturing forces are required to disintegrate the clayey agglomerates and/or other impurities, the special washers with greater disintegrating forces are needed.

Preliminary washers disintegrate and float adhering clay, slimes and colloidal particles from the coarser particles of ore. The disintegrating forces used are of gentle nature, not intended to fracture the ore particles but simply to tumble particle against particle until the slimes arc detached from the surface of coarser particles. A water jet, or simple stirring, may suffice to detach slimes from the lump material so that it flows out of the washer as a dirty effluent while the cleaned ore particles leave from the other end of the washer. Some machines do only the work of disintegration and require a subsequent machine to make the separation. Attrition scrubber followed by classifier is an example. Scrubbers of this type are extensively used in the washing of iron ores, phosphate rock, and sometimes coal and various other materials.

The following two most common types of washers are used in mineral industries:

1. Washers using hand tools for stirring, e.g. trough washers
2. Washers using some form of rotating stirrers driven by power, e.g. log washers, drum washers and attrition scrubbers

Trough washer: Trough washer is a wooden trough which may or may not be lined with iron or steel plate. The bottom usually slopes from the head end, which is closed, to the tail end, which is open. The angle of slope varies between 0 and 10 degrees. Sometimes, this slope is downward toward the tail end. Sometimes the slope is upward toward the tail end. The latter form is more desirable because it allows larger charges to be worked and prevents the loss of certain small sizes of rich ore which would be carried off in the case of a downward slope. Water is supplied at the head end often by means of a feed box. The water, falling from a height, exerts a considerable washing force. The ore, which may be run-of-mine ore or the product of some breaker or concentrator, is fed in at the tail end, shovelled over, and worked toward the head end until the slimes are removed. It is then shovelled to a gravel screen. Sometimes by skilful turning over of the mass, not only is the slimes washed away, but the waste rock is separated from the mineral.

Log washer: Log washer (Fig. 98) consists of an inclined tank in which normally two logs rotate. The logs are armed with stirring paddles set to form part of two interrupted spirals. Feed material is normally entered between two to four feet from the feed end of the box and is subjected to an aggressive washing action resulting from the intermeshing paddle design. The paddles perform two functions: they help convey material and are the catalyst for particle on particle scrubbing. After deleterious material is liberated, clean material move upward between the shafts and discharges out the bottom of the box after a final rinse. The slope of the log washer varies from 0 to 14 degrees depending upon the severity of the washing action. As the percentage of deleterious material increases, the slope must be raised to increase retention time. Deleterious materials are absorbed into the water and carried over a weir on either side of the box. In high capacity applications, an optional rising current system helps lift waste fractions to the top of the water level and over the weir. Log washers are designed to remove tough, plastic clay contaminants and other deleterious materials from gravel, stone and various ores.

Fig. 98 Log washer (Courtesy of JXSC Mining, China)

Drum washer: Drum washer consists of a rotating steel cylinder, equipped with internal lifters and blades to thoroughly scrub the feed material in an intense action. The wet or dry raw material to be treated is fed in at one end of the drum and the clean water enters at the opposite end. The feed raw material combined with water for pre-washing and pulping to obtain a homogenised mixture. The lifting bars and blades, specifically configured to suit the duty, lift and drop the aggregate in a cascade manner and advance it within the drum, releasing the fine particles and clay adhering to the coarse fractions. The stresses to which the material is subjected inside the drum, a combination of friction generated within the aggregate itself, and between the drum wall and the internal components, produce a self-grinding effect in the coarser fractions, which helps break down agglomerates, especially when the drum is full. The counter current arrangement ensures that the entering ore is first washed by the dirtiest water as the latter runs out. After traveling through the machine, it receives its final washing from the cleanest newly-entered wash-water. The washing or breakdown effect is directly related to the duration of time the material remains inside the drum, with the throughput capacity decreasing as the retention time increases, being adjusted depending on the condition of the material to be processed. Washing drums are designed as heavy-duty equipment, for the vigorous washing of coarse/medium-size materials such as ores, rocks and aggregates with the presence of clay-based agglomerates and/or contaminated by clay or other contaminants.

Attrition scrubber: Attrition means to wear away a surface, typically by friction or abrasion. An attrition scrubber does exactly this, using impellers to force particles against each other, causing intense scrubbing and disintegration. Attrition scrubber consists of two opposing axial turbine impellers mounted on a shaft inside a tank. Shafts are coupled to either gearbox or spindle bearing drives. Feed and outlet boxes are integral or flanked with the tank. Attrition scrubbers work well if they have a good mixing action, with fast blend time, multiple turnovers and high product churn. Better mixing within a scrubber results in particles contacting more often and at a greater force. The impeller positioning is also an important design consideration because, if they are too far apart, the particles will lose energy as they travel between them and impact velocity will be reduced. The object of the attrition scrubber is to achieve the strongest possible attrition of particle surfaces without reduction of particle size. This is done by introducing the feed at high density of 70% to 80% solids in the case of sand. The most important functions of the attrition scrubber are the scrubbing of particle surfaces to remove slime coatings and disintegration of clay agglomerates.

12.2 HAND SORTING OR PICKING

Hand sorting is a common practice which is adopted when mining minerals such as asbestos, barites, manganese ore, diamond, mica, coal etc. In case of crysotile asbestos, the run of mine containing long staple fibre is sorted out manually from the mill grade fibre. Hand sorting is a universal practice. In India, in the case of barites mining, differentiation of three colour grades—namely, snow-white, white and off-grade—is done entirely by hand sorting. In the case of manganese ore, battery grade ore is sorted out from other grades from the ROM ore by manual operations. Similarly, in case of diamond, the gem and industrial varieties are sorted out by hand-picking only. In the case of mica, hand sorting is the only means known so far for grading mica. Hand picking was the oldest method in case of coal and performed based on the colour difference between the clean coal and waste mineral particles. The workers are trained to identify waste pieces which contain shale, slate, silica, etc. and remove them manually. The workers standing by the side of a conveyor belt knock out the rejected pieces in a bin located below. The conveyor belt is illuminated to make the identification easy with dust free pieces using water showers.

13 Gravity Concentration

Gravity concentration is a beneficiation operation performed on ore to separate the valuable mineral particles from the ore particles by using gravity force. Heavy medium separation, jigging, spiralling and tabling are the chief gravity concentration operations used in mineral processing plants.

Some idea of the type of separation possible can be gained from the ratio, called concentration criterion, and is defined as

$$\frac{\rho_h - \rho_f}{\rho_l - \rho_f}$$

where ρ_h is the density of the heavy mineral, ρ_l is the density of the light mineral and ρ_f is the density of the fluid medium. This ratio is a comparison of the buoyancy forces which are at work.

In general, when this ratio is greater than 2.5, whether positive or negative, then gravity separation of particles down to 75-micron size is possible. As the value of this ratio decreases, the efficiency of separation decreases, and if it is below 1.25, gravity concentration is not commercially feasible in general. Following are the values of ratios and particle sizes based on experimental data:

Concentration criterion	**Suitability to gravity separation**
> 2.50	easy down to 75 μm (200 mesh)
–2.50	possible down to 150 μm (100 mesh)
1.50–1.75	possible down to 1.7 mm (10 mesh)
1.25–1.50	possible down to 6.35 mm
< 1.25	no separation is possible in water at any size Another (heavier) fluid or a pseudo fluid medium heavier than water is needed. An alternative is to alter the effective particle density by causing it to adhere to an air bubble as in flotation.

Some concentration criterion ratios for minerals that are treated by gravity separation are given in Table 33.

Table 33: Concentration criterion of minerals separated by gravity separation from a gangue of density 2.65 g/cc

Mineral	Fluid	Concentration Criterion
Gold	Water	10.3
Gold	Air	6.8
Cassiterite	Water	3.5
Coal	Water	3.4
Hematite	Water	2.5

The motion of a particle in a fluid depends not only on its specific gravity but also on its size; large particles will be affected more than smaller ones. The efficiency of gravity concentration operation, therefore, increases with particle size. Smaller particles respond poorly because their movement is dominated mainly by surface friction. In practice, close size control of feeds to gravity concentration processes is required to reduce the size effect and make the particles to move depending on their specific gravities.

13.1 SINK AND FLOAT

Sink and float, also called as Heavy Liquid separation (HLS), is an operation where particles of different specific gravities are separated by using suitable heavy liquid. The principle of float and sink is

> *"When two particles of different specific gravities are immersed in a liquid having specific gravity intermediate between that of two particles, heavier particle would sink and lighter particle would float."*

In principle, it is the simplest and standard laboratory method for separating minerals of different specific gravities. Solutions of inorganic salts like calcium chloride and zinc chloride are used as heavy liquid. Table 34 shows some of the organic liquids used for separation of mineral particles.

Table 34: Organic liquids and their specific gravities

Liquid	Chemical formula	Specific gravity
Benzene	C_6H_6	0.80
Tetra Chloro Ethylene	C_2Cl_4	1.62
Pentachloro ethane	C_2HCl_5	1.68
Methylene Bromide	CH_2Br_2	2.48
Bromoform	$CH\ Br_3$	2.89
Tetrabromo ethane	$C_2H_2Br_4$	2.96
Methylene iodide	CH_2I_2	3.31
Thallous formate solution	H COOTl	3.39

Solutions of required specific gravity are prepared by adding two or more liquids at different proportions. When the ore particles are introduced in this solution, mineral particles of more specific gravity will sink and mineral particles of less specific gravity will float. Then two products are removed from the solution.

Separation of minerals, Ilmenite, Monazite, Rutile, Zircon, Garnet from Beach Sands, separation of raw coal into different specific gravity fractions are the two important examples for the application of sink and float.

The buoyant forces acting on the light particles in a heavy medium make them to rise to the surface and the heavy particles sink to the bottom because of gravity force. In a static bath, force balance equation is written as

$$F_g = (m_p - m_f)\, g \qquad ...13.1$$

where

F_g = gravitational force

m_p = mass of the solid particle

m_f = mass of the fluid displaced by the particle

For the particles which float, F_g will have a negative value and for the particles which sink. This value is positive.

Nowadays, Heavy Liquid Separation is not industrially performed in mineral or coal processing plants owing to the following reasons:

1. Some liquids are toxic, some others are corrosive
2. Organic liquids are expensive
3. They absorb on particle surfaces and contaminates the mineral particles
4. Loss of liquid is more due to volatility
5. These liquids cannot be recovered economically
6. It is not possible to get required heavy liquids always

An alternative process known as Heavy Medium separation (HMS) or Dense Medium Separation (DMS) is adopted in processing plants.

13.2 HEAVY MEDIUM SEPARATION

Heavy Medium Separation or Dense Medium Separation is a process similar to that of Heavy Liquid Separation. However, a pseudo liquid is used instead of heavy liquid. A pseudo liquid is a suspension of water and solids which behaves like a true liquid. The solids used are known as medium solids, and they must meet the following requirements:

1. They must be sufficiently high in density
2. They must be physically strong
3. They must be chemically inert
4. They must be readily available at a low cost
5. They must be easily recoverable
6. They must not enter the cracks of cleaned ore lumps
7. The resulting viscosity of mixture of solids and water should be low

To produce a stable suspension of sufficiently high density, with a reasonably low viscosity, it is necessary to use fine, high specific gravity solid particles. Agitation is necessary to maintain the suspension and to lower the apparent viscosity.

The fraction of medium solids (C_w) to be added to maintain the suspension at required specific gravity can be calculated from the equation 5.5

$$C_w = \frac{\rho_p\,(\rho_{sl} - 1)}{\rho_{sl}(\rho_p - 1)}$$

Some of the medium solids employed for making suspensions are given in Table 35.

Table 35: Medium solids

Solids	Specific gravity
Sand Loess Shale	2.6
Barytes	4.0
Magnetite	5.0
Ferrosilicon	7.0
Galena	7.5
Some clays	----

The heavy medium separation is extensively used to clean coal and for the concentration of a wide variety of ores such as those of iron, chrome, manganese, tin, tungsten, lead-zinc, fluorspar, magnesite, garnet, diamonds, gravel, etc. It may be used wherever a significant density difference occurs between two minerals, and commercial separations are typically made in the range of 1.30–3.80 specific gravity. The particle size treated ranges downward from 6–8 inches top size. The minimum size of the ore particles must always be more than the maximum size of the medium particles used. Particles greater than about 10 mesh may be treated in a static bath, though for reasons of separation efficiency, +1/2-inch feed is usually preferred. For particles of less than this size, separation in a heavy medium cyclone is generally used.

Heavy medium separation offers the following potential advantages.

1. Ability to make sharp separations
2. Ability to quickly change the specific gravity of separation to meet changing conditions
3. Ability to continuously remove products
4. Ability to treat a broad size range of products
5. Ease of start-up and shutdown without loss of separation efficiency
6. Relatively low cost of the medium solids and low medium losses
7. Low operating and maintenance costs

8. High capacity with the use of relatively little floor space
9. Relatively low capital investment per ton of capacity

Heavy medium separation may be used to produce a finished concentrate, or a concentrate and a middling of differing quality, or a pre-concentrate by rejection of unwanted gangue. It is widely used for washing coal at coarser sizes. Chance Cone, used for washing coal of 75–13 mm size, is the oldest and efficient HMS unit.

13.2.1 Chance Cone Process

Patented in 1917, the chance cone process is the most popular coal washing process. The first plant was erected in USA in 1921 for washing anthracite. In 1951, the West Bokaro coal washery under the management of TISCO, the first Indian coal washery, employed the chance cone process.

This process employs a suspension of sand in water, the sand having a specific gravity of about 2.60 and a size range of 30–100 B.S. mesh. Such a suspension is unstable as the sand particles at this size settle readily in water. Therefore, the process requires some method of maintaining the sand in suspension. This is accomplished by stirring the sand-water mixture and using rising currents of water of sufficient velocity to hold the sand suspension.

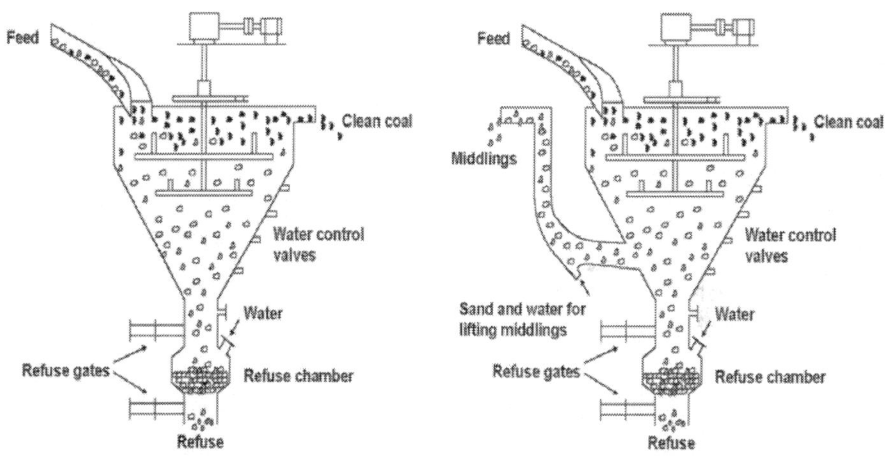

Fig. 99 Two- and three-product chance cones

The separating vessel of a Chance cone process (Fig. 99) consists of an inverted cone with a cylindrical upper section. Three or four water inlets with control valves are arranged in the side of the cone. These water inlets provide the rising currents of water in the vessel and used to control the specific gravity of suspension. Two rotating paddles at the centre of the cone driven by bevel gears revolve at a slow speed of 12 rpm and fulfil three functions.

- They scrap away any sand or refuse accumulating on the sides of the cone and break up any banks of sand

- They impart a rotary motion to the suspension which assists the flow of the coal from the feed chute to the clean coal exit weir
- They also assist in maintaining an intimate mixture of the sand and water

The base of the cone is provided with two air operated slide gates. Between them, a refuse chamber is mounted for receiving the sinking refuse. These gates are automatically controlled and operated.

The chamber is first supplied with water from the filling receiver, via the air operated filling valve, closing a pressure switch when full. The closing of this switch automatically opens the top slide gate which allows the refuse that has accumulated in the bottom of the cone to fall into the refuse chamber. After a given time—usually 15–20 seconds—the top gate closes and automatically opens the bottom gate, allowing sand, water and refuse to fall on to the de-sanding screen. The bottom gate closes after the predetermined time and the cycle is re-commenced by the opening of the air operated filling valve. The cycle is timed to deal with the maximum rate at which refuse is likely to be delivered and is adjustable. Each cycle time is about 1 minute.

Raw coal is fed into the top of the separating vessel at one side by means of a chute. The clean coal, being lighter than the suspension, floats at the surface, and owing to the rotation induced in the separator by paddles, gradually moves around to the discharge point where it overflows with water and sand. Sand is added to the vessel continuously in the form of thick pulp to keep the level of the suspension up to the level of the overflow lip at the top of the vessel. Thus, clean coal floating on the suspension overflows constantly, carrying with it a certain amount of sand suspension. The water required for agitation and maintenance of the suspension is introduced through water inlets at various levels in the cone.

The amount of sand in suspension at any given time and, hence. the specific gravity of the suspension depends directly on the velocity of the upward water current in the cone. Low water velocities give thick suspensions of high specific gravity; high water velocities give dilute suspensions of low specific gravity. Thus, regulation of the water velocity is the means by which the operator is able to adjust the cone to make a separation at any desired specific gravity.

Both the clean coal overflow from the top of the cone and the refuse discharged from the bottom of the cone are de-sanded and de-watered on separate screens. Sand and water is carried to main sand sump where, due to the low rising velocity of the water, the sand is retained at the bottom of the sump and the water overflowed into an outer compartment. From this sump, sand and water are pumped back to the top of the cone and the water control valves to the cone water inlets.

Another function of the sand sump is to hold a reserve of sand. If the cone has been operating at a high gravity and is suddenly changed to a low one, the excess of sand automatically leaves the cone and is retained in the sump. This reserve of sand is likewise, available when more sand is required in the cone.

The sand losses in operating the Chance cone process are around 1.5–2 lb/ton of raw coal treated, and to compensate for this, it is usual to add make-up sand once in a shift. This make-up sand is also stored in the sand sump thus avoiding the inconvenience of adding sand continuously.

A Chance cone may be designed for either two- or three-product separations. When it is desired to obtain middling product in addition to the clean coal and refuse products, a separate middlings column is provided at the side of the cone (Fig. 99). This column is in direct communication with the sand suspension in the cone, and is operated by either increasing the gravity at the bottom of the cone or lowering the gravity at the top by adjusting the water control valves. The middlings, floating at the level of higher specific gravity are carried up the tube by introducing additional sand and water at its base. This type of cone can be operated as a single gravity separator or a two-gravity separator merely by opening or closing the valves controlling the sand and water for lifting the middlings up the column.

The chief advantages of the Chance Cone process are as follows.

1. As the separation depends solely on difference in specific gravity, the size and shape of the particles does not affect the washing efficiency.
2. Within the maximum capacity of the cone, intermittent or variable feed does not affect the washing efficiency.
3. Coals containing high proportions of near-gravity material can be efficiently washed.
4. The operating density can be altered in a few minutes if the products of different seams are to be washed or in accordance with market requirements.
5. Variation in the inherent ash of the raw coal does not affect the efficiency.
6. Breakage in the cone is low.
7. Power consumption is fairly low and is largely a pumping load.
8. A middling product can be separated if required.

Barvoys Process, Dutch State Mines Process, Tromp Process, Drewboy Process, and Wemco Drum Separation Process are the other Heavy Medium Separation Processes employed in Coal Washeries. Barrel Washer, works with self-generated coal slurry as medium, is used in non-coking coal washing plants.

13.2.2 Centrifugal Separator

Heavy medium cyclone is a centrifugal separator similar to the conventional hydrocyclone in principle of operation. It provides a high centrifugal force and a low viscosity of the medium. It has been developed with an included cone angle of approximately 20°, and is used to treat ores and coal in size range of 40–0.5 mm. Instead of water alone, a suspension of heavy particles in water is used in Heavy Medium Cyclone. This suspension, called **heavy medium**, is a mixture of –325 mesh particles of magnetite and water in case of beneficiation of coal.

In a centrifugal separator, equation 13.1 becomes

$$F_C = (m_p - m_f)\frac{v^2}{R} \qquad 13.2$$

where

F_C = centrifugal force

v = tangential velocity

R = radius of the centrifugal separator

From this equation, it is clear that the forces causing the separation of the particles are proportional to $\frac{v^2}{R}$ which is much greater than the gravity forces used in static heavy medium separator. Hence particles down to 0.5 mm size can be separated by centrifugal separators.

As Heavy Medium together with the coal particles are fed through the feed inlet orifice, a vortex with a hollow air core extending from the overflow to the underflow orifice forms in the cyclone. Under the influence of the centrifugal force, high specific gravity coal particles move through the medium to the wall of the cyclone and descend in a spiral flow pattern to the underflow orifice. Those coal particles in the feed stream having the lower specific gravity than the heavy medium follow the major portion of the flow to the centre of the core where they are caught in the high velocity upward central current and are carried out through the overflow orifice. Some particles arriving in the core leave again under the influence of the centrifugal acceleration imparted to them and move to the apex discharge opening along the wall. At the point, near the apex opening, the non-tangential current is direct upward into the core, the particles can again be selected for discharge through either the apex or the overflow opening.

Particles much below the specific gravity of separation will immediately, after entering the cyclone, move rapidly toward the centre and issue through the overflow opening without being recirculated. Heavy particles, on the other hand, will immediately after getting into cyclone move towards the wall and issue through the apex opening.

Both overflow and underflow of HM Cyclone are drained separately on sieve bends. The drained medium of −28 mesh is fed to constant head tank for reuse. The clean coal and refuse are rinsed on the rinsing screens separately with water sprays. The underflow is processed in two stage magnetic separators to recover magnetite for reuse.

Centrifugal force also acts on the very fine magnetite medium and therefore the specific gravity of the medium will increase towards the apex discharge opening. Therefore, the specific gravity of separation is slightly higher than that of the medium.

In this cyclone, floating means moving towards inner air core of the cyclone, while sinking means moving towards the wall of the cyclone. Both movements are rapid as they are under the action of centrifugal force not gravitational force. The specific gravity of separation is related not only to the specific gravity of the heavy medium suspension but also to the inward flow towards the centre. The specific gravity of separation is usually slightly higher than that of the medium. Increased overflow diameter increases the difference between the specific gravity of the medium and specific gravity of separation. The cyclone is at its most efficient when the specific gravity of separation is as close as possible to the specific gravity of the medium. To achieve this, the diameter of the vortex finder should be slightly greater than that of the apex.

The capacity of Heavy Medium Cyclone increases with its diameter. Heavy medium cyclone diameters normally range from 200 mm up to 1 metre, with the inlet pressure usually being higher for the smaller diameter units. They are operated with inlet pressures as low as 6–8 psig, but pressures of 20 psig and higher are used in practice. At very low pressures, the separation is much less effective than at higher pressures, especially for the finer sizes. HM Cyclones with a cone angle of 20° is standard and they are commonly installed with axes inclined at 10°–15° to the horizontal (Fig. 100) to enable the unit to be fed at a comparatively low inlet pressure, preferably from a steady head tank and to allow drainage after shutdown. The optimum static head required is approximately 9 times the diameter of the cyclone for washing coal.

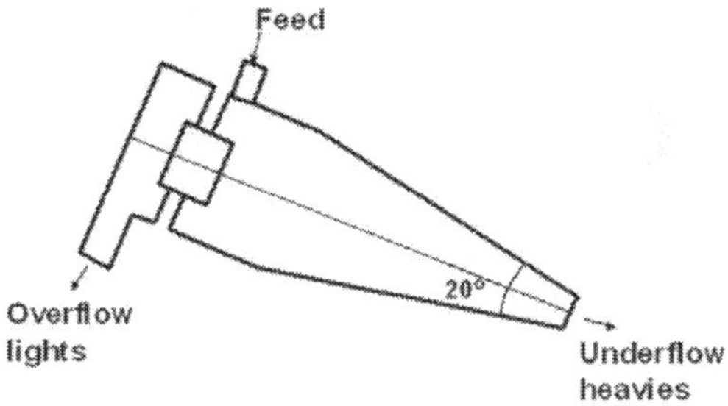

Fig. 100 Typical Heavy Medium Cyclone

The heavy medium cyclones provide efficient and sharp separations in treating coals at separating specific gravities ranging from 1.42 to 1.55. Considering the very short residence time within the cyclone, the sharpness of separation is quite remarkable. The top size of the coal particles to be treated is commonly 13 mm. The bottom size of the coal particle is generally related to the medium recovery system. 0.5 mm is the lowest accepted size for efficient separation and it is due to the difficulty in reclaiming magnetite from very fine size coal.

By giving a special shape to the cone and to the cylindrical part of the cyclone, it is possible to affect separation without using heavy medium. Such cyclone is known as water-only cyclone or water washing cyclone. Vorsyl separator, LARCODEMS, Dyna Whirlpool and Tri-Flo separator are some other centrifugal separators.

13.3 JIGGING

Jigging is the process of separating the ore particles by introducing them on a perforated surface (or screen) in a layer or bed with a depth many times the thickness of the largest particle and is subjected to an alternating rising and falling flow of fluid with the objective of causing all the particles of high specific gravity to travel to the bottom of the bed while the particles of lower specific gravity collect at the top of the bed. The fluid may be either water or air, water being the more commonly used and it is called **hydraulic jigging**. If the air is used as fluid, it is called **pneumatic jigging.**

In the history of mining it was found that heavy minerals could be separated from lighter ones by mixing with water in a tub and stirring or agitating the mixture until the heavier particles found their way to the bottom. It was later found that if the tub had a porous bottom and was moved up and down in water, the alternating upward and downward pulses produced a separation much more efficiently and quickly. This was the jigging in fundamental form and it was so effective and simple, both in equipment and operation. It remained the principal method of mineral concentration.

In jigging, the particles are allowed to settle only for short period and the particles will never attain their terminal velocities. It means that separation will depend on the initial settling velocities of the particles. The particles will settle during their accelerating period. The initial settling velocity is extremely low and the drag force due to frictional effects is not developed. Under these circumstances, the drag force is practically zero and the two principal forces acting on the particle are the gravitational and the buoyant forces. Then the force balance equation becomes

$$m_p \frac{dv}{dt} = (m_p - m_f)\, g \qquad 13.3$$

$$\frac{dv}{dt} = \frac{m_p - m_f}{m_p} g = \left[1 - \frac{\rho_f}{\rho_p}\right] g \qquad 13.4$$

Equation 13.4 shows that the initial acceleration of the particles during settling depends on the force of gravity, density of the particle and the fluid and does not depend on the size or shape of the particle. Initial acceleration is maximum for the denser particles. This situation indicates that light and heavy particles can be separated by providing short duration of settling. It can also be explained by considering three heavy particles and three light particles of different sizes. The six curves in Fig. 101 shows the settling velocities of these particles as a function of time.

If the particles are allowed to settle for time t_3 small heavy and coarse light particles settle equally. Similarly, fine heavy and small light particles settle equally. Coarse heavy particle settles faster than all other particles and fine light particle settles slower than all other particles. Hence only these two particles can be separated.

If the time of settling is reduced to t_2, it is not possible to separate fine heavy and coarse light particles as they fall to an equal distance in a settling period of t_2. However coarse and small heavy particles can be separated as they settle faster than all other particles. Similarly, small and fine light particles can be separated as they settle slower than all other particles.

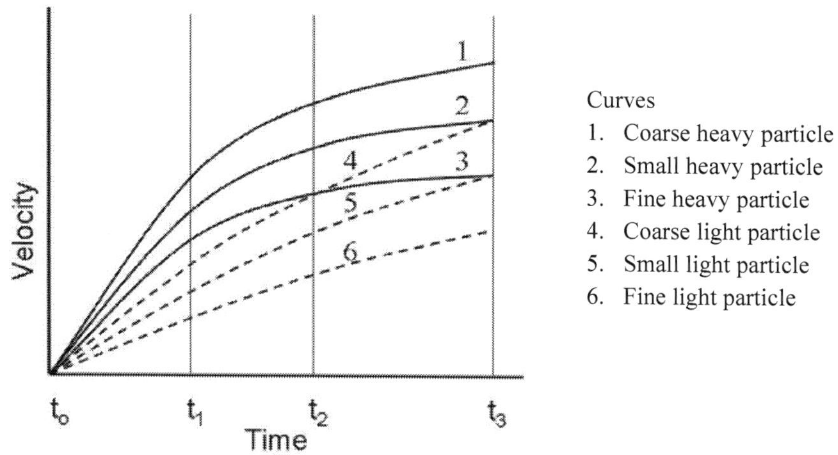

Fig. 101 Settling velocities of six particles

If the time of settling is still reduced to t_1, all heavy particles will fall to a greater distance than all light particles and hence all heavy and light particles can be completely separated irrespective of their size.

The short settling time t_1 is achieved in jigging by giving frequent pulsion and suction strokes of a fluid to a bed of mixed particles of different densities and sizes. After repeated pulsion and suction strokes of the fluid, the particles are rearranged in a bed according to their densities and sizes.

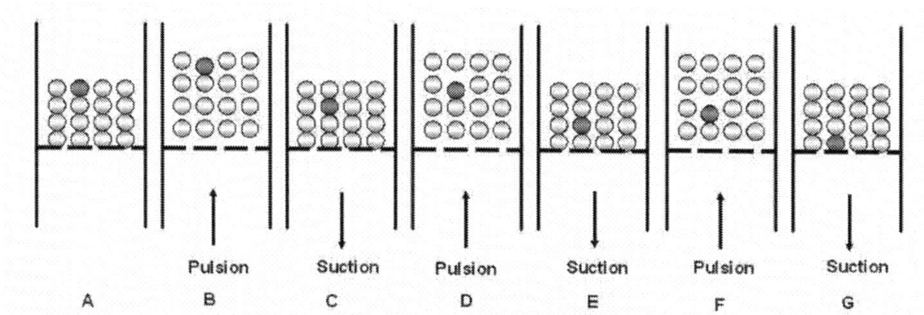

Fig. 102 Jigging process

Let a simple experiment with a circular screen fixed inside a vertical hallow cylinder, which can be called a Jig, be considered. A group of light particles, all of the same size, with one heavy particle at the top of light particles are placed on a screen as shown in Fig. 102A. If water is introduced from the bottom to create a pulsion stroke, energy of the water acts to lift all the particles in the bed, expanding the contents to the teeter stage. Light particles are pushed more the distance while heavy particle is pushed less the distance as shown in Fig. 102B. Now, if the water is start withdrawn from the bottom to create a suction stroke, light particles settle down less the distance while the heavy particle settle down more the distance. Due to the suction created, heavy particle is pulled down more the distance and light particle is pulled down less the distance. This result rearrangement of the particles as shown in Fig. 102C where heavy particle is placed one layer below from the top. During the further repeated pulsion and suction strokes, the same process is repeated as shown in Fig. 102D to G and at the end, the heavy particle is placed in the bottom layer. The particles settle by **hindered settling** as less quantity of water is present so that heavy particle is always placed at the bottom.

If a group of both light and heavy particles, all of the same size, are taken in equal number as shown in Fig. 103A and do the same experiment, all the particles are rearranged as shown in Fig. 103B after several pulsion and suction strokes are given. Rearrangement of particles is called as **stratification**.

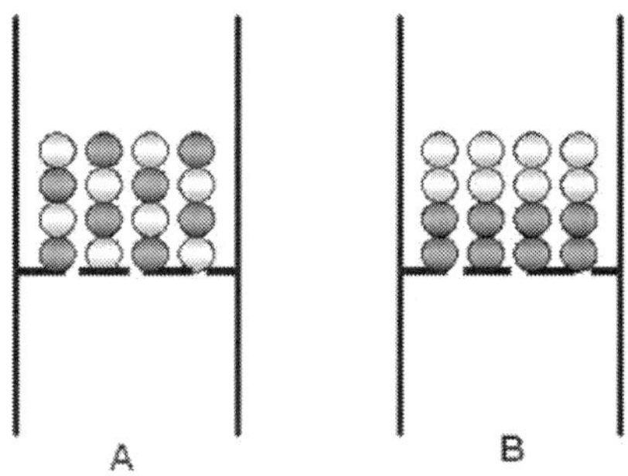

Fig. 103 Stratification of equal size particles due to Jigging action

If a group of both light and heavy particles of different sizes are taken and do the same experiment, all the particles are rearranged in similar fashion. However, small particles pass through the interstices between coarse particles as shown in Fig. 104. This phenomenon is called **consolidation trickling**.

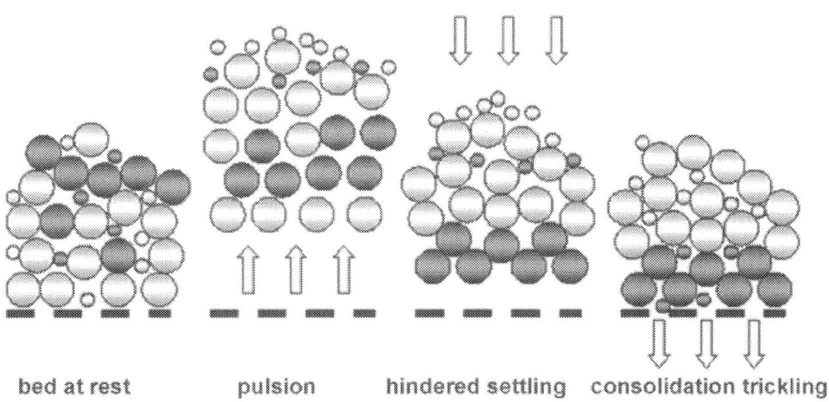

Fig. 104 Stratification of different size particles due to jigging action

If repetition of fall is frequent enough, and the duration of fall is short enough, the difference in distance travelled by dissimilar particles will be more due to the difference in their initial acceleration. Under these conditions, stratification would take place on the basis of density alone. This is the reason why the particles having little density difference are stratified effectively in jigging.

Following three effects contribute to the stratification in jigs:

1. Hindered settling classification
2. Differential acceleration at the beginning of fall
3. Consolidation trickling at the end of fall

During the pulsion stroke, solid bed is opened and expanded. When pulsion ceases, the particles settle into more homogeneous layers under the influence of gravity during the suction stroke. Stratification during the stage that the bed is open is essentially controlled by hindered settling classification as modified by differential acceleration, and during the stage that the bed is tight, it is controlled by **consolidation trickling.** The first process puts the coarse-heavy particles on the bottom, the fine-light particles at the top, and the coarse-light and fine-heavy particles in the middle. The second process puts the fine-heavy particles at the bottom, the coarse-light particles at the top, and the coarse-heavy and fine-light particles in the middle. By varying the relative importance of the two, and by varying the importance of differential acceleration, an almost perfect stratification according to density alone can be obtained. The frequency of pulsations usually varies from 50 to 300 cycles per minute.

The pulsation of the water in a jig can be effected by means of a reciprocating piston. The Fig. 105 shows the displacement of the water in a piston type jig as well as its velocity, considering the simple harmonic motion of the jigging cycle.

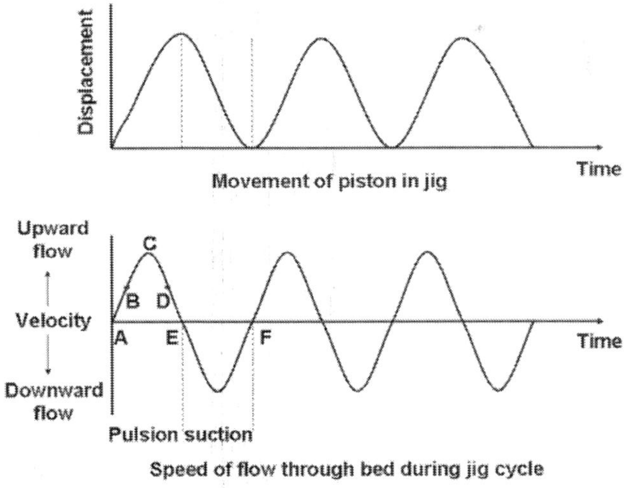

Fig. 105 Harmonic motion cycle of piston Jig

At the beginning of the cycle, the upward flow of water increases from point A. As the velocity increases, the particles will be loosened and the bed will be forced open, or dilated. At point B, the particles are in the form of hindered settling in an upward flow. From B to C, the light fine particles move upward more rapidly than the coarse heavy particles. At point C, the chance of the finest particles, both heavy and light, being carried along with the top flow is the greatest.

Between C and D, the upward force of the rising water on the particles will have been spent acting against gravity, and the velocity of the particles will reach zero by about point D, even though the water velocity is still upward. At this point, differential acceleration prevails, as each particle reverses its direction and begins to fall. The particles continue to fall during the period D to E by the combination of initial acceleration and hindered settling.

During the suction period, E to F, the water flow is downward and the particles are pulled down. The coarse heavy particles arrive first. They are followed by the coarse light particles to form a locked bed, while many slower moving particles continue to settle by the mechanism of consolidation trickling through the interstices of the coarser particles. One complete jigging cycle is shown in Fig. 106 depicting how the particles stratify.

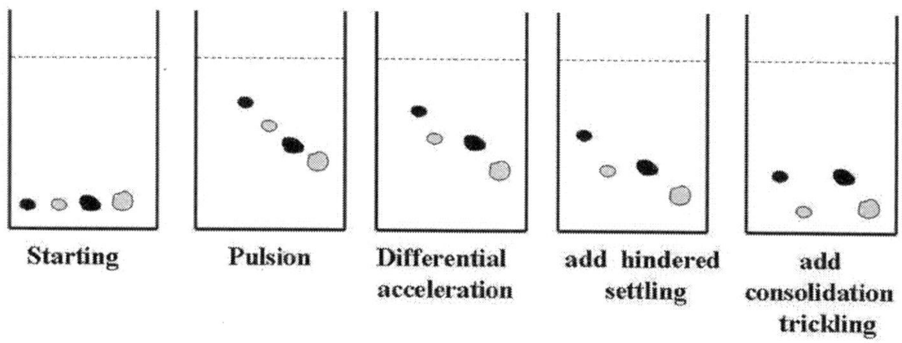

Fig. 106 Particles stratification in one jig cycle

The device used for the jigging process is called **Jig**. The basic construction of a hydraulic jig is shown in Fig. 107.

It consists of a rectangular open top container with the screen at the bottom which supports the bed of solids. Pulsion and suction strokes of water are given through the screen. The feed is introduced at one end of the jig and products are taken out by two methods. In one method, the size of the feed particles is more than the aperture size of the jig screen, and both concentrate and tailing are removed from the jig bed from the end opposite to the feed end. This method is called as **Jigging on the screen**. In another method, called **Jigging through the screen**, the size of the feed particles is less than the aperture size of the jig screen, and light product is removed from the jig bed whereas heavy product is removed from below the jig screen i.e. from the bottom of the jig box known as **hutch**. In such a case, heavy product is known as **hutch product** which passes through the jig screen. A mineral bed, called as **ragging** material, usually having the specific gravity intermediate between the light and heavy particles, is sometimes placed on the jig screen to permit the heavy product to pass through the screen and to prevent the passage of light product into the hutch.

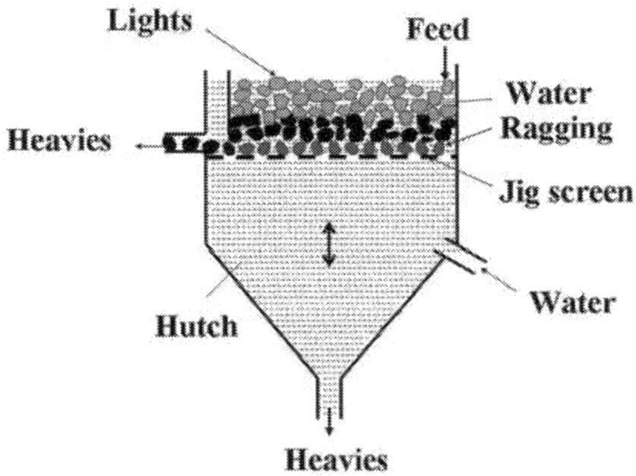

Fig. 107 Basic construction of a Hydraulic jig

13.3.1 Types of Hydraulic Jigs

Hydraulic jigs are primarily of two types namely **Fixed screen jigs** and **Movable screen jigs**. In Fixed screen jigs, water is made to pulsate while screen is stationary whereas in Movable screen jigs, screen moves downward and upward in a stationary water and produces pulsion and suction strokes. All modern jigs are of the fixed screen jigs. There are many variations of the modern jigs. Since the fundamental principles of all jigs are essentially the same, the basic differences between the various types of jigs are a matter of practical engineering to optimise the operating performance, materials handling, maintenance and control. The basic design features of any jig are as follows:

1. A screen to support the mineral bed
2. A hutch or tank containing the water beneath the screen
3. A means of creating a jig stroke
4. A method of modulating the jig-stroke wave form
5. A method of regulating water up-flow
6. A method of supplying feed to the bed
7. A method of removing products from above the screen and from the hutch

Different types of fixed screen jigs are given in Table 36.

Table 36: Types of fixed screen jigs

Type and machine	Pulsating mechanism	Stoke modification
Harz	Mechanical piston	Differential piston action
Cooley, Collum		
Woodbury		
Denver-Harz		
Denver MS	Mechanical diaphragm	Water valve pulsion
Richards	Water pulsion	Water valve control
Pan American		
Pulsator		
Baum	Air pulsion	Air discharge control
Jeffrey		
Link Belt		
Batac		
Bendelari	Mechanical diaphragm (internal)	None
Yuba	Mechanical diaphragm	Differential stroke, hydraulic or mechanical
Jeffrey	(side)	
Rouss		
Cleveland-IHC		
Pan American Placer	Moving hutch	Mechanical
Kraut		
Remer		
Russian	Pneumatic piston	Air release control
MOBK		

The oldest form of mechanical fixed screen jig is Harz jig where the pulsation is effected by a reciprocating piston or plunger and it is a typical of jigging on the screen.

In **Harz jig** (Fig. 108), the piston moves up and down creating the necessary pulsations of water in the jig compartment. It consists of number of independent, concomitant rectangular hopper shaped compartments and are placed in series for successive stages of separation of products of different qualities. The number of compartments varies depending on the size of material being treated and the difficulty of separation. Regardless of the number of compartments, the jig as a whole has a fixed number of strokes per minute, but the stroke of each individual piston may be varied. Water is usually added to the piston compartment to replace that removed with the products, and each compartment has its independent adjustable supply. This water has a marked influence upon the pulsation and suction strokes, reinforcing or opposing the strokes according to the quantity added, and to the amount removed with the products. The sized feed is introduced at one end of the first compartment. The feed should be distributed evenly over the full width of the box. In the first compartment the controls are manipulated to produce a high-grade concentrate as the bottom layer. The light fraction stratifies upward and moves over the far end of the compartment into the second compartment and so on down to the final discharge.

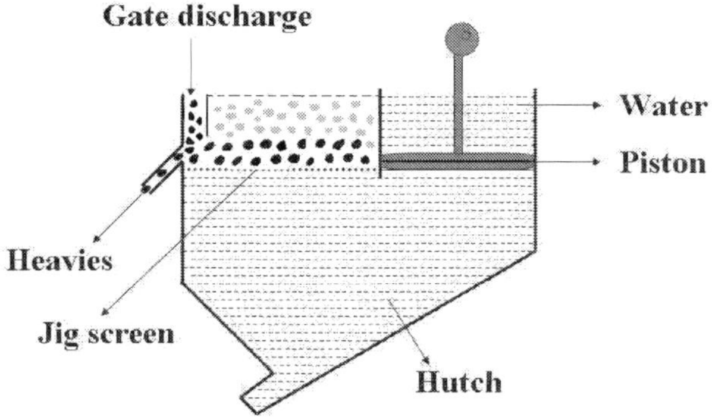

Fig. 108 Harz Jig

The final light fraction overflows from the last compartment as tailing. The heavy fractions are removed continuously through a discharge gate at the side of each compartment. A high-grade concentrate is produced in the first compartment and successively lower grades are produced in the other compartments. Although the Harz jig is essentially on the screen jig, some fine concentrate will inevitably find its way to the hutch, where it is periodically removed by manual methods. Generally, the piston in this type of jig is actuated by a simple eccentric on a drive shaft giving harmonic wave form. However, some manufacturers developed variants of this waveform to increase or decrease a particular part of the jigging cycle. Harz jigs are simple to fabricate, either out of steel, wood, or concrete and lend themselves to shop fabrication. Harz jigs are used in processing of metallic ores in the size range of 37 mm to 0.5 mm.

The **Denver Mineral Jig** (Fig. 109) consists of a conventional divided box with a top driven diaphragm on one side and screen box on the other. It is widely used, especially designed for removing heavy minerals from closed grinding circuits preventing over grinding.

The Denver Mineral Jig incorporates a rotary water valve synchronised with the diaphragm through timing belts. The rotary water valve can be adjusted to open at any desired part of the jig cycle. By suitable adjustment of the valve, any desired variation can be achieved, from complete neutralisation of the suction stroke with hydraulic water to a full balance between suction and pulsion. The water addition can be set such that water enters only during the diaphragm up-stroke, or suction stroke to essentially neutralise the suction and permit a lengthening of the hindered settling portion of the separation, reducing interstitial trickling.

The **Baum jig** is the basis of most of the jigs commonly used in coal washing industry. It is originally designed by Fritz Baum in 1890. The Baum jig (Fig. 110) consists of a U-shaped steel container divided into two in cross section and longitudinally divided into two or more compartments. On one side of the jig is a perforated screen which supports the particle bed and on which the particle separation is affected. The region below the supporting screen, forming the bottom of the U, is referred as the hutch compartment. Usually a screw conveyor is located at the bottom to remove fine particles which have passed through the screen with the flowing water.

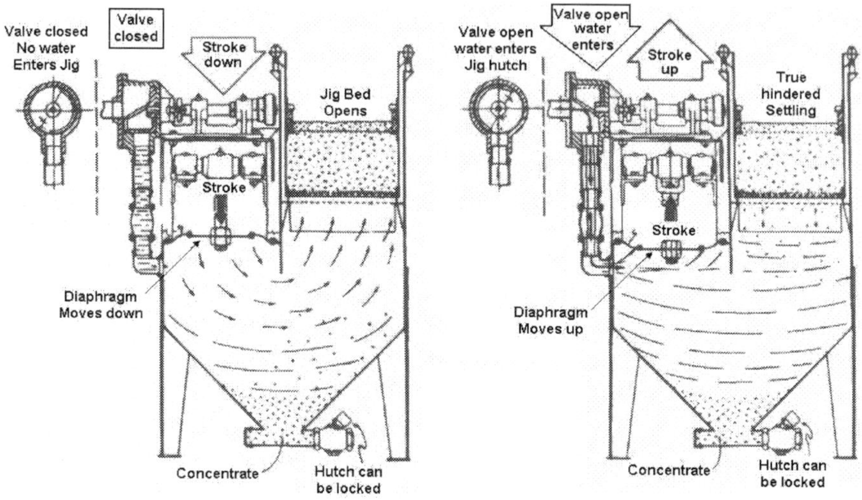

Fig. 109 Denver Mineral Jig

On the opposite side of the screen plate is a pulsion chamber in which the water pulsations are initiated. A sealed air chamber above the hutch compartment is fitted with an air valve, connected to a compressed air supply. The air valve is actuated mechanically to admit air over the hutch compartment forcing water upward through the perforated screen to produce pulsion stroke. As the valve continues to rotate, the air is released at a controlled rate to atmosphere, and makes the water to move down wards to produce suction stroke. In Baum jig, water pulsation is caused at the rate of 30–60 cycles/min.

In its longitudinal phase the U-shaped jig box is divided vertically into compartments. Each compartment, which is essentially a complete jig in itself is separated by fixed weir and contains the screen and means to remove heavy refuse material either by a gate and dam discharge or a rotary valve mechanism (dirt extractor) at a controlled rate. The heavy refuse drops into the hutch compartment from which a bucket elevator raises the heavy refuse out of the jig. In this jig, primary discharge of heavy dirt is at the feed end and the secondary discharge of lighter dirt, middlings and inferior coal is at the discharge end. It is suited to either steady or fluctuating loads of raw coal, containing at times large proportions of dirt which are quickly disposed at the feed end. An unsized coal is washed with this arrangement. The extraction of dirt must be regulated to correspond exactly with the rate at which it separates, otherwise the dirt will either accumulate in the washer (if the extraction rate is too slow) or coal will pass out with the dirt (if the extraction rate is too fast).

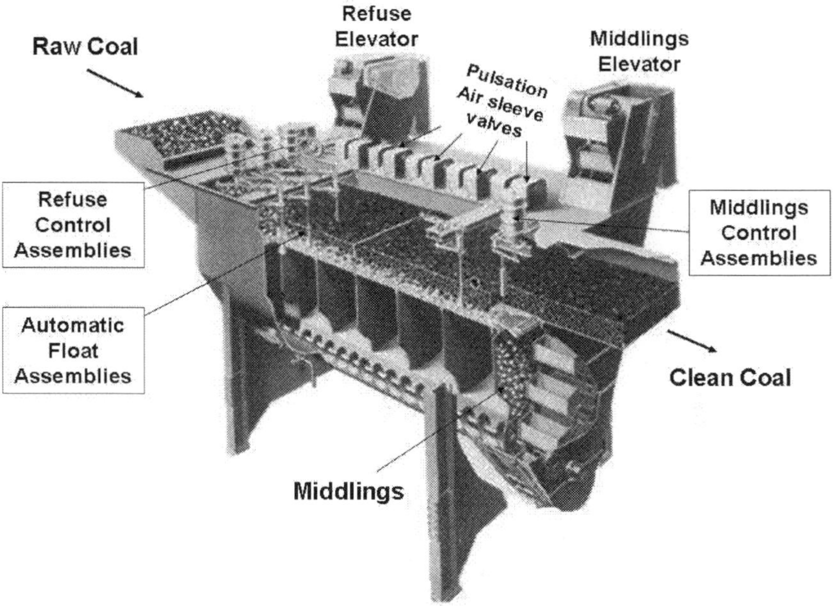

Fig. 110 Baum Jig
(Courtesy of MBE Coal & Mineral Technology India Pvt Ltd, Kolkata)

Originally the regulation was by hand, but in modern machines several systems are available for automatic control of refuse material. The most successful system removes the refuse continuously, stabilises the bed and does not impair the stratification. In some systems of automatic control, a float is immersed in the bed in such a way that it takes up a level intermediate between that of dirt and the coal in the stratified bed. If the dirt layer increases in thickness, the float is raised and accelerates the rate of extraction of dirt by increasing the rotation of dirt extractor.

Baum jig has an ability to treat coal of a wide size range of a coal below 20 cm and handles large tonnages up to 1000 tons/h. Because the air chamber in a Baum jig is on the side of the unit, unequal water forces are applied along the width of the jig screen and uneven stratification tends to develop across the bed with a resulting loss in separation efficiency. This tendency is not so important in relatively narrow jigs. As jig size increases, the U-shaped hutch no longer gives an even flow across the whole bed. The efficiency of the jigs is usually good when the NGM is less than 20%.

In **Batac jig**, water pulsations are produced by valve controlled compressed air acting on the water from air chambers arranged underneath the jig bed. This allows the air to be uniformly distributed across the width of the jig. This design promotes more even stratification and gives both improved efficiency and greater processing capacity. Batac jig is extensively used in coal washing plants.

A **ROM Jig** is a movable screen jig and used for de-shaling of coal. Separation at relatively high density to remove nearly pure rock is referred as **de-shaling**. In a ROM Jig, jigging takes place in a water bath. Pulsion and suction strokes required for separation is achieved by lifting and dropping of jig screen to which a hydraulically moved rocker arm is fixed (Fig. 111). The so induced jigging action causes the separation of rejects from the product, the rejects being on the bottom of the bed, the products on top. The rocker arm movements and the slope pressure result in material transport. The rejects are discharged by a discharge-roll that has the effect of a retaining edge. The hydraulic pressure applied during the upward movement of the rocker arm reflects the reject layer thickness that has accumulated on the screen jig. This value is used as a controlled variable for the discharge-roll velocity. The separated coal is transported over a chute incorporated in the rocker arm. The two products, i.e. rejects and coal are discharged and at the same time de-watered by a twin-type bucket elevator.

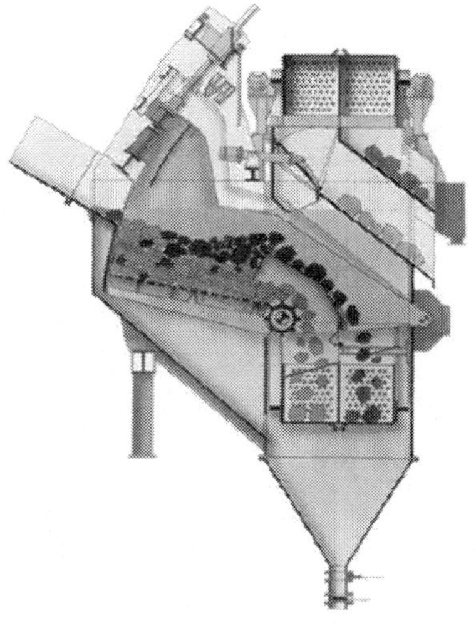

Fig. 111 ROM Jig
(Courtesy of MBE Coal & Mineral Technology India Pvt Ltd, Kolkata)

The fines dropped through the jig screen in the hutch are directed to fine-grain separation over a gate. The water discharged through the gate and water removed from the coal and rejects are recycled.

ROM Jig is a single-cut machine developed for primary separation of rejects from the coarse raw coal. The main benefit of the application of the ROM Jig is the reduction of the material to be treated in the next separation steps. Since the material removed is predominantly high mineral content material, the abrasive tendencies of the product will be reduced and the downstream processes will suffer less mechanical wear, thus reducing maintenance costs. The rejects separated in ROM Jigs can be stored in mined-out cavities to alleviate numerous impacts on the environment.

The development of the ROM Jig opened the possibility, to use a very simple process for the beneficiation of the lump size coal of −400+30 mm using only one technologically improved machine and some standard auxiliary equipment.

Jigs were the chief concentrating devices for many centuries. Presently, they are mainly used in coal washing plants and to a lesser extent in iron and manganese ores and non-metallic ores.

13.4 FLOWING FILM CONCENTRATION

Sorting of mineral particles on flat inclined surface in accordance with the size, shape and specific gravity of the particles moved by a flowing film of water is called as **Flowing Film Concentration**. The concentration in thin film (Fig. 112) where the film thickness and size of the particles are of similar magnitude and the rate of shear on the fluid is relatively low is called as **Thin Film Concentration**. Another type of concentration, called as **Thick Film Concentration**, takes place in much thicker films (Fig. 113) under the conditions of substantial fluid shear.

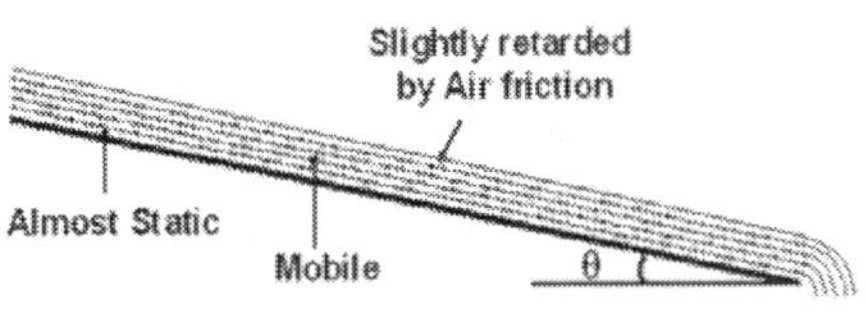

Fig. 112 Thin film flow over the deck

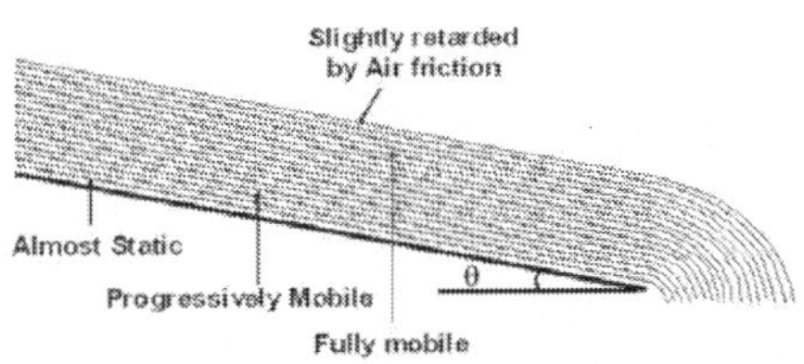

Fig. 113 Thick film flow over the deck

When the water is made to flow over a bare sloping deck, the velocity of the water adjacent to the deck is zero and increases as the distance from the deck increases reaching maximum at the top surface of the water. This increase is not uniform but at a constantly decreasing rate. The velocity distribution is parabolic. However, velocity at the top surface of the water is slightly less than the maximum due to air friction. Such laminar flow conditions occur at $N_{Re} = \dfrac{\delta v_w \rho_w}{\mu_w} < 500$, where δ is the thickness of the water film, v_w is the velocity of the water, ρ_w is the density of the water and μ_w is the viscosity of the water. If y is the distance within the flowing film measured at right angles to it from the deck surface (Fig. 116), the velocity of the water is

$$v_w = \frac{\rho_w g \sin\theta}{2\mu_w}(2\delta - y)y$$

where θ is the angle at which the deck makes with the horizontal.

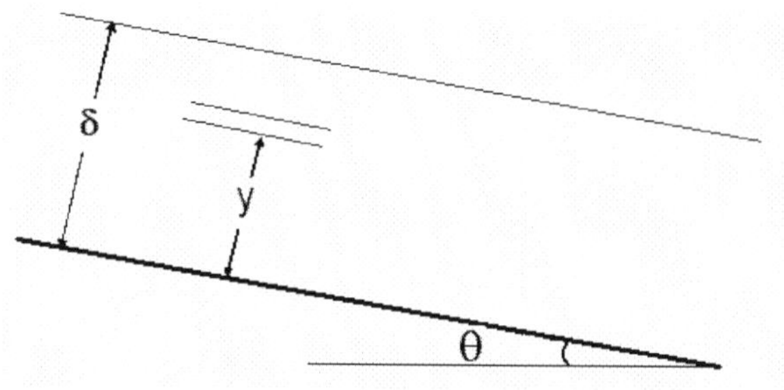

Fig. 114 Water film flow over the deck

The thickness of the water film δ increases with increase of rate of flow at constant slope and increases with decrease of slope at constant rate of flow.

Consider sorting of spherical particles in flowing film of water, the behaviour of particles at the bottom of a flowing fluid film is affected by the following factors:

1. The slope of the deck.
2. The thickness of the fluid film (or the rate of flow of the fluid).
3. The viscosity of the fluid.
4. The coefficients of friction between the various particles and the deck.
5. The specific gravity of the particles.
6. The shape of the particles.
7. The roughness of the deck.

On a horizontal deck, there is no motion of particles. As the deck is tilted, particles begin to move, until all are moving. On a perfectly smooth deck, the critical angle below which motion of a particle does not occur increase with the size and specific gravity of the particle. On a rough deck conditions are different.

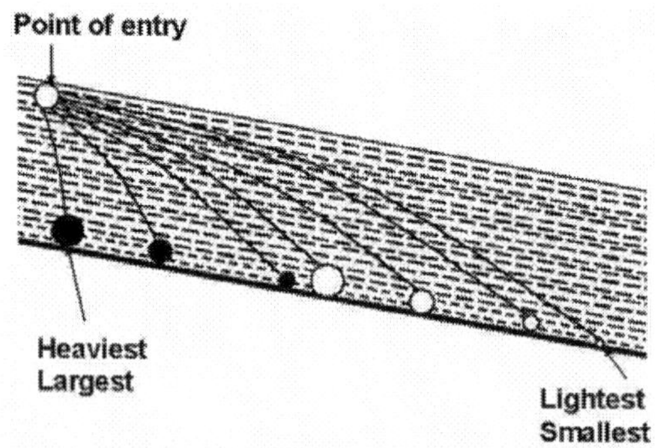

Fig. 115 Particles drift in flowing water

If number of spheres, composed of two kinds of mineral, one heavy another light, and are of different sizes, are introduced into a layer of water, they will be separated during their fall through this layer. The biggest heavy sphere falls faster on to the deck through water and least effected by the current and lies nearest to the point of entry. The smallest light sphere will be drifted farthest downstream. The others will be drifted to different distances (Fig. 115).

The flowing water presses the sphere and makes to move downstream. The differential rate at which the water is flowing over the deck causes low pressure on the bottom of the sphere tending to slide on the deck and causes high pressure at the top of the sphere tending to roll on the deck (Fig. 116).

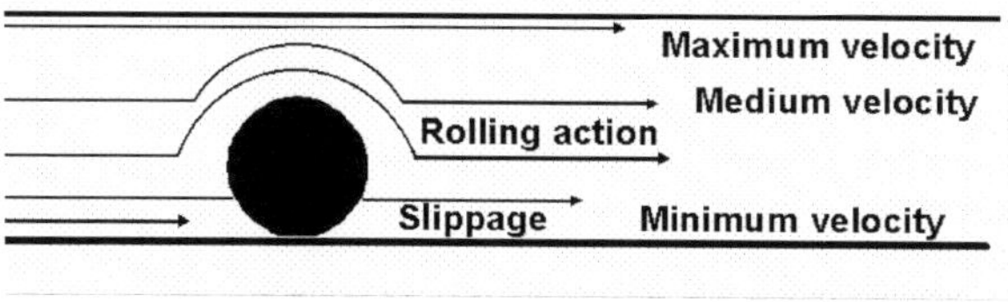

Fig. 116 Forces of flowing water on particle

Since small particles are submerged in the slower-moving portion of the film, they will not move as rapidly as coarse particles. If the combined influence of deck slope and streaming velocity is sufficient to keep all the spheres in rolling movement, they rearrange themselves in the following downslope sequence (Fig. 117)

1. Small-heavy particles
2. Coarse-heavy and small-light particles
3. Coarse-light particles

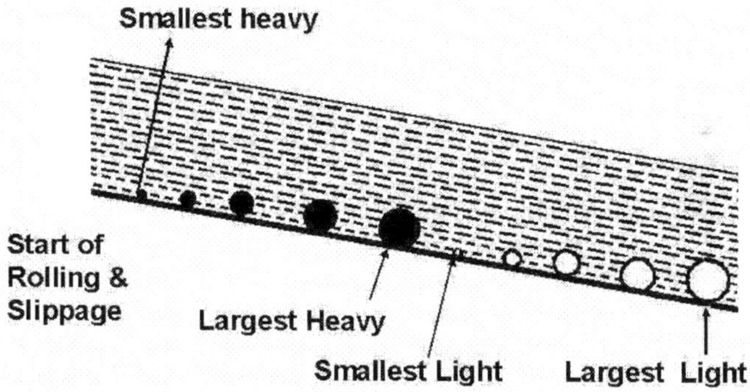

Fig. 117 Arrangement of particles over deck of flowing film

It is to be noted that in flowing film concentration coarse-heavy particles are placed with small-light particles which is reverse of the stratification takes place in classification.

Pinched sluice (Fig. 118) is an inclined stationary trough of 60–90 cm length, narrowing from about 24 cm in width at the feed end to 3 cm at the discharge end. Feed consisting of 50%–65% solids enters the sluice and stratifies as the particles flow through the sluice and crowds into the narrow discharge area. The crowding causes the bed to dilate allowing heavy minerals to migrate and move along the bottom, while lighter particles are forced to the top. The resulting mineral bands are separated by a splitter at the discharge end. Pinched sluices are simple devices and are inexpensive to build and run. Pinched sluices are mainly used for separation of heavy mineral sands.

Reichert Cone (Fig. 119) is based on the pinched sluice concept but employs an inverted cone instead of a rectangular channel. The crowding and dilating effect of the bed is produced by a reduction in perimeter as the material approaches the centre discharge point. Reichert cones are more efficient than pinched sluices because there are no sidewalls to interfere in the separation process.

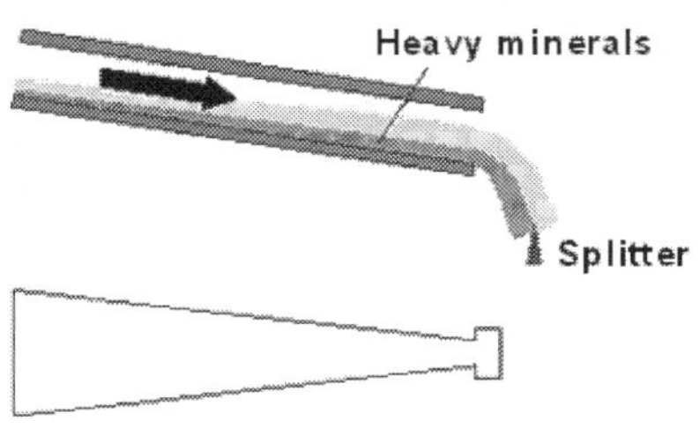

Fig. 118 Pinched Sluice

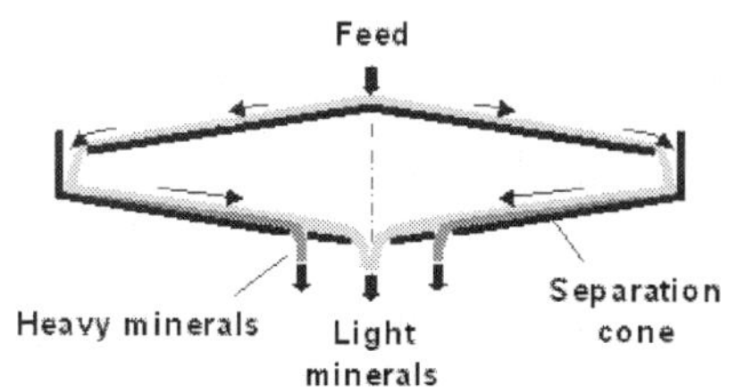

Fig. 119 Reichert Cone

Inclined stationary tables are the simplest means of flowing film concentration. These consist of inclined rectangular tables. The pulp is fed at the upper end which flows down over the table surface. The slope is so adjusted that the lighter particles get washed off, while the heavier ones settle and accumulate on the surface.

13.4.1 Spiral Concentrator

Humphrey Spiral (Fig. 120) is a typical unit of spiral concentrator making use of a several separating forces—gravity and centrifugal action, film flow and frictional drag. It consists of a descending spiral launder of a helical conduit with a modified semi-circular cross section wrapped around a central supporting column which also acts as a discharge pipe for collected concentrate. Wash water runs down a small trough alongside the inner horizontal part of the cross section and can be deflected into the portion carrying the pulp whenever required. At intervals, withdrawal ports are provided at the lowest points in the cross section down the spiral. Adjustable splitters are arranged at the ports to control the removal of part of the inner pulp stream which is further discharged to the central collecting pipe.

Feed is introduced through the feed box, which reduces its velocity and establishes the correct pattern of flow. The feed enters the spiral channel as an almost homogeneous slurry. As the pulp flows around the helix of the spiral concentrator, stratification occurs in a vertical plane. This stratification is usually considered to be the result of a combination of hindered settling and interstitial trickling. In the vertical plane, the heavies proceed to the lower velocity zones near the trough surface, while the lights tend to stratify above them in the higher velocity zones.

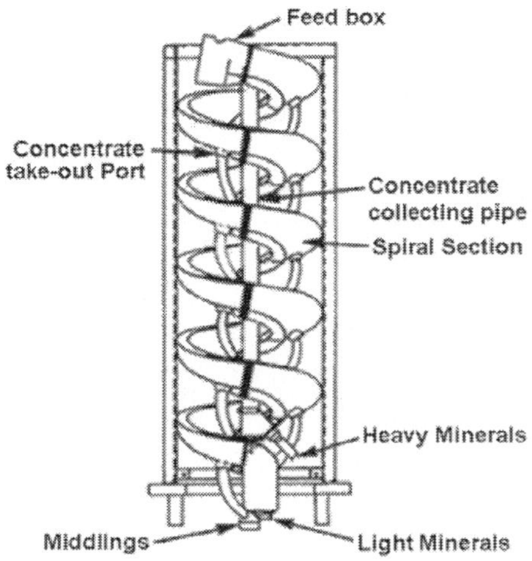

Fig. 120 Humphrey Spiral

The helical twist of the spiral concentrator causes not only a flowing film velocity gradient to be set up in a vertical plane, but also a radial or centrifugal velocity gradient in a horizontal plane. The difference in centrifugal forces acting upon the varying stream components causes a cross-sectional rotation to develop. The portion of the stream nearest the surface moves outward to the point of maximum stream velocity. From there it moves down into the stream to near the spiral surface. It then follows the spiral surface inward to the inner margin of the stream. This cross-sectional rotation of the stream serves to shift the heavy minerals inward toward the collecting ports, or outlets, and the light, faster flowing, but slower settling minerals, outward away from the ports. The heavier and lighter mineral components of the streams are thus shifted laterally in opposite directions so that one is separated from the other. Cross-sectional rotation in the spiral makes it unique as a gravity concentration device.

Wash-water is added from the inner radius of the trough in between the discharge ports to assist in washing light particles from the surface of stratified bed at the inner radius back towards the outer radius of the trough.

A cross section of a spiral concentrator (Fig. 121) can be divided into various regions, with each region describing the effect it has on the slurry traveling through it. On the outer most region (1) (perimeter), will have mostly water, with fine particles, trapped by the high velocity of the moving water. Moving inward towards the centre of the spiral, the next region (2) would consist of a very small area where the maximum water velocity exists, and prevents any separation to occur. This region is defined since it separates the next region (3) from the first region.

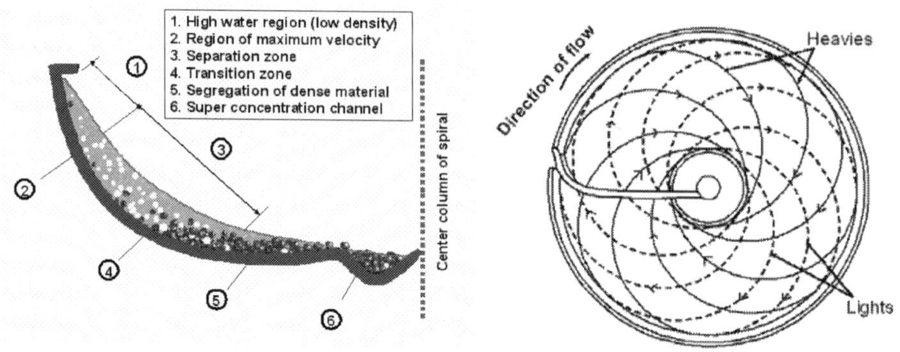

Fig. 121 Cross section of a Spiral showing particle movement

Region 3 is a very active region where the velocity begins to slow down and most of the separation occurs, as more dense particles settle to the bottom and the water velocity keeps the light density particles in the stream near the surface, where they eventually wind up in the outer regions (2 and 1). The next region is actually where two regions overlap (region 3 and 4), and is a very narrow region (like region 2). Next to the last region (region 5) is where the heavy density concentrates collect. The remaining low density particles in this region find their way to the top of the slurry surface and are carried off by the fast-flowing water to the perimeter of the spiral, with the bulk of the water and the low-density solids. Thus, high-density particles get thrown towards the centre of the spiral, medium density particles stay in the centre and the low-density particles collect towards the periphery of the spiral.

Some spirals have a wash water section, where additional water is added to free any trapped light density material in the concentrates, and on a wash water spiral, the innermost portion is where this water is added, and it is called region 6. From the innermost region of the spiral, the concentrates then flow to the bottom section of the spiral, where splitter 'bars' actually make a cut of the material, channelling the inner most material to the heavy concentrate port, a 'middlings' splitter can be used to channel the intermediate to a separate discharge port, and most of the water and the light density material is cut to a low-density port. These cutter bars are adjustable, and are usually set up during the start up. They can be changed easily, to accommodate differing feed material.

The most important in spiral operation is steady feed rate and density of the slurry. It should be as close as design conditions. Grade of the ROM ore and size range of the particles in feed slurry are important factors that affect the separation. Spirals are effective for particles in the range of 3 mm to 75 μm and the minerals with specific gravity differences greater than 1.0. The spiral is used for the recovery of heavy minerals such as chromite, cassiterite, zircon, scheelite, garnet, rutile, monazite and iron ores.

13.4.2 Concentrating table

Concentrating table is a mechanical version of hand pan. Inclined stationary tables are the simplest means of flowing film concentration. They are inclined rectangular tables. The pulp is fed at the upper end which flows down over the table surface. The slope is so adjusted that the lighter particles get washed off, while the heavier ones settle and accumulate on the surface.

If an obstruction like cross-riffles (bars set across the stream) is placed on a bare sloping deck, this obstruction introduces turbulent flow, and vortices; the smallest and heaviest particles occupy the bottom place and are covered by heavy particles of increasing size and then small lighter particles. At the top of the bed are coarsest light particles which may pass over the obstruction (Fig. 122).

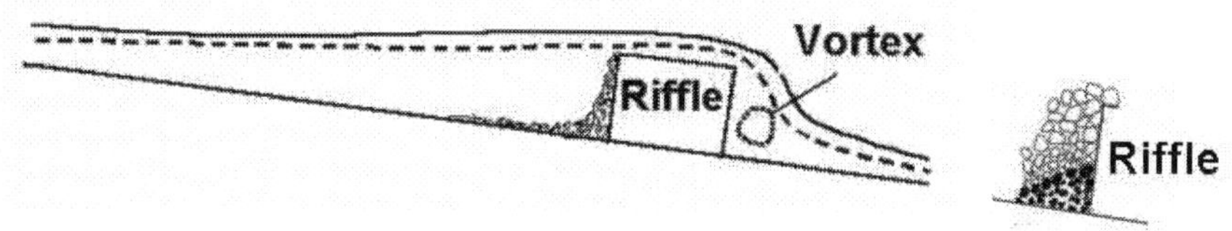

Fig. 122 Effect of Riffle over deck of flowing film

If a reciprocating motion is given to the deck, these particles in between the successive riffles are subject to hindered settling and consolidation trickling and move to one side of the deck whereas lighter particles are washed off across the riffles and along the streaming current. This operation is called as **tabling**. Fig. 123 shows how the particles are stratified in between the riffles as a result of shaking motion in a tabling operation assuming all the valuable minerals are completely liberated.

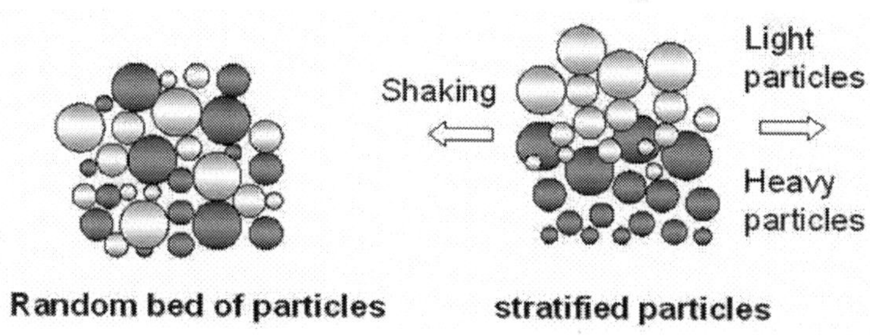

Fig. 123 Particles stratification due to shaking motion in a Table

Fig. 123 clearly indicates that coarse lights and fine heavies are separated as individual fractions (products) and fine lights and coarse heavies cannot be separated as individual fractions (products) by tabling and they are obtained as middling fraction.

Shaking table is a device utilises reciprocating motion and cross riffles in addition to flowing film. **Wilfley table** is most widely used shaking table. It consists of (Fig. 124) a slightly inclined deck on to which feed, at about 25% solids by weight, is introduced at the feed box and is distributed along the feed launder, wash water is distributed along the balance length of the feed launder. The table is given a shaking motion longitudinally using a slow forward stroke and a rapid return, which causes the mineral particles to move along the deck parallel to the direction of motion. In this table, the riffles run parallel with the long axis and are tapered from the maximum height on the feed side (nearest the shaking mechanism) till they die out near the opposite side, part of which is left smooth. Where the riffles stand high, a certain amount of eddying movement occurs, aiding the stratification and jigging action between the riffles. As the load of material is jerked across the table, the upper most layer is swept or rolled over into the riffled gap. In this way, the upper most layer of particles is repeatedly move over the riffle after riffle until it leaves the deck. As the water film is thinnest while climbing over the riffle, any suspended solids drop into the trough and move crosswise as a result of the mechanical shaking movement.

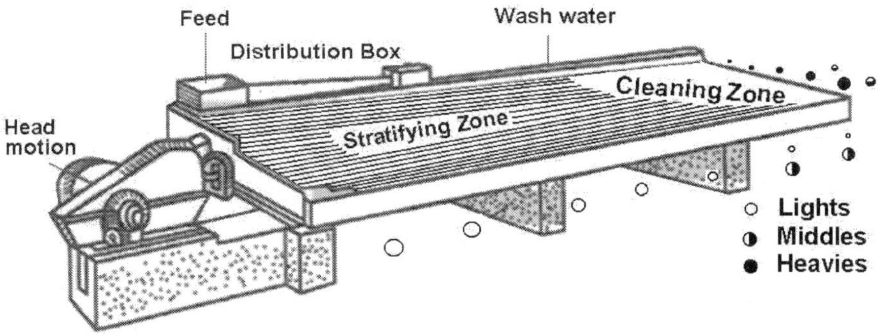

Fig. 124 Wilfley Table

The high-density particles settle into the troughs between the riffles and the motion of the table throws them toward the right side of the table and finally discharge them at the corner diagonally opposite to the feed corner. These particles are stratified while they travel. The light particles which remain in suspension are washed across the table by the water owing to the inclination of table. The other particles (middling) discharge at intermediate places between lights and heavies. From the Fig. 123, it can be observed that the middling fraction produced contains middling particles i.e. locked particles of valuable mineral and gangue, as well as coarse heavies and fine lights. In actual practice of tabling operation, middling fraction is reground for further liberation of valuable minerals from middling particles. If these are reground in a grinding mill, coarse heavies and fine lights discharged along with middling particles will be needlessly reground.

In an assumed idealised situation where all the valuable mineral particles are completely liberated and the table feed contains wide range of particle sizes, middling fraction from the table contains fine lights and coarse heavies. Shaking tables effectively separate coarse light from fine dense particles as shown in Fig. 123.

Hence it is a common practice to classify the feed, since classifiers put such particles into the same product on the basis of their equal settling rates. To feed as narrow a size range as possible on to the table, classification is usually performed in multi-spigot hydrosizers; each spigot product, comprising a narrow range of equally settling particles, is fed to separate tables. Sorting classification followed by tabling gives complete separation. A comparison between the tabling alone and sorting classification followed by tabling is shown in Fig. 125.

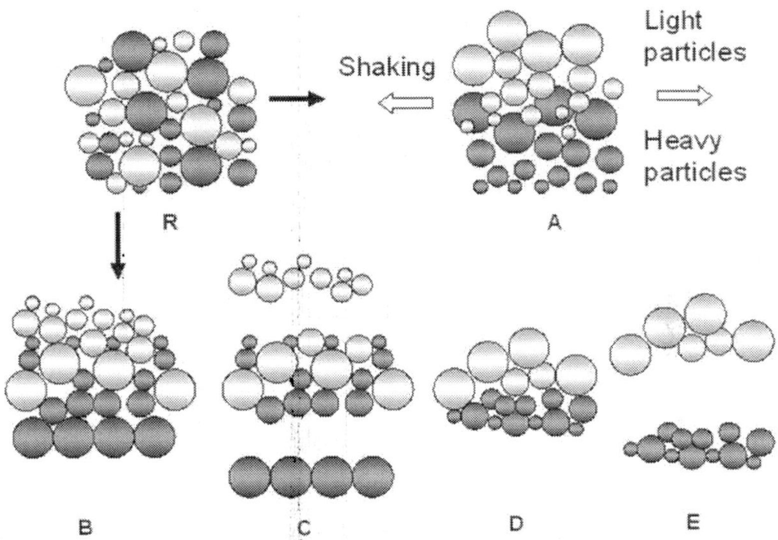

Fig. 125 Comparison between tabling and sorting followed by tabling

In Fig. 125, R is the mixture of particles. A is the stratification after tabling operation as shown in Fig. 123. B is the Stratification after sorting classification wherein fine lights are separated as overflow and coarse heavies are separated as first spigot product of a classifier, as shown in C. The middle fraction (second spigot product of a classifier) as shown in C is subjected to the tabling operation. The resulting stratification is as in D. The stratified layers can then be separated as in E. With this example, it can be concluded that sorting before tabling can result complete separation. It is to be reiterated that this separation takes place due to stratification in tabling is reverse to that of the stratification in classification.

Tables are successfully used for the recovery of heavy minerals from beach sands, chromite beneficiation and the like. These normally operate on feed sizes in the range of 3 mm to 100 microns. The particles of size less than 100 microns are treated in **slimes tables** whose decks have a series of planes rather than riffles.

Vanners, GEC Duplex concentrator, Bartles cross-belt separator, Bartles–Mozley table are the other stirred bed flowing film concentrators. Knelson concentrator, Falcon Concentrator and Multi-Gravity Separator (MGS) are the Centrifugal separators. The Knelson and Falcon concentrators are high-speed centrifugal separators that set a centrifugal force to the particles in the slurry against a fluidisation water flow. The MGS combines the centrifugal motion of an angled rotating drum with the oscillating motion of a shaking table, to provide an enhanced gravity separation, particularly suited to fine particles.

13.5 DRY GRAVITY CONCENTRATION

Separation of valuable minerals from the ore is generally a wet process where separation takes place in water medium. The separation of valuable mineral particles in a stream of air has been practiced for centuries. This separation is called as Dry separation or Pneumatic separation.

Various dry gravity concentration devices have been developed for desert use. They are inherently inferior to their water counterparts, except where water cannot be tolerated or is not available. Dry processing of dredged ilmenite, rutile, or cassiterite, upgrading of asbestos and mica by dry gravity concentration are the few commercial examples. One of the largest users of dry gravity concentration processes is the coal industry, which have the advantage of producing a dry product. Air heavy medium fluidised bed separator, all air jig and FGX separator are the notable examples of dry gravity concentration devices used in coal industry.

■■■■

14 Froth Flotation

Froth Flotation is a method of separating the minerals in a relatively finely divided state by utilizing the differences in physico-chemical surface properties of particles of various minerals.

In a beaker of water, if air is introduced from the bottom, air bubbles are produced and rise to the surface of water (Fig. 126A) as the density of air bubble is much less than that of water. Similarly, air bubbles also rise if air is introduced in a pulp containing mineral particles. If a mineral particle of high density adheres to the air bubble, air bubble along with mineral particle rise to the surface because the apparent density of air bubble and the adhered mineral particle is less than that of water. If many number of mineral particles are adhered to the air bubble, still air bubble rise to the surface (Fig. 126B) as the apparent density of air bubble and adhered mineral particles is less than that of water due to relatively large volume of air bubble. This concept is the basis for froth flotation operation.

To use this concept or phenomenon, the following two are required:

1. Method to make the mineral particles adhered to the air bubble.
2. Method to keep the air bubble alive when it reaches the surface of water. It is known that the air bubble collapses when it reaches the surface of water. When it is collapsed, the adhering mineral particles are dropped into the water. Therefore, the air bubble must be kept alive on the surface of water for sufficient length of time till it is removed from the surface.

Following are the methods adopted in practice:

1. To make the mineral particles adhered to the air bubble while it is rising through the pulp, the mineral particles are treated with suitable chemicals to acquire adhering property for the mineral particles.
2. For making the air bubble not to collapse, some other chemicals are used to prevent the collapsing of air bubble or in other words to increase the life of the air bubble.

The required mineral particles after adhering to the air bubble float to the surface along with the air bubble. The aggregation of several such mineral adhered air-bubbles forms the froth on the surface of the pulp (Fig. 126C). Hence, this operation of mineral separation is named as **Froth flotation**.

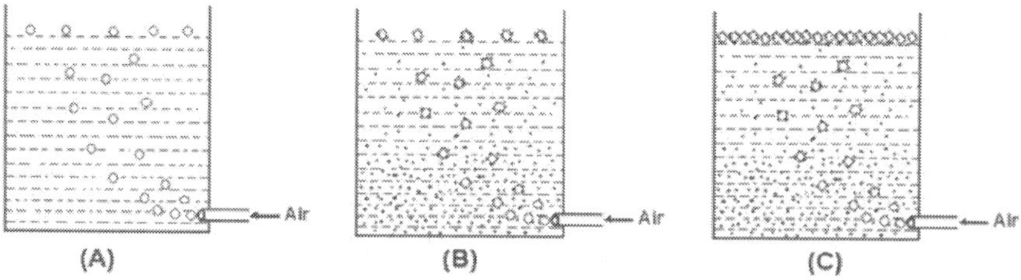

Fig. 126 Process of rising air bubbles and forming froth

Froth flotation can only be applied to relatively fine particles (less than 150 μm). Particles smaller than 400 mesh (37 micron) often will not attach to the air bubble. If the particles are too large, the adhesion between the particle and the bubble is less than the weight of the particle and the bubble drops the mineral particles. The air bubbles can only stick to the mineral particles if they can displace water from the mineral surface. This can only happen if the mineral is water repellent or hydrophobic. Air bubbles, after reaching the surface, can continue to hold the mineral particles if they can form a stable froth. If not, air bubbles will burst and drop the mineral particles. To achieve the favourable conditions for froth flotation, the pulp is treated with various chemical reagents known as flotation reagents. The chemicals used for treating the mineral particles to make them to adhere to air bubble are called as collectors and the chemicals used to increase the life of air bubbles are called as frothers.

The ordinary simple froth flotation comprises the following steps:

1. The ore is ground in water to at least 48 mesh
2. The pulp thus formed is diluted with water to a consistency between 25% and 45% solids by weight
3. Small quantities of surfactant chemicals called modifiers are added to the pulp to modify the surfaces of specific minerals
4. Another reagent called collector specifically chosen to affect the mineral, to be recovered by froth flotation, is added. It coats the mineral particle with an aerophilic surface (i.e. water repellent)
5. Another reagent called frother is then added which imparts persistence to bubbles when they reach the surface and assists in establishing a stable froth at the surface.
6. The chemically treated pulp is aerated either by agitation or by air injection through the porous bottom of the containing tank, or through pipes, during which the coated mineral particles become more or less firmly attached to air bubbles.
7. The mineral bearing froth which rises to the surface is skimmed off. The pulp passes on through a series of containers or cells to provide both time and opportunity for mineral particles to come in contact with air bubbles and to be recovered in the froth.

These steps frequently follow the sequence stated. Dilution is usually affected in the classifier. Modifiers and the collector are added in a separate tank called conditioning tank. Sometimes they may be added in the grinding mill. The frother is normally added in flotation machine where aeration is going on.

The crux of the froth flotation process is the existence of a selective tendency for some particles to adhere to air and for other particles to adhere to water. Adhesion to air in preference to water is a property peculiar to hydrocarbon groups of atoms, whether they are part of the mineral itself or part of a coating on the mineral surface.

Most ores mined at present require fine grinding for high degree of liberation of valuable minerals, and thus the froth flotation becomes the only possible means of beneficiation to higher grades and recoveries. Froth flotation treatment is extensively applied to the concentration of metalliferous minerals both sulphides and oxides. About 90% of the world's important ores of lead, zinc and copper are upgraded by froth flotation operation. Froth flotation technique is also extensively used in upgrading of non-metallic ores like graphite, limestone, fluorite, fluorspar, clay, rock phosphate and coal. All in all, it has been estimated that at least 100 different minerals are currently being recovered by froth flotation.

Physicochemical understanding of the mechanism of froth flotation requires the use of various concepts such as atomic bonding, surface energy and surface tension, adsorption, floatability, and contact angle.

14.1 ATOMIC BONDING

An **Ionic Bond** (or electrovalent bond) is a chemical bond resulting from the transfer of electrons from one bonding atom to another. In this type of bond, one atom gives up one or more electrons to another atom. A **covalent bond** is a form of chemical bonding wherein each atom contributes one or more electrons to a pool. The atoms can be of the same element or different elements. From flotation standpoint, the important covalent bond is that displayed by hydrocarbons as in methane and by the hydrocarbon portions of large molecules. In each molecule, the bonds between the atoms are strong but the bonds between molecules are usually weak. A **coordinate covalent bond** is a description of covalent bonding between two atoms in which both electrons are shared in the bond by the same atom. A **metallic bond** can be understood as the sharing of 'free' electrons among a lattice of positively charged ions (cations). A **hydrogen bond** is the attractive force between one electronegative atom and a hydrogen covalently bonded to another electronegative atom. It results from a dipole-dipole force with a hydrogen atom bonded to nitrogen, oxygen or fluorine. The hydrogen bond is a very strong fixed dipole-dipole force of attraction called Van der Waals force, but weaker than covalent, or ionic bonds.

Polar and non-polar bonds are types of covalent bond. A covalent bond formed between two different atoms is known as Polar covalent bond. For example, when a Covalent bond is formed between hydrogen H and chlorine Cl, it is polar in nature because Cl atom is more electronegative than H atom. Therefore, electron cloud is shifted towards Cl atom. Due to this reason a partial negative charge appeared on Cl atom and an equal positive charge on H atom. A covalent bond formed between two like atoms is known as Non-polar bond. Since difference of electro negativity is zero therefore, both atoms attract electron pair equally and no charge appears on any atom and the whole molecule becomes neutral. Hydrogen molecule H_2, Chlorine molecule Cl_2 and Fluorine molecule F_2 are the examples. None of the bonds in hydrocarbon molecules, like hexane, C_6H_{14}, are significantly polar, so hydrocarbons are nonpolar molecular substances.

A molecule is said to have a permanent di-pole when the centres of its positive and negative charges do not coincide. The **dipole moment** is the product of either the positive or negative charge (regardless of sign) and the distance between charge centres. Many molecules have such dipole moments due to non-uniform distributions of positive and negative charges on the various atoms. A gaseous HCl molecule has a permanent di-pole. The electrons are considered to be distributed around the two atoms such that it appears as if the centre of the negative charge is located close to the chlorine atom and the centre of the positive charge is close to the hydrogen atom. A molecule with a permanent dipole moment is called a **polar** molecule.

A water molecule is polar because (1) its O-H bonds are significantly polar, and (2) its bent geometry makes the distribution of those polar bonds asymmetrical. The Oxygen and Hydrogen atoms are not arranged in a line, but bent in a special way. As a result, part of the molecule is negatively charged and part positively charged. It makes the water molecules highly polar. One region of the water molecule, its oxygen, is negative, while another region, its two hydrogen atoms, is positive. Polar molecules tend to line up with each other, the positive pole of one molecule closely aligned to the negative pole of another molecule. Therefore, polar water molecules create a network that is held together by special weak bonds, called hydrogen bonds.

Hydrogen bonds between water molecules give water its special properties. For example, the presence of hydrogen bonds is responsible for water being a liquid at normal room temperature, and its ability to dissolve ionic and organic compounds effectively. The hydrophobic effect is another unique property of water caused by the hydrogen bonds. Hydrogen bonds are very weak, and can be easily broken, giving water its flexible, fluid nature. Water molecules tend to make as many hydrogen bonds as possible.

Mineral hydration is an inorganic chemical reaction where water is added to the crystal structure of a mineral, creating a new mineral, usually called a **hydrate**.

Water occurs in both the covalent and electrovalent (ionic) forms as shown below

$$\mathrm{H:\ddot{\underset{\cdot\cdot}{O}}:H} \quad \text{and} \quad [\mathrm{H.\,HOH}]^{+} \quad [\mathrm{:\ddot{\underset{\cdot\cdot}{O}}:H.HOH}]^{-}$$

covalent **ionic**

Approximately 46% of the bonding is covalent and the balance of 54% is ionic. Thus, water consists of both di-polar H_2O molecules and hydrogen and hydroxyl ions.

Under the influence of hydrogen bonds, covalent form of the water forms chains such as H_4O_2, H_6O_3, H_8O_4, $H_{10}O_5$ etc.

By ionisation or reaction with dissolved substances and the surfaces of solid particles, water molecules can change to the electrovalent forms

$$3\,H_2O \rightleftharpoons [H.\,HOH]^+ + [:\ddot{\underset{\cdot\cdot}{O}}:H.HOH]^-$$

Being polar, these electrovalent forms are attracted to the oppositely charged lattice points of the solid surface.

At the particle surfaces, the hydrogen ion concentration controls the ionisation in the pulp. Hydrogen ion concentration is expressed as number of grams of hydrogen ions per litre of solution. The pH of a solution is defined as the negative logarithm of the hydrogen ion concentration.

$$pH = \log_{10}\frac{1}{[H^+]}$$

As pure water at ordinary temperature dissociates slightly into hydrogen ions and hydroxyl ions [$H_2O \rightarrow H^+ + OH^-$], the concentration of each type of ion being 10^{-7} mol per litre, the pH of pure water will be $\log_{10}\frac{1}{10^{-7}} = 7$. This is taken to represent neutrality on the pH scale.

At pH	7,	a solution is neutral
At lower pH	(1–6),	a solution is acidic
At higher pH	(8–14),	a solution is basic

14.2 SURFACE ENERGY AND SURFACE TENSION

A molecule in the interior of a liquid is under attractive forces in all directions and the vector sum of these forces is zero. However, a molecule at the surface of a liquid is acted by a net inward cohesive force that is perpendicular to the surface. It requires work to move molecules to the surface against this opposing force, and surface molecules have more energy than interior ones. The surface energy may be defined as the excess energy at the surface of a liquid compared to the interior of the liquid. The manifestation of this excess free surface energy is termed as surface tension of the liquid. Surface tension, then, is a force acting parallel to the surface, which opposes any effort to extend the interfacial area. Surface energy is the product of surface tension and area (ergs/cm^2). The surface tension of a liquid can also be defined as the work that must be done to bring enough molecules from inside the liquid to the surface to form one new unit area of that surface (ergs/cm^2). Surface energy is numerically equal to the surface tension. Surface tension of a liquid is the same at all points on the liquid surface. It is, however, affected by the temperature. It decreases with increase in temperature.

The surface tension of aqueous solutions of organic substances is intermediate between that of water and of the solute, but much closer to that of the solute due to greater concentration of solute molecules at the surface than within the liquid and also due to the orientation of the organic molecules on the surface. Organic substances such as fatty acids and alcohols contain a polar hydrophilic group and a non-polar hydrophobic group. The hydrophilic group, –COOH of fatty acids, makes the whole molecule reasonably soluble. The non-polar hydrocarbon group tends to come to the surface from the interior of the liquid. At any given instant, the proportion of organic molecules that are presenting the non-polar end to the gas phase (away from the water) is enormously greater than average. As a result, the surface tension decreases.

14.3 ADSORPTION

Adsorption is a process of increase in the concentration of a dissolved substance at the boundary of the phase in which it was dissolved. The substance, which accumulates on the surface is called adsorbate and the adsorbing phase is called adsorbent. Adsorption process creates a film of the adsorbate (the molecules or atoms being accumulated) on the adsorbent's surface.

Adsorption is a surface phenomenon. In this process, accumulation or concentration of a substance takes place at the surface or interface as compared to the bulk phases. This happens because the molecules of the bulk phase at the surface are attracted only from below and from the sides as compared to the molecule inside the bulk of the phase. In the interior, molecules experience forces of attraction from all sides. Hence, the molecules at the surface experience a net attraction downward and tend to move towards the interior to increase attractions. On account of the unbalanced forces, the surface of a liquid or solid is in a state of strain. This implies that the surface molecules have higher energy than those present in the interior. So, when the surface of liquid or solid attracts other molecules, this energy is reduced. In other words, adsorption results in decrease of surface energy.

The adsorption process is generally classified as physical adsorption and Chemisorption. In physical adsorption, the forces of attraction between the molecules of the adsorbate and the adsorbent are of the weak Van der Waals forces. Since the forces involved are of a molecular nature, multimolecular layer adsorption can take place. In physical adsorption, the reagent (adsorbate) maintains its chemical identity. Mineral surface holds the adsorbates (liquids in froth flotation) adhesively at its surface without penetration. In chemisorption, the forces of attraction between the adsorbate and the adsorbent are very strong; the molecules of adsorbate form chemical bonds with the molecules of the adsorbent present in the surface. Compared to physical adsorption, the bonding forces in chemisorption are from 20 to 50 times as great as those of physical adsorption. In chemisorption, the adsorbate forms new compounds in reactions of the following types:

1. Chemisorption of either an anion or cation is promoted when they form insoluble complexes with corresponding cations or anions of the lattice of the adsorbate.

   For example, (a) sulphur ions adsorb readily on minerals containing heavy metal cations to form insoluble metal sulphides. (b) Cyanide anions adsorb equally well on the surfaces of some heavy metal sulphides since they form with the metal cations, compounds and/or complexes of very low solubility

2. Ions which have an effective radius approximately equal to the radii of the crystal lattice adsorb better than ions, which differ appreciably in size from the lattice of the adsorbate. Exchange adsorption is a process where ions in solution displace specific ions in the mineral surface. This type of adsorption proceeds in the direction in which either more stable compounds or less soluble compounds are formed.

14.4 FLOATABILITY

Solid paraffin is an excellent natural floatable substance. Natural hydrocarbons and soft coals are naturally floatable and give substantial or even large contact angles. All minerals are classified into **polar** or **non-polar** types according to their surface characteristics. The surfaces of non-polar minerals are characterised by relatively weak molecular bonds. The minerals are composed of covalent molecules held together by van der Waals forces, and the non-polar surfaces do not readily attach to the water dipoles, and in consequence are hydrophobic. Minerals of this type, such as graphite, sulphur, molybdenite, diamond, coal, and talc, thus have high natural floatabilities with contact angles between 60° and 90°. Although it is possible to float these minerals without the aid of chemical agents, it is universal to increase their hydrophobicity by the addition of hydrocarbon oils. Creosote, for example, is widely used to increase the floatability of coal.

Table 37: Classification of polar group of minerals

Group 1	Group 2	Group 3(a)	Group 4	Group 5
Galena	Barite	Cerrusite	Hematite	Zircon
Covellite	Anhydrite	Malachite	Magnetite	Willemite
Bornite	Gypsum	Azurite	Gothite	Hemimorphite
Chalcocite	Anglesite	Wulfenite	Chromite	Beryl
Chalcopyrite		**Group 3(b)**	Ilmenite	Feldspar
Stibnite		Fluorite	Corundum	Sillimanite
Argentite		Calcite	Pyrolusite	Garnet
Bismuthinite		Witherite	Limonite	Quartz
Millerite		Magnesite	Borax	
Cobaltite		Dolomite	Wolframite	
Arsenopyrite		Apatite	Columbite	
Pyrite		Scheelite	Tantalite	
Sphalerite		Smithsonite	Rutile	
Orpiment		Rhodochrosite	Cassiterite	
Pentlandite		Siderite		
Realgar		Monazite		
Native Au, Pt, Ag, Cu				

Minerals with strong covalent or ionic surface bonding are known as polar types, and exhibit high free energy values at the polar surface. The polar surfaces react strongly with water molecules, and these minerals are naturally hydrophilic. The polar group of minerals is subdivided into various classes (Table 37) depending on the magnitude of polarity, which increases from groups 1–5. Minerals in group 3(a) can be rendered hydrophobic by sulphidisation of the mineral surface in an alkaline aqueous medium.

Apart from the native metals, the minerals in group 1 are all sulphides, which are only weakly polar due to their covalent bonding, which is relatively weak compared to the ionic bonding of the carbonate and sulphate minerals. In general, therefore, the degree of polarity increases from sulphide minerals, through sulphates, to carbonates, halites, phosphates, etc. then oxides, hydroxides, and, finally, silicates and quartz.

14.5 CONTACT ANGLE

Contact angle (θ) is an angle of contact of an air bubble with the surface of a solid measured across the water (Fig. 127). It is a convenient measure of the forces of adhesion between the bubble and the solid surface. The contact angle marks the position of equilibrium between the solid-water and water-air surfaces on a wetted surface i.e. it is the position of equilibrium between three tension forces; the surface tension of water T_{WA}, surface tension of solid mineral T_{MA} and interfacial tension T_{MW} between the solid mineral and water.

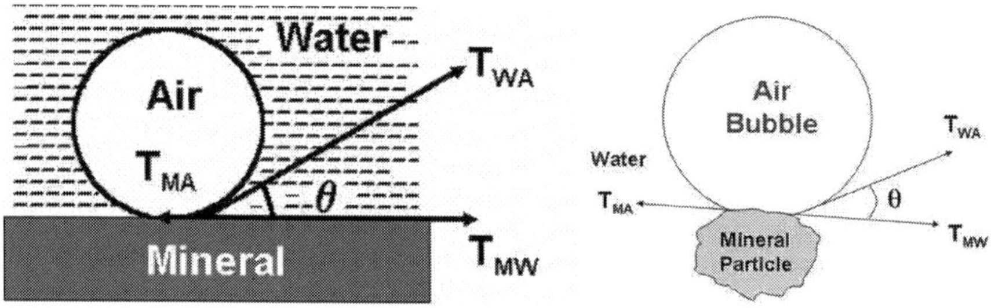

Fig. 127 Contact angle

If the surface tension of the solid mineral T_{MA} is more, the water is pulled over the solid till an acute angle θ is reached, when the component of water tension T_{WA} together with the interfacial tension T_{MW} is sufficient to bring about equilibrium. Under these conditions, the solid shows a preference for water. If the interfacial tension T_{MW} is greater, the water will be drawn back and an obtuse angle will form. Under these conditions the solid has a preference for air.

When solid shows affinity for water $T_{MA} = T_{MW} + T_{WA} \cos \theta$

When solid shows affinity for air $T_{MW} = T_{MA} + T_{WA} \cos \theta$

Therefore, $T_{WA} \cos \theta$ is a measure of degree of wetting. When the contact angle is nil, $\theta = 0$, $\cos \theta = 1$, the degree of wetting is maximum. When the contact angle is 180°, $\theta = 180°$, $\cos \theta = -1$, the water will contract its extent and the degree of wetting is at a minimum. Since there is always some adhesion between solids and liquids in contact, there is no such thing as complete non-wettability, i.e. a contact angle of 180°. Adherence of the mineral particle to the air bubble depends on contact angle. As the contact angle increases, adherence increases and hence floatability increases. Minerals with high contact angle are called aerophilic (hydrophobic), i.e. they have higher affinity for air than for water. Most minerals are aerophobic (hydrophilic) in their natural state. To make the valuable mineral particles aerophilic, reagents called collectors are added to the pulp which adsorb on mineral surfaces, increases contact angle and facilitates bubble attachment. Many freshly formed mineral surfaces exhibit a natural contact angle of a few degrees. Graphite and some coals have a high contact angles to float without aid of a collector. They are said to have natural floatability.

14.6 FLOTATION REAGENTS

Flotation reagents (surfactants) are substances added to the ore pulp prior to or during flotation to make possible to float valuable mineral particles and not to float the gangue mineral particles. The interfacial tensions normally existing between minerals and water are generally insufficient in the magnitude and range to be used effectively in flotation, and it is the function of the flotation reagents to intensify the characteristics of the interfaces in each of the required directions to separate the mineral components from the associated constituents. Since flotation surfactants are, in general, supplied to the interfaces through the aqueous solution phase, mainly those reagents which are somewhat soluble in water are used in flotation. In some cases, it is necessary to use insoluble hydrocarbons or other oils; these liquids are dispersed in the aqueous phase as emulsions, with the help of soluble surfactants, to facilitate their reaching the interfaces in a reasonably short time period. Important flotation reagents are collectors, frothers, depressants, activators and pH regulators.

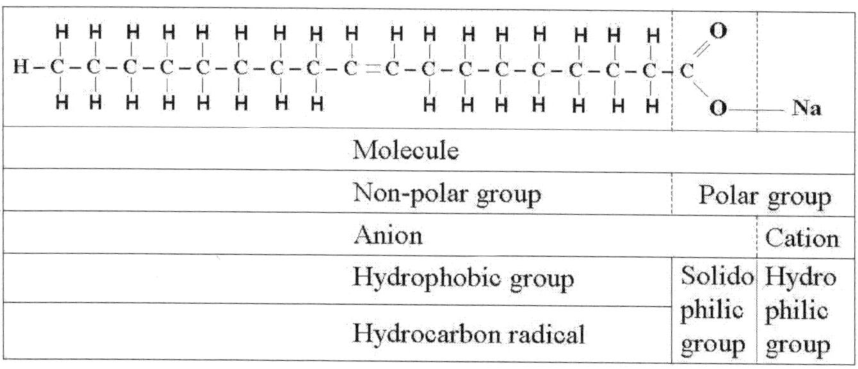

Fig. 128 Structure of Sodium Oleate

14.6.1 Collectors (Collecting agents)

Collector is a chemical reagent, an organic compound, either an acid, base or salt, and is hetero-polar in nature; polar part of it has an affinity towards a specific mineral and non-polar part has an affinity towards air bubble. The non-polar portion of the collector molecule is a hydrocarbon radical, which does not react with water and is therefore water-repellent. In contrast to the non-polar part of the molecule, the polar part can react with water. A typical example of such a hetero-polar molecular structure is sodium oleate (Fig 128).

Small amount of collector is added to the pulp and agitated long enough so that the polar part is adsorbed on to the mineral to be floated while the non-polar part is oriented outwards (Fig. 129) and makes the surface of mineral particles hydrophobic.

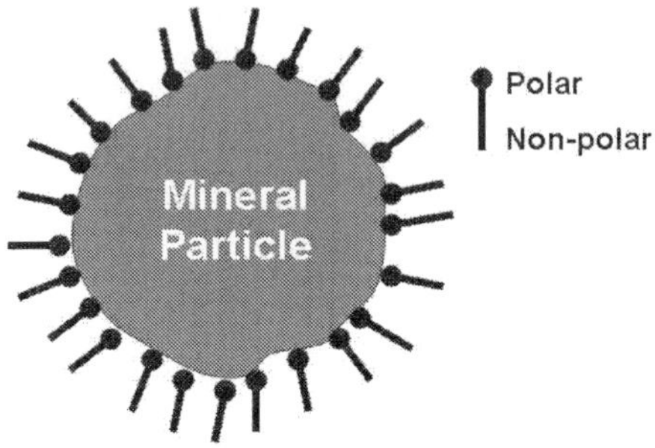

Fig. 129 Collector Adsorption on mineral surface

Collector increases the contact angle of the valuable mineral particles.

The collectors must be soluble in water to at least a small degree. Collectors are classified according to their ability to dissociate into ions in an aqueous solution and with regard to the type of ion which produces the water-repellent effect. When a collector dissociates into an anion and cation, the ion which is the direct cause of the water-repellent action is considered as the active repellent ion and the other as non-active (non-repellent) ion. The water-repellent ion structure always includes a hydrocarbon radical, the presence of which ensures that the mineral is water repellent. However, hydrocarbons do not exist in free state and are unable to attach themselves directly to a mineral surface. Therefore, the repellent ion structure includes a second group of atoms (in addition to the hydrocarbon) to form a link between the hydrocarbon and mineral surface. This linking group is called the solidophil group. Thus, a water-repellent collector ion includes a hydrocarbon radical linked to a solidophil group as shown in structure of sodium oleate. The water-repellent effect is directly associated with the length and the structure of the hydrocarbon radical, whereas the effect of the solidophil group depends on (a) the nature of the reaction with the mineral surface (b) the strength of the collector attachment, and (c) selectivity.

Collectors are broadly classified into the following two groups:

1. Ionizing collectors, which dissociate into ions in water.
2. Non-ionizing collectors, which are non-polar compounds, principally hydrocarbon compounds, insoluble in water. It is believed that these collectors render the mineral water-repellent by covering its surface with a thin film to cause bubble attachment.

The larger group of collectors are ionizing, and are divided into

1. Anionic collector, where the anion renders the mineral surface water-repellent.
2. Cationic collector, where the cation renders the mineral surface water-repellent.

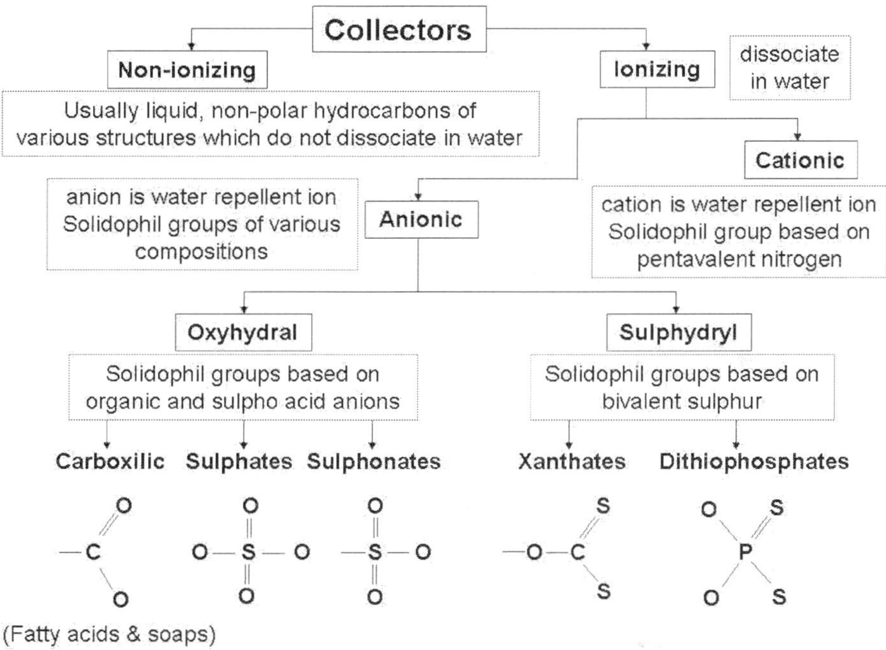

Anionic Collectors: Many of the anionic collectors in use are noted for their remarkable selectivity and strong attachment. The most widely used collectors are of the sulphydryl type, called as thiols, in which the solidophil group contains bivalent sulphur (thio compounds). Xanthate is an example. Xanthates, chemically termed as dithiocarbonates, are prepared by reacting an alcohol, carbon disulphide and an alkali hydroxide:

$$ROH + CS_2 + KOH = RO.CS.SK + H_2O$$

Sodium or Potassium xanthate is 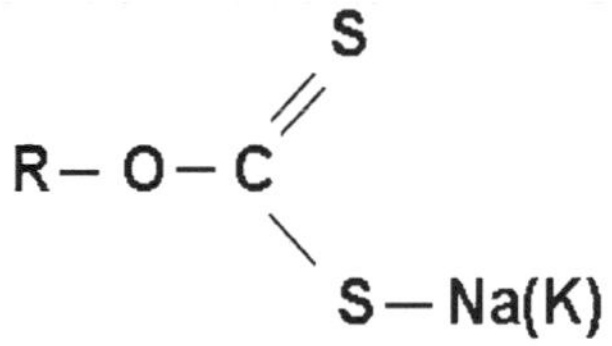

The active anion is 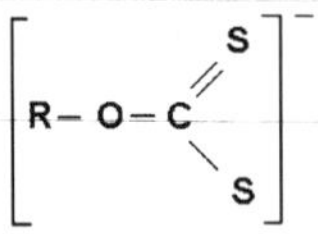 connected solidophil group is 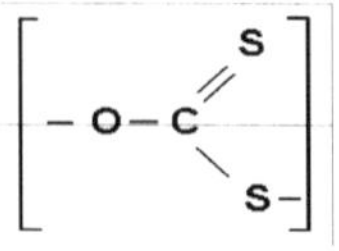

where R is the hydrocarbon group and contains normally one to six carbon atoms. The most widely used xanthates are ethyl, isopropyl, isobutyl, amyl, and hexyl. Following are the general formulae for ethyl and isopropyl xanthates:

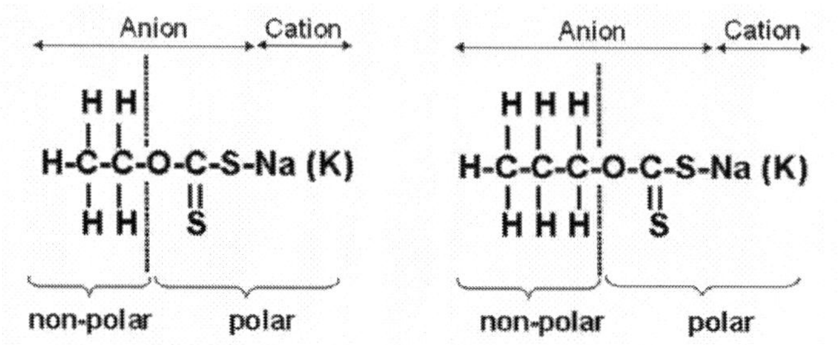

Ethyl Xanthate **Isopropyl Xanthate**

The anion consists of a hydrocarbon non-polar radical and a connected polar group. The cation (sodium or potassium) plays no part in the reactions leading to mineral hydrophobicity. Xanthates are very powerful and highly selective in the flotation of sulphide minerals.

Xanthates are assumed to adsorb on sulphide mineral surfaces due to chemical forces between the polar group and the surface, resulting in insoluble metal xanthates, which are strongly hydrophobic.

The Chemisorption mechanism consists of the following steps:

1. Oxidation of surface sulphide to sulphate (or thiosulphate) by dissolved oxygen in the pulp

$$PbS_{(S)} + 2O_{2\,(g)} \quad \leftrightarrow \quad PbSO_{4\,(S)}$$

2. Ion exchange replacement of surface sulphate ion by carbonate ion

$$PbSO_{4\,(S)} + CO_3^{\;2-} \quad \leftrightarrow \quad PbCO_{3\,(S)} + SO_4^{\;2-}$$

   In aerated pulps, carbonate ions will be present. Solubility products of the lead sulphate and carbonate indicate this reaction will occur.

3. Ion exchange replacement of surface carbonate and sulphate ions (and hydroxyl ion, depending on pH) by xanthate ion forms lead xanthate

Xanthate ion is indicated by X^- 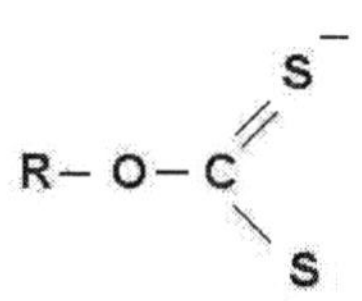

$$PbSO_{4(S)} + 2X^- \leftrightarrow PbX_{2(S)} + SO_4^{2-}$$
$$PbCO_{3(S)} + 2X^- \leftrightarrow PbX_{2(S)} + CO_3^{2-}$$
$$Pb(OH)_{2(S)} + 2X^- \leftrightarrow PbX_{2(S)} + 2OH^-$$

Since lead xanthates are more stable than the carbonate or sulphate, lead xanthates form at the galena surface.

4. Bulk precipitation of lead xanthate at the mineral surface by greater than monolayer.

Xanthates are preferably used in weakly alkaline solutions. Bubble attachment to collector coated particles is difficult to establish in highly alkaline solutions. In acid solutions, the salt goes over to the relatively insoluble acid which decreases dispersion and requires more reagent and time for coating.

Another anionic collector with a bivalent sulphur in the solidophil is the Dithiophosphate group (known as Aerofloat). Dithiophosphates differ from Xanthates in that they have pentavalent phosphorous in the solidophil group in place of carbon.

Dialkyl dithiophosphate of Sodium or Potassium is 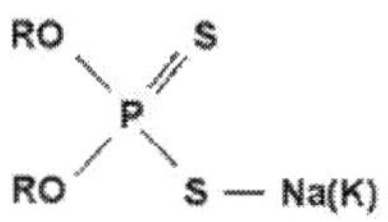

The anion is 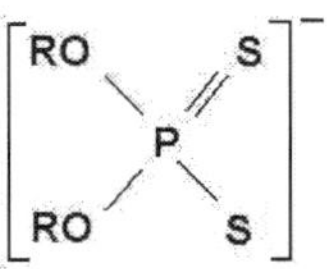

and the solidophil is 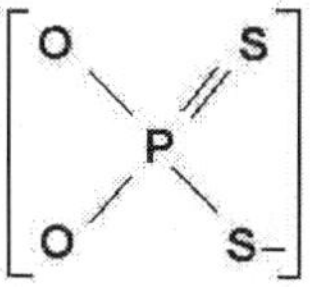

Dithiophosphates are not as widely used as the xanthates, but are still important reagents in practice. They are comparatively weak collectors, but give good results in combination with xanthates. They are often used in separation of copper sulphides from lead sulphides as they are effective selective collectors for copper sulphide minerals.

Dithiophosphates are the most easily affected by depressants. Since the iron sulphides are the most easily depressed, the dithiophosphates are useful for differential work involving iron depression. On the other hand, in pulps of low alkalinity or slightly acid, they are effective to float iron sulphides.

Thiocarbanilide is the commercial term for diphenyl-thiourea, $CS(NH.C_6H_5)_2$. Its commercial form is a solution (about 15%) in orthotoluidine, known as 'T-T mixture'. This reagent was formerly used, but, owing to its insolubility in water, its cost, and its somewhat limited applicability, it has now been largely replaced by the xanthates and the aerofloats. It can be applied in the flotation of galena, where it does not to float as much sphalerite and pyrite as the xanthates.

Oxyhydryl collectors have organic and sulpho-acid anions as their polar groups and the cation takes no significant part in the reagent-mineral reaction. Typically, oxyhydryl collectors are organic acids or soaps. The carboxylates are known as fatty acids. The salts of oleic acid, such as sodium oleate and linoleic acid, are commonly used collectors.

Sodium Oleate

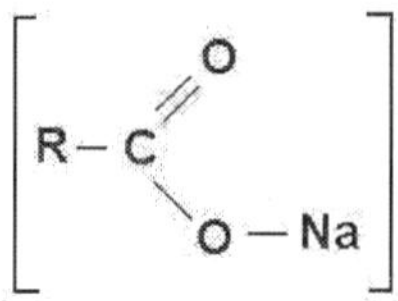

Active anion is

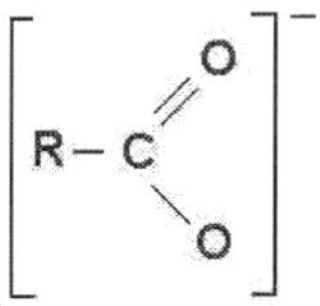

Solidophil is

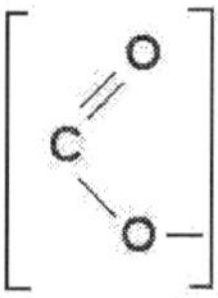

The carboxylates are strong collectors, but have relatively low selectivity. They are used for the flotation of minerals of calcium, barium, strontium, and magnesium, the carbonates of non- ferrous metals, and the soluble salts of alkali metals and alkaline earth metals.

The soap is produced by substituting a sodium ion for the hydrogen of the carboxyl, i.e. COONa instead of COOH. Oleic acid and its sodium or potassium soaps are used in the flotation of oxides, halides, silicates, and oxygen-salt minerals, e.g. fluorite, apatite and garnet.

The sulphates and sulphonates are used more rarely. They possess similar properties to fatty acids, but have lower collecting power. However, they have greater selectivity and are used for floating barite, celestite, fluorite, apatite, chromite, kyanite, mica, cassiterite, and scheelite.

Cationic collectors: The characteristic property of Cationic collectors is that the water-repulsion is produced by the cation where the polar group is based on pentavalent nitrogen. Amines are most common cationic collectors. The anions of such collectors are usually halides, or more rarely hydroxides, which take no active part in the reaction with minerals.

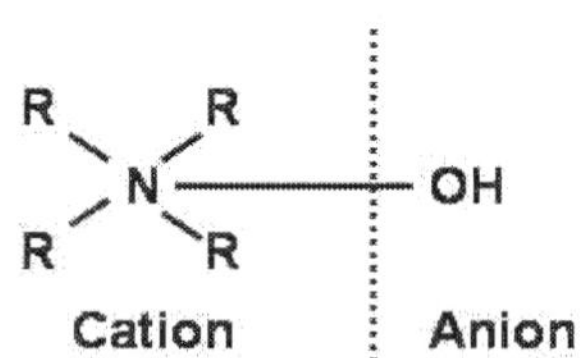

Amines are considered to adsorb on mineral surfaces primarily due to electrostatic attraction between the polar head of the collector and the charged electrical double layer on the mineral surface. These collectors are relatively weak in collecting power. They are used for oxide and silicate minerals including quartz. Cationic collectors are very sensitive to the pH of the medium, being most active in slightly acid solutions and inactive in strongly alkaline and acid media.

Non-ionizing Collectors: Non-ionizing collectors are chemical substances that cannot dissociate to form ions or are insoluble in water. They are non-polar hydrocarbon oils and are obtained from crude oil (distillation, fractionation etc.) or from coal. The reagents from this group do not have a defined chemical composition because of differences in the characteristics of the crude oil or coal from which they are obtained. These reagents do not interact with water dipoles or mineral surfaces. Note that the molecules of the non-polar hydrocarbons have covalent bridging. Because non-polar reagents do not have a solidophilic group, they do not form orientated absorption layers on the mineral surface. It is believed that their adsorption on the mineral surface is imparted by adhesion. They readily adsorb on minerals that are naturally hydrophobic such as graphite, coal, elemental sulphur and on some of the molybdenite found in vein deposits.

Apart from the use for flotation of naturally hydrophobic minerals, non-polar reagents have an application as froth modifiers during fatty acid and amine flotation. They are useful in modification of fatty acids to improve selectivity of the collector.

The most widely used non-polar reagents are kerosene, transformer oil and synthetic hydrocarbon oils. Performance of the non-polar reagents is significantly improved when they are emulsified before use. It has been reported that use of the xanthate-fuel oil system improved the recovery of galena and also resulted in a reduction of xanthate consumption.

Oils are used in froth flotation with carboxylic collectors to escape the necessity for accurate control of pH. With cationic collectors, oil is a marked aid, since it makes a much larger contact angle with the chemically coated surfaces than air does and therefore aids air-bubble attachment. Oils should not be used when differential flotation is sought between two or more minerals all more or less reaction coated by the same collector.

14.6.2 Frothers

Frother is a surface active organic reagent and is heteropolar in nature; polar part of it has an affinity for water and non-polar part has an affinity for air or repulsion for water. Frother acts upon the air water interface. When surface-active molecules react with water, the water dipoles combine readily with the polar groups and hydrate them, but there is practically no reaction with the non-polar hydro-carbon group. The result is to force the latter into the air phase. The frother molecules are adsorbed on the air-water interface and arranged in a way such that the hydrophilic or polar groups are oriented into the water phase, and the hydrophobic or nonpolar hydrocarbon chain in the air phase (Fig.130).

The addition of a frother decreases the surface tension of water and increases the life of bubbles produced. The main objective of a frother is to permit the production of a sufficiently stable froth to hold the mineral particles that form a network around the bubbles until they are removed from the flotation unit.

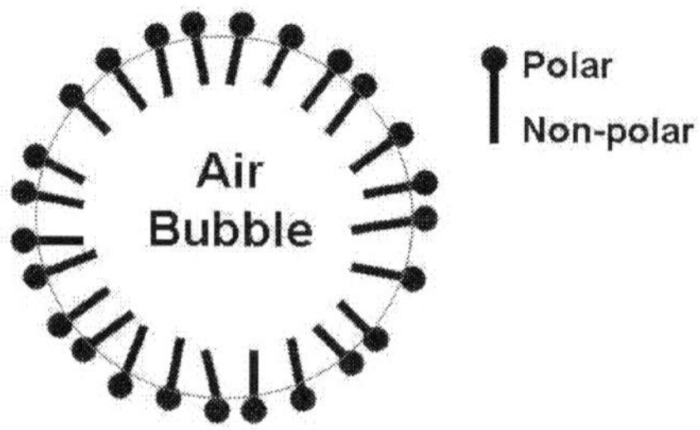

Fig. 130 Action of the frother

In fact, the frother creates conditions for froth formation. The flotation froth is a three-phase system. The frother concentrates at the interface of water and air bubbles, forming an envelope around the bubbles, which prevents them from colliding or touching. The forces created around the air bubble in the presence of a frother prevent the bubbles from collapsing.

Requirements of a frother

1. Ideal frother should have no collecting property. The frother must be controllable independently of collection.
2. Frothers must be to some extent soluble in water, otherwise they would be distributed very unevenly in an aqueous solution and their surface-active properties would not be fully effective.
3. Frothing properties of a frother should not be affected by collectors and salts even when both are present in high concentrations.
4. A frother should be intensely surface active, so that it produces frothing in low concentrations.
5. A frother should produce a copious, lively, non-persistent but non-brittle froth.
6. A frother should froth equally well in alkaline or acid medium.
7. A frother should be readily dispersible in an aqueous medium.
8. A frother should not ionise materially.
9. A frother should not absorb soaps or other organic colloids to the exclusion of solids.
10. A frother should be liquid, cheap and stable against storage and high atmospheric temperatures.

As a result of the addition of a frother, the gas bubbles, formed under the surface of water, are more or less completely lined with monomolecular sheath of frother molecules which allows each bubble to come in contact with other bubbles without coalescing. This forms a froth. Thus, a froth is simply a collection of mineral bearing bubbles. The effect produced by the addition of a frother is proportional to the amount of frother. Past a certain point, the effect of further additions of frother is less than that of preceding additions and addition of further amount results in a decrease of the frothing action and finally in the total absence of frothing. The point at which total absence of frothing is obtained corresponds to the saturation by the dissolved substance of the solution.

Overfrothing is a condition of a froth which involves an uncontrollable amount of froth. Over-oiling produces an overfrothing condition in which the froth contains small bubbles, the froth is highly fluid and carries a heavy solid load with little or no selectivity. Overfrothing can be corrected by a reduction in quantity of frother.

The most effective frothers include one of the following polar groups in their composition.

Hydroxyl **– OH**

Carboxyl 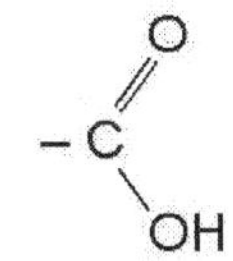

Carbonyl **= C = O**

Amino **– NH_2**

Sulfo **– OSO_2.OH** or **– SO_2.OH**

The hydrophobic group consists of a hydrocarbon chain or a cyclic hydrocarbon radical. The most common fothers are those containing the OH group, i.e. cresylic acid $CH_3C_6H_4OH$ and pine oil $C_{10}H_{17}OH$. The OH group has strong hydrophilic properties and is only sparingly adsorbed on minerals, thus producing a minimum collecting effect. Frothers with a hydroxyl polar group (– OH) alcohols have no collector properties and for this reason they are preferred over other frothers.

A wide range of synthetic frothers are now in use in many plants. Methyl Iso-Butyl carbinol (MIBC) $C_5H_{13}OH$ is most important among the synthetic frothers. The synthetic frothers have a guaranteed composition which makes for ease of control in the plant. Following are the chemical formulae of most used organic reagents.

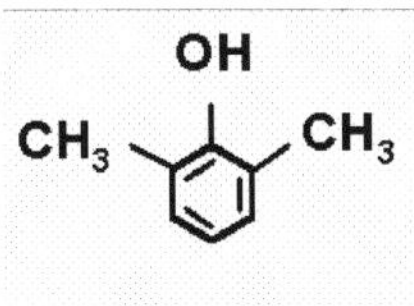

Cresylic Acid

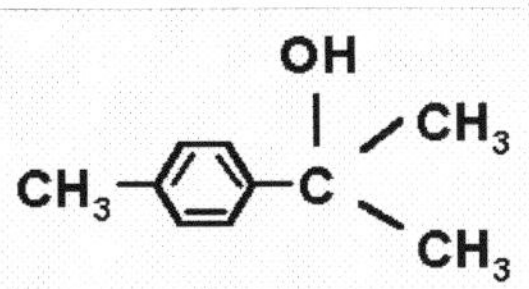

Pine oil

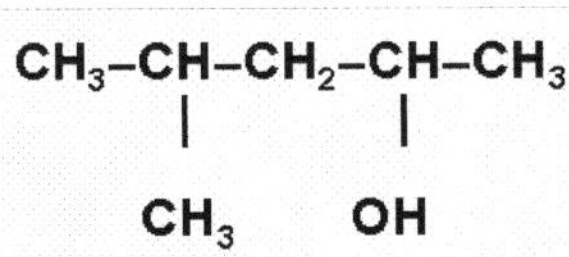

Methyl Isobutyl Carbinol (MIBC)

Eucalyptus oil, camphor oil and sagebrush oil are used when they are more cheaply available than the common frothing agents.

While the most important function of a frother is to form a stable froth which permits removal of a concentrate, frothers have other valuable uses in a flotation circuit such as

1. result in the formation of finer bubbles, i.e. improve air dispersion in a flotation cell.
2. Prevent coalescence of separate air bubbles.
3. Decrease the rate at which bubbles rise to the surface of the pulp.
4. Affect collector action.
5. Increase bubble wall strength and the stability of the froth formed when mineralised bubbles rise to the surface.

The effect of collector on froth properties

The volume of froth generated in the flotation machine under standard operating conditions is often referred to as **froth power**. The froth power is influenced by the type of collector and frother used. The compounds like fatty acids, sulphonates and amines which are in use as collectors have also frothing properties. The reagents having both frothing and collecting properties are known as **frother-collectors.** Frothers with the carboxyl group (COOH) are both frother and collector. Kerosene is a frother-collector used in coal flotation. In fatty acid flotation, certain alcohol-type frothers, such as pine oils, cresylic acid, etc. act as froth modifiers. In cationic flotation with amines, addition of frother to the amine reduces collector consumption and improves the rate of flotation. To have an independent control, the frother should not have a collecting property.

14.6.3 Modifiers (Regulators)

Modifiers are defined as reagents used in flotation to control the action of a collector on minerals, to intensify or reduce its water-repellent effect on the mineral surface. Modifiers make collector action more selective with regard to minerals, thus ensuring a reasonably sharp separation from each other in flotation. When such a modifier is added to the pulp, the collector is then able to render only certain minerals water repellent without perceptible result on other minerals. The function of the modifier involves reaction with the mineral and/or ions present in the pulp; in most cases the reaction is chemical in nature.

Resurfacing of mineral particles to improve the selectivity of the flotation process is the most important and most complex function performed by any modifying agent. The term resurfacing is used in the broad sense to include both adsorption and chemical interaction of modifying agents at or near the mineral surface. Depressants, Activators and pH regulators are most important regulating agents.

Depressants are inorganic chemicals. They react chemically with the mineral particle surfaces to produce insoluble protective coatings of a wettable nature making them non-floatable even in the presence of an appropriate collector. Thus, the formed protective coatings prevent the formation of collector film. Sodium or potassium cyanide is a powerful depressant for sphalerite and pyrite.

Sodium cyanide (NaCN) hydrolyses in aqueous solution to form free alkali and relatively insoluble hydrogen cyanide:

$$NaCN + H_2O \leftrightarrow HCN + NaOH$$

The hydrogen cyanide then dissociates:

$$HCN \leftrightarrow H^+ + CN^-$$

An increase in pulp alkalinity reduces the amount of free HCN, but increases the concentration of CN^- ions. The available CN^- ions depends on the pH value, which must therefore be held within controlled limits to maintain the desired degree of depression. Research shows that for a specific mineral and collector there is a critical value for CN^- ions concentration above which flotation is not possible. CN^- reacts with metal cations in two stages.

$$Me^+ + CN^- \leftrightarrow MeCN$$

$$MeCN + CN \leftrightarrow Me(CN)_2^-$$

The end product is complex soluble salt. Silver, Copper and Gold form the complex $Me(CN)_2^-$ and Zinc, Nickel, and Cadmium form the complex $Me(CN)_4^=$. It has been observed that minerals with metallic ions which do not form stable complex compounds with cyanide i.e. antimony, arsenic, bismuth, lead and tin, are not depressed by cyanide.

Apart from the reactions of cyanide with metal ions in solution, it can react with metal xanthates to form soluble complexes, preventing xanthate adsorption on the mineral surface. The greater the solubility of the metal xanthate in cyanide, the less stable is the attachment of the collector to the corresponding mineral surface. It has been shown that lead xanthates have very low solubilities in cyanide, copper xanthates are fairly soluble, while the xanthates of zinc, nickel, gold, and iron are highly soluble. Iron and zinc can, therefore, be very easily separated from lead in complex ores.

The depressive effect of cyanide depends on its concentration and on the concentration of the collector and the length of the hydrocarbon chain. The longer the chain, the greater is the stability of the metal xanthate in cyanide solutions and the higher the concentration of cyanide required to depress the mineral. Relatively low concentrations of xanthates with short hydrocarbon chains are therefore used for selective flotation where cyanides are used as depressants.

Zinc sulphate ($ZnSO_4$) is used as a sphalerite depressant in the separation of lead-zinc and lead-zinc-copper ores. Zinc sulphate alone is not very effective depressant for copper-activated sphalerite and is often used together with sodium cyanide. Combination of sodium cyanide and zinc sulphate ($ZnSO_4$) is more effective in depressing zinc sulphide minerals, sphalerite and marmatite. Cyanide reacts with zinc sulphate to form zinc cyanide. This is relatively insoluble and precipitates on the sphalerite surface, renders it hydrophilic and prevents collector adsorption.

$$ZnSO_4 + 2NaCN \leftrightarrow Zn(CN)_2 + Na_2SO_4$$

In an alkaline pulp, zinc hydroxide, which adsorbs copper ions, is also formed and it is precipitated on to the sphalerite surface, and prevents collector adsorption.

The use of zinc sulphate reduces the cyanide consumption. Combination of sodium cyanide and zinc sulphate also have depressing action on pyrite.

Lime is used sometimes to depress pyrite in sulphide flotation. It is known that the presence of free hydroxyl (OH^-) ions is detrimental to pyrite flotation. Iron tends to form a very insoluble hydroxide indicates the possibility of the formation of a protective hydroxide-containing film or coating. In using lime as a depressing agent or as a pH control agent, care should be taken to prevent its depressing any of the valuable sulphide minerals. For example, a large excess of lime may prevent the flotation of galena with xanthates. Lime also has a depressing effect on free gold. High lime reduces the effectiveness of aerofloat collectors in the flotation of copper minerals.

Potassium dichromate ($K_2Cr_2O_7$), sodium dichromate ($Na_2Cr_2O_7$) are used to depress galena. The depressive action is due to the chemical reactions between the galena surface and CrO_4^- anions, which produces insoluble lead dichromate that increases wettability and prevents flotation. It is used primarily in floating copper from lead, also to some extent in cleaning zinc concentrates and molybdenum concentrates. Neither sodium dichromate nor sodium cyanide was found to be a very effective depressant when used alone, but when added together they significantly enhanced the depression of both chalcopyrite and galena. Sodium sulphide (Na_2SO_3) and sulphur dioxide (SO_2) have also been used for the depression of sphalerite. They are less powerful depressants than cyanide and thus find a few applications where this may be of advantage. Sodium sulphide (Na_2S) has a depressing action on sphalerite, and laboratory work indicates that under certain conditions it may act as a depressant for many other minerals.

Oxidizing agents such as sodium hypochlorite (NaOCl) and potassium permanganate ($KMnO_4$) are effective depressants for copper, iron and cobalt sulphides in lime pulps. As an example, the separation of copper and molybdenum sulphides can be carried out by depressing the copper sulphide with sodium hypochlorite. Sulphuric acid is used to depress quartz in soap flotation. Lactic acid is a powerful depressant in iron sulphide flotation. Metaphosphates are used as depressants for non-silicates like barite, fluorite and calcite.

Depressants are normally added either in conditioner or in flotation cell. Sometimes it is found necessary to add cyanide directly to the ball mill to achieve the desired selective action, while on the occasions conditioning time of a few minutes to an hour may be required.

Activators, generally inorganic compounds, can modify the surface of non-floatable or poorly floatable mineral particles by adsorption on particle surface so that the collector may film the particle and induce flotation. Activators are generally soluble salts which ionise in solution, the ions then reacting with the mineral surface. An example of this is the use of copper sulphate in the flotation of sphalerite. Sphalerite is not floated satisfactorily by a xanthate collector, since the collector products formed, such as zinc xanthate, are relatively soluble in water, and so do not provide a hydrophobic film around the mineral.

Floatability can be improved by using large quantities of long chain xanthates, but a more satisfactory method is to use copper sulphate as an activator, which is readily soluble and dissociates into copper ions in solution. Activation is due to the formation of molecules of copper sulphide at the mineral surface because copper is more electronegative than zinc and therefore ionises less readily. Copper sulphate dissociates into copper ions in solution and copper sulphide is formed at the surface of sphalerite.

$$CuSO_4 + ZnS \quad \leftrightarrow \quad ZnSO_4 + CuS$$

Thereafter, it reacts with xanthate and forms insoluble copper xanthate which makes the sphalerite surface hydrophobic. In the flotation of lead-zinc ore, galena is floated by depressing sphalerite. After galena flotation, the sphalerite is activated with copper sulphate. Copper sulphate forms copper sulphide coating on the zinc sulphide thus giving it the floatable characteristics. To some extent, copper ions can also activate galena, calcite, and pyrite. When sphalerite is associated with pyrite or pyrrhotite, selectivity is usually ensured by the high alkalinity (pH 10.5–12) of the pulp, lime being added in conjunction with the copper sulphate activator. Copper sulphate also activates previously depressed pyrite when added in sufficient quantity.

Sodium sulphide is used to activate oxide minerals of lead, zinc and copper such as cerussite, smithsonite and malachite. As sodium sulphide imparts sulphide surface to the mineral particles to facilitate for collector coating, this activator is also known as **sulphidiser**. The amount of sodium sulphide added to the pulp must be very strictly controlled, as it is a very powerful depressant for sulphide minerals and will, if in excess, depress the activated oxide minerals, preventing collector adsorption. Sodium sulphide is also used to float previously depressed pyrite. It has dispersing effect on sulphide minerals when added in large quantities.

pH regulators are used to modify the alkalinity or acidity of a flotation circuit or in other words to control the pH of the pulp. With few exceptions, the effectiveness of all flotation reagents depends greatly on the concentration of hydrogen or hydroxyl ion in the pulp solution. pH determines the driving force for various interactions between the mineral and the reagents in solution, and therefore it is an important control parameter for regulating flotation. The flotation of almost all minerals can be suppressed by sufficient addition of H^+ or OH^- ions through the addition of acids or alkalis. The exact role of H^+ and OH^- ions depends on the mineral, the collector and other modifying agents present in the flotation slurry.

Flotation where possible is carried out in an alkaline medium, as most collectors, including xanthates, are stable under these conditions, and corrosion of cells, pipework etc. is minimised. Most commonly used pH regulators are lime (CaO) and sodium carbonate (soda ash) (Na_2CO_3). Sodium hydroxide (NaOH) can also be used but is more expensive. On account of availability and low cost, lime is used in all circuits where the calcium ion is not objectionable, including most zinc and copper circuits. Some minerals, notably pyrite and galena tend to adsorb calcium ions in preference to a collector and are thus not so easily floated in a lime pulp. This difficulty is eliminated by using soda ash which precipitates naturally occurring calcium ions as well as producing the necessary alkalinity. Sulphuric or sulphurous acids are used where a decrease in pH is required.

pH regulators are often used in very significant amounts in almost all flotation operations. Although they are cheaper than collectors and frothers, the overall cost is generally higher with pH regulators per ton of ore treated than with any other processing chemical. For example, the cost of lime in sulphide mineral flotation is roughly double that of the collector used, so significant operational cost savings can be achieved by the appropriate choice and use of pH regulators.

In flotation with ionic collectors, free H^+ and OH^- ions may compete with the collector for surface sites. Thus, high concentrations of H^+ or OH^- ions may lead to depression through a decrease in collector adsorption.

With xanthates as collectors, sufficient alkali will depress almost any sulphide mineral, and for any concentration of xanthate there is a pH value below which any given mineral will float, and above which it will not float. This **critical pH** value depends on the nature of the mineral, the particular collector and its concentration, and the temperature.

Quantity of reagents used in flotation varies from ore to ore and day to day or hour to hour for one ore. Small quantities of reagents are normally required. Low and high quantities of the reagents are of no use. In certain cases, increasing quantities results other effects. For example, use of increased quantities of collector tends to float other mineral particles (other than required) also. Table 38 shows the approximate quantities of various reagents used in flotation.

Table 38: Quantities of flotation reagents

Reagent	Quantity in pounds per ton
Collectors	0.05–2.5
Frothers	0.025–0.25
Depressants	0.05–1.0
Activators	0.5 –2.0
pH regulators	0.5–10.0
Dispersants	0.1–0.7

14.7 TYPES OF FLOTATION

As a mineral is selected and floated in flotation, it is called **selective flotation**. As required mineral is selected and floated, it is also called **direct flotation**. When an unwanted mineral is selected and floated, in which case, sink is the required product, it is called **reverse flotation**. When an ore contains two or more valuable minerals, one valuable mineral is selected and floated first and second valuable mineral is floated from the tailings. This flotation is called as **differential flotation**. Alternatively, when all the valuable minerals are selected and floated, it is called **bulk flotation**. The individual minerals are selected and floated one after the other from the floats.

14.8 FLOTATION MACHINES

A flotation machine is an equipment used to carry out flotation operation. It provides the hydrodynamic and mechanical conditions which effect the separation. Basically, the flotation machine must include

1. means for receiving and discharging the pulp
2. means for agitation and mixing the pulp
3. means for air introduction and dispersion
4. means for settling the pulp away from the froth
5. means for discharging the froth and conveying it to the next processing step

Regulation of rates of intake and discharge of pulp, of rate of air introduction and dispersion, and of the pulp level is essential for better control of flotation operation.

The most commonly used flotation machines are of two types namely mechanical type and pneumatic type. In accordance with the mode of introduction of the air, the flotation machines can be divided into four types

1. Agitation machines in which air is drawn by a vortex caused by a rotating impeller.
2. Sub-aeration machines in which air is introduced by suction or by a blowing through or to the base of a rotating impeller.
3. Cascade machines in which air is introduced by tumbling of the pulp.
4. Pneumatic machines in which air is introduced directly by blowing.

In a mechanical type flotation machine, mechanically driven impeller agitates the pulp and disperses the incoming air into small bubbles. The air may be drawn in by suction created by the impeller or may be introduced to the base of the impeller by an external blower. These flotation machines are often composed of several identical cells arranged in series in such a way that one cell receives the de-frothed pulp (tailing of preceding cell) as feed. Such a series of cells called as bank. The Denver sub-aeration machine (Fig. 131) is well known cell-to-cell machine.

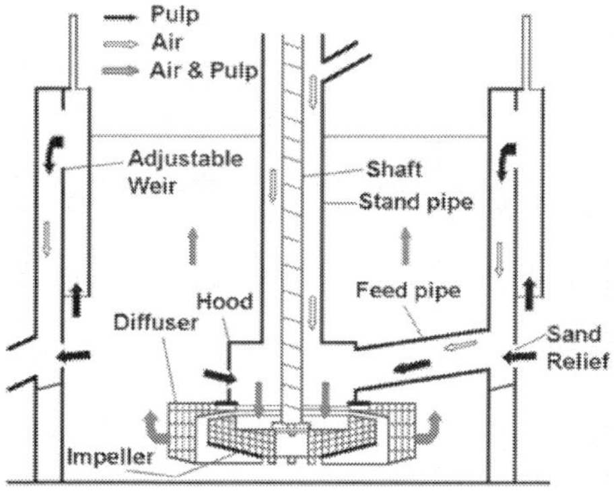

Fig. 131 Denver Sub-aeration Flotation Cell

The pulp from the weir of the preceding cell flows through the feed pipe on to the rotating impeller. The positive suction created by the impeller draws air through hallow standpipe, sheared into fine bubbles by impeller and intimately mixed with the pulp. The diffuser arranged around the impeller prevents the agitation and swirling of the pulp above the impeller. The stationary hood above the impeller prevents the pulp in the cell to have direct contact with the impeller. The mixture of bubbles and pulp leaves the impeller and bubbles ascend with their mineral load to form the froth. As the bubbles move to the pulp level, they are carried to overflow lip by crowding action of succeeding bubbles and removed by froth paddles. Pulp from the cell flows over the adjustable weir on to the impeller of the next cell. Particles which are too heavy to flow over the weir are by-passed through sand relief ports.

In an 'open-flow' or 'free-flow' type machine, intermediate partitions and weirs between cells are eliminated. The impeller speed is normally in the range of 7–10 m/s peripheral. By increasing the impeller speed, more air can be drawn. More air over-agitates the pulp as well as increase impeller wear and energy consumption. In such cases, supercharging may be applied by introducing additional air down the stand-pipe by means of an external blower. Supercharging is required with the Denver D-R flotation machine. This machine was developed as a result of the need for a machine to handle large tonnages in the bulk flotation circuits. Intermediate partitions and weirs between cells are absent in these cells (Fig. 132). The pulp is free to flow through the machine without interference. The pulp level is controlled by a single tailings weir at the end of the trough. Flotation efficiency is high, operation is simple, and the need for operator attention is minimised.

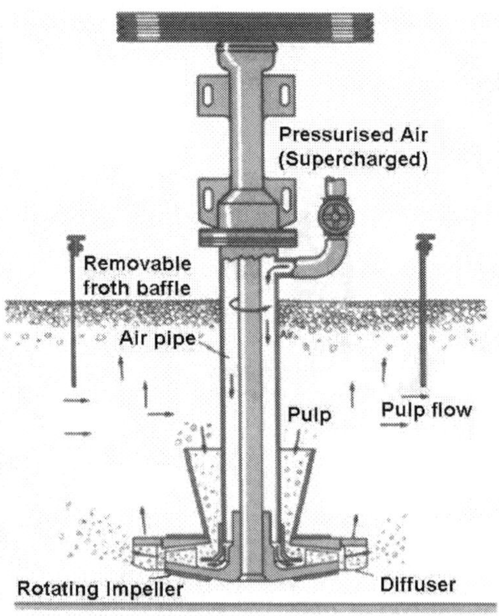

Fig. 132 Denver D-R Flotation machine

The Wemco Fagergren flotation cell is an open-trough type, self-aerating mechanical flotation cell (Fig. 133), and is most commonly used in flotation of coal. The mechanism consists of a star-like rotor, which rotates inside the shroud or disperser. When operating, the rotor creates a fluid vortex inside the standpipe and in the draft tube, this vortex being at a sufficient vacuum to induce air into the standpipe through the air inlet from above the cell surface. The induced air is mixed inside the rotor-stator zone with the pulp that has been sucked into the rotor from below.

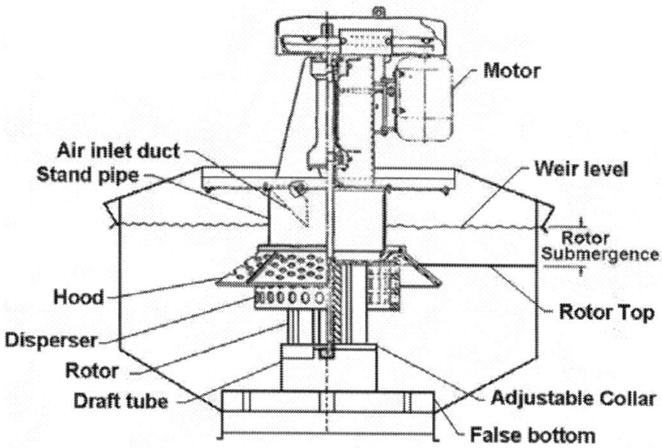

Fig.133 Wemco Fagergren cell

The large capacity Wemco flotation cells have a false-bottom draft-tube system that enhances the circulation of the pulp. It permits a low rotor submergence, even for the large-capacity cells. The three-phase mixture leaves the shroud (disperser) through the disperser apertures in radial direction. The new disperser fulfils the shrouding effect perfectly, since the tangential velocity of the three-phase mixture imparted on it by the rotor is converted to an entirely radial direction. The particle-loaded air bubbles rising to the froth surface are separated outside the disperser, and the remaining pulp, which flows down the tank walls and returns to the rotors, is separated in the large cells through the false bottom and the draft tube. The conical stator hood is to quieten the cell surface and to keep any turbulence created by the rotor away from the froth layer.

The function of the mechanism in a Wemco flotation cell is therefore threefold: to circulate pulp, to mix pulp with air, and to draw in air. These functions depend on the shape and geometric dimensions of the rotor and stator, on the speed and submergence of the rotor, on the vertical distance between the rotor and the real tank bottom.

Galigher Agitair machine (Fig. 134) is a supercharged machine and is often used in large-capacity plants. In each compartment, is a separate impeller rotating in a stationary baffle system. Air is blown into the pulp through the hollow standpipe surrounding the impeller shaft, and is sheared into fine bubbles, the volume of air being controlled separately for each compartment. Pulp depth is controlled by means of weir bars or dart valves at the discharge end of the bank, while the depth of froth in each cell can be controlled by varying the number and size of froth weir bars provided for each cell. Agitair machines produce copious froths and have found favour in mills handling ores of poor floatability, which require large froth columns to help weakly aerophilic particles to overflow.

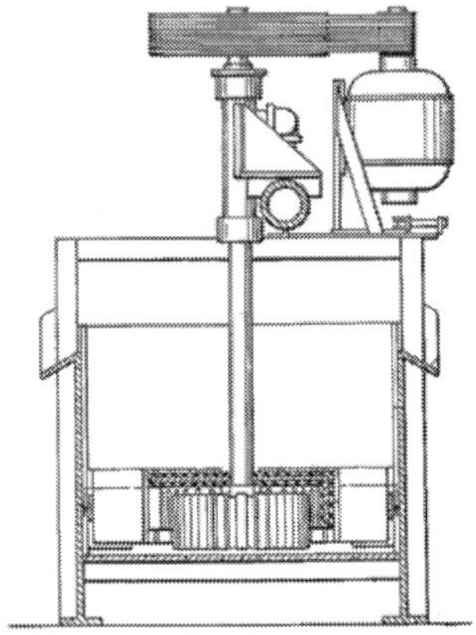

Fig. 134 Galigher Agitair machine

In pneumatic machines, air is introduced through the porous bottom of the cell. The air used in these machines not only creates aeration and produce froth but also maintain suspension by circulation. Pneumatic machines are now rarely seen, except in a few old concentrators. Mechanical machines are better suited to difficult separations, particularly where fines are present.

14.9 AERATION AND FROTH FORMATION

Aeration may be defined as the introduction of air bubbles in the pulp in a manner to result in the stable froth formation.

Frothing means the maintenance of a body of bubbles at and above the upper surface of the flotation pulp. The liquid walls of these bubbles contain ore particles either at rest or in motion or in both states. Collection of mineral bearing bubbles forms the froth. Coalescence interferes with froth building. Bubbles in pure water coalesce immediately on contact. Frothing agents and the solids in the bubble interfaces reduce coalescence markedly. The functions of the froth are

1. Froth changes the effective density of the minerals included in it so that bubble-mineral aggregate will rise to the surface
2. It permits differential draining back of gangue minerals into the body of pulp
3. It maintains a body of ore particles above the pulp long enough to permit such draining to affect concentration
4. It transports floated material horizontally to the concentrate discharge point.

A froth is simply a collection of bubbles. Prolongation of bubble life is, therefore, an essential element in frothing. Sufficiently long life is necessary to give time for the desired draining and horizontal transport. Beyond such time, however, further life is a nuisance, since froth is difficult to transport and dewater. The draining function is best served by a relatively large bubble froth, comparatively thick walled, with plenty but not an excessive amount of internal movement.

A good froth

1. should have sufficient buoyancy to carry the mineral load without collapse or excessive showering to the surface of the pulp
2. should be stable enough to exist until the bubbles with their mineral load can pass over the lip of the flotation machine
3. should be sufficiently small bubbled to support fairly heavy load of solids
4. should not be so tight that gangue is carried in it
5. should not be so tough that it will be difficult to break up after passing from the machine
6. should also be fluid enough to flow over the lip of the machine without clogging.
7. should be lively and actively coalesce in its upper portion to build large bubbles (several inches to a foot or more in diameter)

Coalescence transfers the solid load to underlying bubbles and affords opportunity to shake out solid particles. If coalescence occurs at low point, mineral particles showers back into the pulp. If bubbles are too fine and too much crowded and does not coalesce frequently, coarse gangue is carried over mechanically. In such case, fine spray of clean water is applied on to the surface of the froth to drain the gangue. Persistence of the froth decreases as temperature increase due to decrease in viscosity of frothing agent and consequent ease of dispersion of frother. Hence more frothing agent is required in winter than in summer.

14.10 FORMATION OF MINERALISED BUBBLES

Mineral particle attachment to air bubbles is the basis of the froth flotation process. Bubble attachment is obtained by one of the following two methods:

1. particles are come in contact with bubbles already formed in the pulp
2. gas bubbles are precipitated from the solution on to the particle surface

In the first method, where particles are come in contact with bubbles, the amount of surface energy is determined by the sum of the free energies on the corresponding interfaces. The Fig. 135 shows bubble and mineral particle before and after attachment.

In the second method, a small bubble initially forms from the solution on a mineral particle. A larger bubble with sufficient lifting power to carry the particle to the surface attaches itself to the smaller bubble. Coalescence of the two bubbles takes place and the large bubble adheres to the grain. It has been found that the surface energy loss is greater in this type of bubble formation. Coalescent type bubbles is more rapid in flotation.

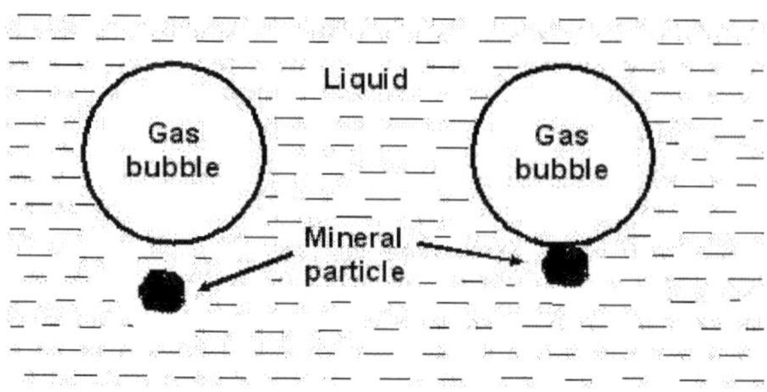

Fig. 135 Mineral particle–bubble attachment

Mineral particles will occupy from 2% (in low grade separations) to 20% (in coal flotation and re-concentrating stages in sulphide ores) of the total bubble surface available. In certain cases, large single particles may be floated by a group of bubbles. In other cases, groups of particles and bubbles coalesce to form air flocs, which occurs when the mineral surface is highly water repellent and there is plenty of aeration but not too much agitation.

14.11 FLOTATION OPERATION

The flotation operation is generally carried out in three stages namely roughing, scavenging, and cleaning called as flotation circuit (Fig. 136). Each stage consists of a bank of cells and the number of cells in a bank is primarily depends upon the residence time of the pulp in the cells and the required throughput. The reagent conditioned feed pulp is treated in a first bank of cells called roughers. The tailing from the rougher cells, which may still contain some valuable mineral particles, is treated in another bank of cells called scavengers. The concentrate from the rougher cells is further treated in a bank of cells called cleaners to obtain high grade final concentrate.

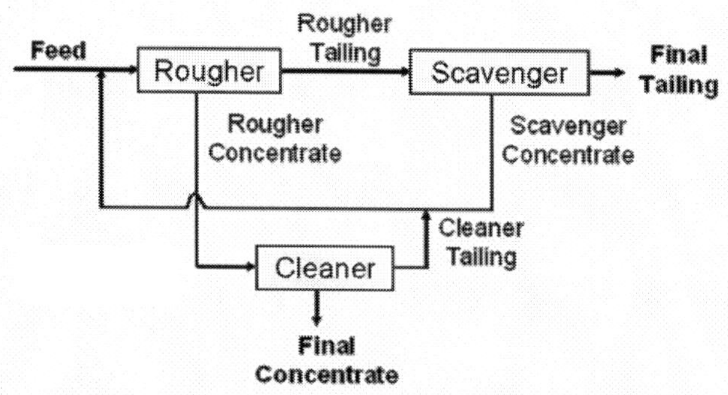

Fig. 136 A typical flotation circuit

The scavenger concentrate and cleaner tailing are re-fed to rougher bank to be treated with the fresh feed pulp. The scavenger tailing is the final tailing. It is to be noted that the pulp is treated in a conditioner, called **conditioning**, with necessary reagents like depressant and collector prior to flotation to convert the mineral particles to respond readily in a flotation cell. Conditioning the pulp reduces its residence time in the flotation cell and hence the capacity of the cell increases.

Commercial flotation is a continuous process. Cells are arranged in series forming a **bank**. Pulp enters the first cell of the bank and gives up some of its valuable minerals as froth; the overflow from this cell passes to the second cell, where more mineralised froth is removed, and so on down the bank, until barren tailings overflow the last cell in the bank. In the case of flotation cells that use weir-type level control, the height of the froth column for each cell is determined by adjusting the height of the tailings overflow weir, the difference in height between this and the cell overflow lip determining the froth depth. In modern tank cells, pulp level is often maintained by adjusting the cell's tailings discharge with a rubber sleeve pinch valve.

New feed enters the first cell of the bank, the froth column in the first few cells being kept high, since there are plenty of hydrophobic mineral particles to sustain it. The pulp level is raised from cell to cell, as the pulp becomes depleted in floatable minerals, by progressively raising the cell tailings weir height. The last few cells in the bank contain relatively low-grade froths, comprising weakly aerophilic particles. These are the **scavengers**, usually containing middlings particles, which are often recirculated to the head of the system.

The flowsheet for the basic system is shown in Fig. 137A. This flowsheet can be successfully operated only when the gangue is relatively unfloatable, and it requires extremely careful control to produce an even grade of concentrate if there are any fluctuations in the head grade. A preferable system is to dilute the concentrate from the first few cells of the bank, known as **rougher** concentrates, and re-float them in **cleaner** cells, where the level of the pulp is kept low to maintain a deep froth and produce a high-grade concentrate. In this rougher-scavenger-cleaner system (Fig. 137B), the cleaning cells receive a comparatively high-grade feed, whilst the scavenging section can be run with an excess of air to obtain maximum recovery. Tailings from the cleaner cells, usually containing aerophilic mineral particles, can be recirculated to the rougher cells, along with the scavengers. This type of circuit, besides being useful for ores that need the maximum amount of aeration at the end of the bank to produce profitable recovery, is often employed when the gangue tends to float and is difficult to separate from the mineral. In such cases, it may be necessary to utilise one or more recleaning banks of cells (Fig. 137C).

It is worth noting that the diluting water used to lower the pulp density of the cleaning bank passes to the roughing cells and dilutes the primary feed, which should therefore contain a correspondingly smaller portion of water as it leaves the grinding section in order that the dilution of the cleaner tailing may bring it to the correct pulp ratio in the roughing cells.

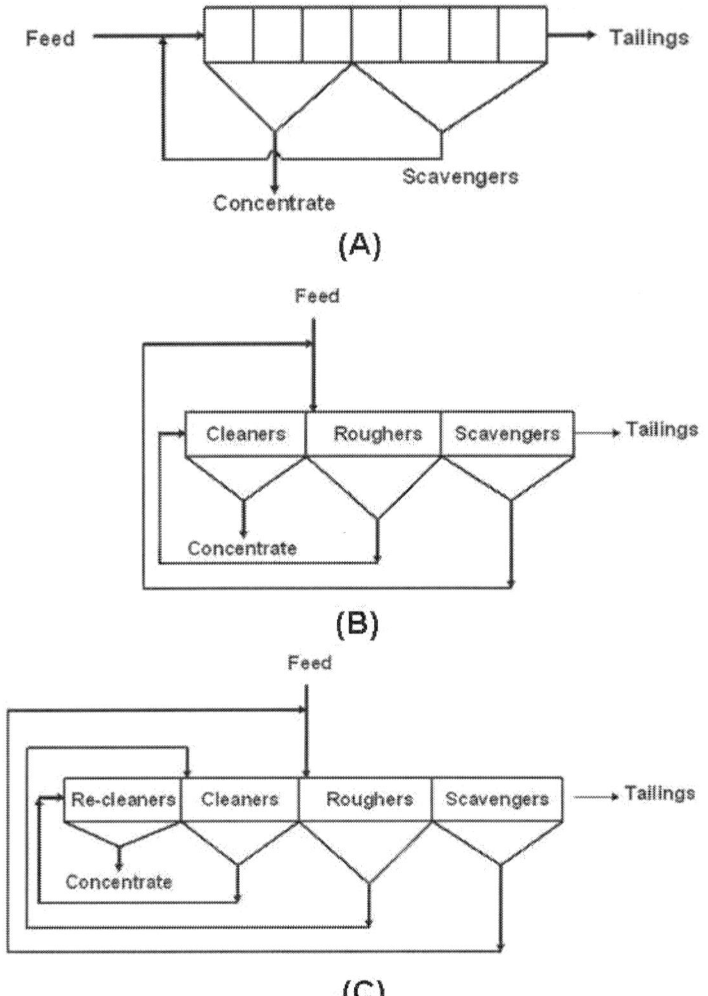

Fig. 137 Flotation Circuits

For successful operation of the flotation, variables effecting the flotation and need to be controlled are ore characteristics, quality of the water, particle size distribution, pulp density, conditioning time, quantity of reagents and their feeding, place of addition of reagents, effect of length of hydrocarbon chain in the reagents, and the time of flotation.

14.12 COLUMN FLOTATION

The column flotation technique uses the principle of counter current wash-water flow for better separation particularly when operating on fine materials. The flotation column is a simplest form of pneumatic type flotation machine. It consists of a tall cylindrical column (Fig. 138) having the height to diameter ratio of more than 10. The reagent conditioned feed pulp is admitted into the upper portion, usually to the depth of one-thirds to one-fifths of the total column height and flotation tailings are discharged in the bottom part of the column. Compressed air is admitted into the sparger (diffuser or aerator) at the bottom of the column so that air bubbles are generated and move upward in counter direction of downward flow of slurry. Hydrophobic mineral particles are attached in the lower enrichment section, also called recovery section, of the column between the feed point and air inlet known as collection zone or flotation zone. Valuable mineral particles attached to the air bubbles are transported to the top part of the column, called washing section, also referred as cleaning zone or frothing zone. This frothing zone is a mobile packed bubble bed that is contacted counter currently with wash water introduced from the top of the column in the form of spray. Gangue mineral particles that are loosely attached to the bubbles are washed down by water sprays and sent back to the flotation zone. Only clean froth rises and discharged through the annular trough.

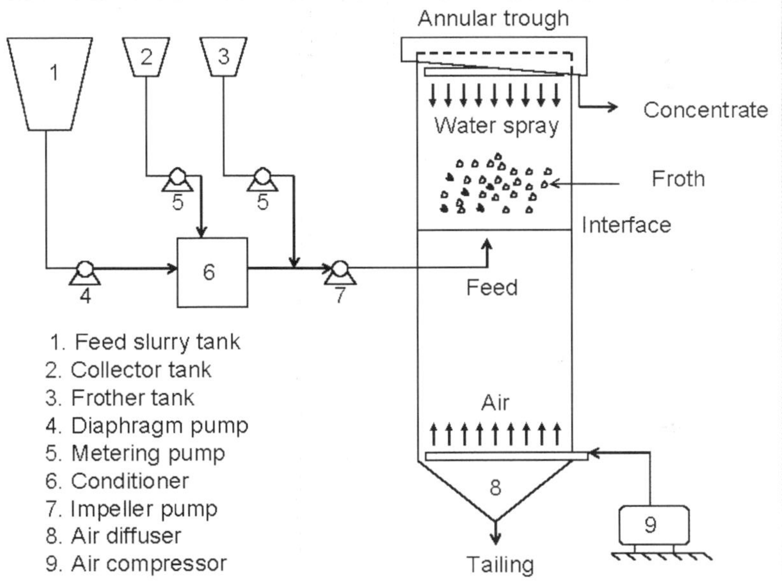

Fig. 138 Flotation Column

Design and operating philosophy of flotation column are totally different in comparison with conventional flotation cell. Although the basic concept of flotation column looks relatively simple, the fundamental principles related to performance of flotation column are quite complex.

The type of the relative motion of particles and bubbles is a major factor governing the probability of bubble/particle attachment, bubble loading, flotation rate and power requirement of the process. Counter current flow of bubbles and particles provide better conditions for bubble-particle attachment and aggregate stability.

The probability of bubble-particle collision is particularly determined by the value of the normal component of relative velocity, the time of contact and the magnitude of inertia forces. The relative velocity of bubbles and particles in the counter current is 10–12 cm/s at a slurry flow rate of 2 cm/s and mean bubble size 1.5–2.5 mm. Such relative velocity provides optimum conditions for collision. Counter current of slurry and air causes a reduction in the rise velocity of the swarm of bubbles, which increases their retention time in the slurry, reduces the compressed air requirement, and increases specific throughput of the column. Without stirrers and at a low turbulence of slurry flows in a column, inertia forces causing the detachment process are insignificant. Furthermore, in a counter current, the probability of bubble-particle collision is higher because of the large aerated volume of the cell and the long way the particles and bubbles have to travel along the column height. And lastly, intensity of longitudinal slurry mixing is low in a column. All this shows that the flotation rate in the counter current is higher than with other directions of slurry and air flows.

Advantages of Flotation Column are

- production of high grade concentrates
- decrease in the reagent dosage
- effective flotation of heavy loaded slurries
- high selectivity of separation
- reduction in process time and number of stages required
- high aerated volume
- possibility of air flow rate and bubble size distribution control
- possibility for creating various aero-hydro-dynamic conditions
- possibility of beneficiation of fine and coarse particles
- low intensity of slurry mixing
- absence of moving parts
- low energy consumption
- Three to four times less floor space requirement than that of conventional cells
- simple and compact in design
- low capital costs
- simplification of process flowsheets
- A single flotation column can replace five to six stages of operations involving conventional cells and yet achieve better performance

■■■■

15 Magnetic Separation

Magnetic separation is a physical separation of particles based on the magnetic property of the mineral particles. The property of a material which determines its response to a magnetic field is the magnetic susceptibility.

According to Weber's theory of magnetism, molecules of all substances are basically magnets in themselves, each having a north and south pole. In a magnet, these molecular magnets lie in line so that north poles of molecular magnets are on one end and south poles on the other. Thus, two definite poles N and S are developed at the ends.

Magnetic field (or force) of a magnet is the surrounding space through which its influence extends. The magnitude of force between two magnetic poles by coulomb's law is

$$F = \frac{m_1 m_2}{4\pi\mu r^2} \qquad 15.1$$

where the force F is expressed in newtons, m_1 and m_2 are the strengths of the poles expressed in weber, r is the distance between the poles in meters, and μ is a constant depending on the medium.

$$F = \frac{m_1 m_2}{4\pi\mu_o \mu_r r^2} \quad \text{in a medium} \qquad 15.2$$

$$F = \frac{m_1 m_2}{4\pi\mu_o r^2} \quad \text{in air (or vacuum or free space)} \qquad 15.3$$

μ_o = **Absolute permeability** of vacuum or air, Henries/metre (H/m), weber/A-m

μ_r = **Relative permeability** of the medium surrounding the poles, dimensionless

The value of μ_o is $4\pi \times 10^{-7}$ H/m

The value of μ_r is different for different media.

If, in the equation 15.3, $m_1 = m_2 = m$ (say), r = 1 metre, $F = \frac{1}{4\pi\mu_o}$ newton

Then, $m^2 = 1$ or $m = \pm 1$ weber (SI unit)

Hence, a **unit magnetic pole** may be defined as that pole which when placed in vacuum at one metre distance from a similar and equal pole repels it with a force of $1/4\pi\mu_o$ newtons.

Magnetic field intensity or **Magnetic field strength** or **Magnetising force** (H) at any point within a magnetic field is the force experienced by a N-pole of one weber placed at that point. This magnetising force induces the lines of force through a material. The unit of H in SI system is Ampere/metre. (1 ampere/metre = $4\pi \times 10^{-7}$ tesla).

The total number of magnetic lines of force in a magnetic field is called **Magnetic flux.** It is represented by ϕ and the unit in SI system is weber. 1 weber = 10^8 lines or maxwells. **Magnetic flux density** (B) at any point is the magnetic flux passing per unit area at that point

$$B = \frac{\phi}{A} \text{ weber/m}^2 \text{ or tesla} \qquad 15.4$$

When an iron rod is placed in the magnetic field of intensity H, it becomes magnetised due to magnetic induction and the number of lines (called lines of induction) passing through iron is much greater. If B is the number of lines of induction per square meter in iron, then B is the **Magnetic induction** or **Magnetic flux density** developed in the rod. It is measured in Weber/m^2, Tesla, or Gauss (1 gauss = 10^{-4} tesla).

$$\therefore \text{ Permeability } \mu = \frac{\text{Magnetic Induction}}{\text{Impressed field intensity or Magnetising force}}$$

$$\mu = \frac{B}{H} \qquad 15.5$$

$$\Rightarrow \quad B = \mu H = \mu_o \mu_r H \text{ weber/m}^2 \qquad 15.6$$

For ferro-magnetic substances, such as iron, nickel, cobalt and alloys such as nickel-iron, and cobalt-iron, relative permeability is much greater than unity whereas for para-magnetic substances, such as aluminium, relative permeability is slightly greater than unity. For diamagnetic materials such as bismuth, the relative permeability is less than unity.

Magnetic flux density is the magnetic force at a point inside a magnetic material and is the resultant of the externally impressed magnetic field and the force due to the induced poles at the ends of the rod. If M is the strength of the induced magnetism, which is called as **magnetisation** or **intensity of magnetisation** and H is the impressed magnetic field, then

$$\text{Magnetic flux density } B = \mu_o(H + M) \qquad 15.7$$

In vacuum, M = 0, and it is extremely low in air.

$$\therefore \text{ Magnetic flux density } B = \mu_o H \qquad 15.8$$

The **Magnetic susceptibility** of a material is defined as the ease with which a magnetic substance can be magnetised. **Magnetic susceptibility** S is the ratio of the intensity of magnetisation produced in the material to the field strength which produces the magnetisation.

$$\text{Magnetic susceptibility} = S = \frac{\text{Intensity of Magnetisation}}{\text{Impressed field intentisy or Magnetising force}}$$

$$S = \frac{M}{H} \quad 15.9$$

Magnetic flux density $B = \mu_o(H + M) = \mu_o(H + SH) = \mu_o(1 + S)H$

$1 + S = \mu_r$ = relative permeability

$\therefore$ Magnetic flux density $B = \mu_o\mu_r H = \mu H$ 15.10

The field gradient is the rate at which the field intensity increases towards surface of the magnet. The force on the particle (F) by the magnet (the capacity of a magnet) to lift a particle is directly proportional to field intensity (H) and field gradient (dH/dl).

$$F \propto H\frac{dH}{dl} \quad 15.11$$

Infinite number of combinations of field and gradient can be used to generate required lifting force. Production of a high field gradient as well as high intensity is therefore an important aspect in designing a magnetic separator.

15.1 ELECTRO MAGNETISM

Magnetic field can be created by electric current. When an electric current (I) flows through a conductor, magnetic field is set up all along the length of the conductor. The magnetic lines of force are circular in a plane perpendicular to the current (Fig. 139). The magnetic field near the conductor is stronger and becomes weaker as the distance increases from the conductor. The magnetic field becomes stronger if current is increased and vice-versa. The direction of the field is reversed if current is reversed.

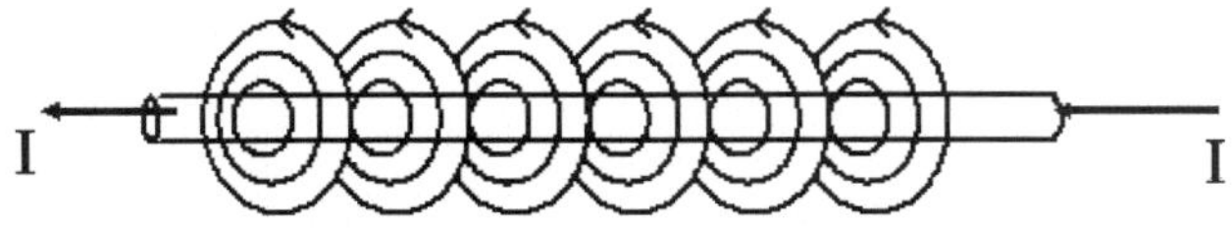

Fig. 139 Magnetic lines of force

If a wire is made into a loop and current is passed through it, the magnetic lines of force are circular round the wire all along its length. When a number of circular coils are arranged co-axially in such a way as same current flows in the same direction in the coils, then this arrangement is called a **solenoid**.

Solenoids are of two types (1) Iron-cored solenoid and (2) Air-cored solenoid. In the iron-cored solenoid (Fig. 140a), the set of circular coils are wound on an iron bar. In the air-cored solenoid (Fig. 140b), there is air inside the coils.

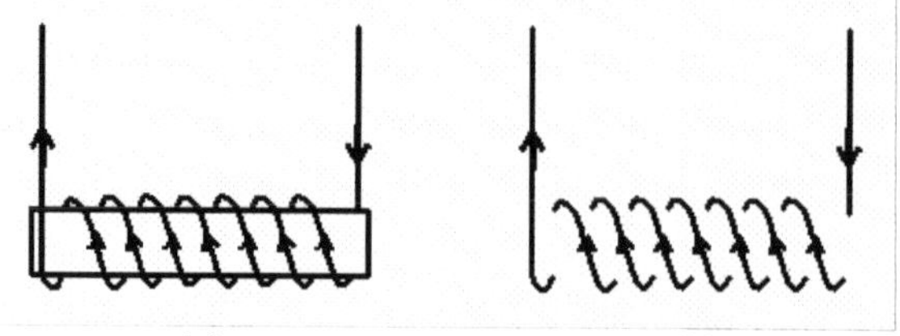

Fig. 140 Types of solenoid

Bars of soft wrought iron, soft steel or cast iron are magnetised by electric currents through the insulated copper or aluminium wire windings around the iron bars. These iron bars form electromagnets as long as the current flows through the windings. The poles of an electromagnet may be reversed by simply reversing the direction of flow of the electric current. The magnetism, or magnetic field, can be obtained of different intensities ranging from indefinitely weak to a certain maximum of strength. The intensity of the magnetic field depends on the size of the magnet, the form of it, the distance between the body to be attracted and the magnet, and the number of ampere turns in the magnet coil, that is, the product of the amperes of current flowing in the coil (I) times the number of turns around the core (N).

One important characteristic of magnetic flux is that each line is a closed loop. The complete closed path followed by group of lines of magnetic flux is known as magnetic circuit. Consider an iron ring wound with a number of turns as shown in Fig. 141. When current is passed through the winding, magnetic lines of force are set up in the ring. The path of these lines is called the magnetic circuit.

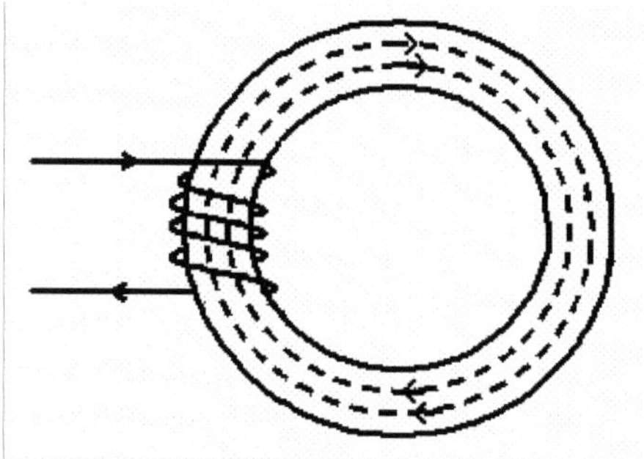

Fig. 141 Magnetic circuit

The magnetic pressure which sets up or tends to set up magnetic flux in a magnetic circuit is called **magneto-motive force** (m.m.f). m.m.f is produced by passing electric current through a wire of number of turns. It is measured in ampere turns, generally written as AT, and is the product of number of turns (N) and the current (I) flowing through these turns. m.m.f. = N×I. The opposition to magnetic flux in a magnetic circuit is known as **reluctance**. The reluctance is directly proportional to the length of magnetic path (l), inversely proportional to the area of cross-section of the material through which flux is passing (A) and depends on the nature of the material.

$$\text{Reluctance} = \frac{l}{A\mu_o\mu_r} \qquad 15.12$$

In any magnetic circuit, m.m.f. sets up magnetic flux and opposing to flux is offered by the reluctance of the magnetic path. Greater the m.m.f., greater is the flux i.e.

$$\text{Flux} \propto \text{m.m.f}$$

$$\frac{m.m.f}{flux} = \text{constant} = \text{reluctance}$$

(or) m.m.f = flux × reluctance

This is called ohm's law for magnetic circuit.

$$\text{Flux, } \phi = \frac{m.m.f}{reluctance} = \frac{AT}{\frac{l}{A\mu_o\mu_r}} \qquad 15.13$$

$$\text{Therefore Ampere-turns required} = \text{AT} = \phi \times \frac{l}{A\mu_o\mu_r} \qquad 15.14$$

$$AT = \frac{\phi}{A} \times \frac{l}{\mu_o\mu_r} = \frac{B}{\mu_o\mu_r} \times l = H \times l \qquad 15.15$$

i.e. Ampere-turns required for any part of magnetic circuit

= Field strength H in that part × length of that part

The practical magnetic circuits have mostly air gap i.e. magnetic flux is intended to pass through a certain distance in air as shown in Fig. 142. Flux through air gap is called useful flux because it is only this flux which can be utilised for various purposes. It is, therefore, desired that total flux produced should pass through air gap. However, as air is not a perfect magnetic insulator, therefore, a part of the total flux returns by such paths as 'a' and 'b' and is called leakage flux. It is so called because it does not pass through the air gap rather it follows a leaking path.

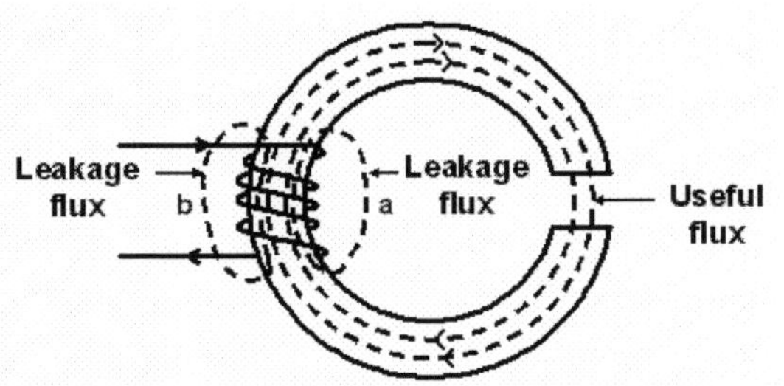

Fig. 142 Magnetic circuit with air gap

In an electromagnet, the induction for a current of given intensity decreases rapidly with increase in air gap and the force of attraction decreases even more rapidly with increase in the air gap. To get a suitable field at the lowest operating cost, it is recommended to reduce air gap to a minimum, to reduce the reluctance to a minimum by using highly permeable core, a large cross section and a short length and to use many turns and small current.

15.2 TYPES OF MINERALS

Based on magnetic susceptibility, minerals are classified into three broad groups as Ferromagnetic, Paramagnetic and Diamagnetic.

Ferromagnetic minerals have very high susceptibility to magnetic forces and retain some magnetism when removed from the magnetic field. They are strongly attracted by the magnetic field. They can be concentrated by low-intensity magnetic separators. The principal ferromagnetic mineral separated is magnetite (Fe_3O_4).

Paramagnetic minerals are weakly attracted by magnetic field. They can be concentrated by high-intensity magnetic separators. Examples of paramagnetic minerals are ilmenite ($FeOTiO_2$), rutile (TiO_2), wolframite (**(**$(Fe,Mn)WO_4$**)**, hematite (Fe_2O_3), siderite ($FeCO_3$), chromite ($FeOCr_2O_3$), pyrrhotite (FeS) and manganese minerals. Hematite and siderite can be roasted to produce magnetite and can be separated by low-intensity magnetic separators.

Diamagnetic minerals are repelled along the lines of magnetic force to a point where the field intensity is smaller. The forces involved here are very small. They cannot be concentrated magnetically. Examples are quartz (SiO_2) and feldspar ($KAlSi_3O_8$ or $NaAlSi_3O_8$).

15.3 MAGNETIC SEPARATORS

The machines used for magnetic separation are called magnetic separators. The operational principle of magnetic separator is

> *"When a stream of ore particles is continuously passed through a field of magnetic force, the magnetic particles are attracted towards the source of magnetic force while the non-magnetics travel unaffected"*

Magnetic separation, in industry, is achieved by simultaneously applying a magnetic force on all particles of an ore which acts on magnetic particles and a combination of forces which acts in a different direction and affects both magnetic and nonmagnetic particles. The most commonly applied nonmagnetic forces are gravitational, centrifugal, and fluid drag. Other forces which usually enter in an incidental manner are frictional, inertial, and attractive or repulsive inter-particle forces.

These forces depend on both the nature of the feed material and the character of the separator. The nature of the feed material includes its size distribution, magnetic susceptibility and other physical and chemical properties that may affect the various forces involved.

Magnets used in magnetic separators are of two types; Permanent magnets and Electromagnets. Permanent magnets are magnetised steel bars which retain magnetism indefinitely. Special-permanent magnet alloys produce a magnetic field at a constant level indefinitely after initial charging. The newer ceramic permanent magnets in the form of barium and strontium ferrites, and rare-earth-cobalt permanent magnet alloys are also being in use.

Bars of soft wrought iron, soft steel or cast iron are magnetised by electric currents through the insulated copper or aluminium wire windings around the iron bars. These iron bars form electromagnets as long as the current flows through the windings. Electromagnets are used for heavy-duty work. The flux is controlled by varying the D.C. current in the coils and the number of turns in the coil surrounding the iron core. The flux can also be controlled by varying the air-gap between the passing ore and the attracting magnet.

The essential elements of design of a magnetic separator are as follows:

1. Production of a suitably converging magnetic field

A field consisting of two flat poles near each other gives a nearly uniform field (Fig. 143), magnetic particles will orient themselves, but will not move along the lines of flux and hence is not suitable. Provision of V-shaped pole above a flat pole produces suitable converging field (Fig. 144). The tapering of the upper pole concentrates the magnetic flux into a very small area giving high intensity. The lower flat pole has the same total magnetic flux distributed over a larger area. Thus, there is a steep field gradient across the gap by virtue of the different intensity levels.

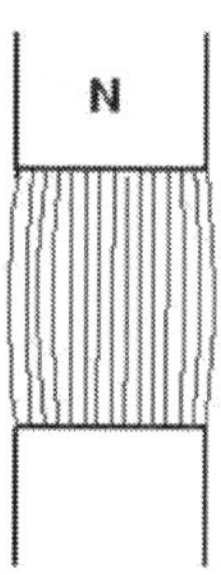

Fig.143 Two flat poles

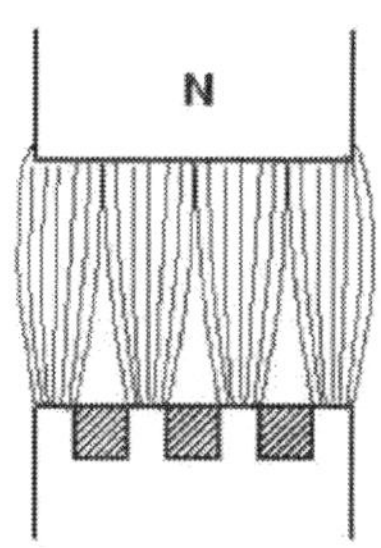

Fig.144 Flat and V-pole

Fig.145 Flat and Laminated pole

If a pole is made in laminations alternately of iron and of a nonmagnetic material, the field in the immediate vicinity of the laminated pole is strongly converging (Fig. 145). This type of magnet is frequently utilised, especially in the form of a laminated drum.

2. Easy regulation of the intensity of the field

Provision must be incorporated in the separator for regulating the intensity of the magnetic field to deal with various types of material. This is easily achieved in electromagnetic separators by varying the current. In case of permanent magnets, the intensity of magnetic field can be varied by varying interpole distance.

3. Even feeding of ore particles as a stream or sheet

Uniform feeding in the case of dry separators means that the feed must indeed be dry and not damp, and is achieved with a suitable feeding gate and bin.

4. Control of speed of passage of ore through the magnetic field

Separation in magnetic separators is carried out on a moving stream of particles passing into and through the magnetic field. Belts or drums having close control of the speed of passage of the particles are very often used to transport the feed through the field.

5. Avoidance and/or correction of occlusion of non-magnetic material between or within magnetic flocs

When the particles are introduced into a magnetic field, a high field gradient is produced, since the lines of force converge to the particles, which causes the particles themselves to behave as magnets, thus attracting each other. Flocculation or agglomeration of the particles can occur if they are small and highly susceptible and if the field is intense. These magnetic flocs can entrain nonmagnetic material and can bridge the gaps between the poles of the magnet which reduces the efficiency of separation.

Occlusion of nonmagnetic material within magnetic flocs is especially serious in dry separators working on fine, highly magnetic material. If the ore is fed as a layer of one particle deep, this occlusion is much less noticeable, but such practice is clearly impossible on fine material, as the capacity of the machine is drastically reduced.

If the magnetic material has low coercive force (e.g. natural magnetite), the occlusion may be reduced by passing the material through consecutive magnetic fields, which are usually arranged with successive reversal of the polarity. This causes the particle to turn through 180°, each reversal tending to free the entrained particles of nonmagnetic material. This is not readily achieved if the magnetised particles have a high coercive force. In that case, it is necessary to break the d-c field in such a way that a damped oscillatory discharge is set up through the magnet windings.

In wet separators, there is no great rush of magnetically susceptible particles toward the region of denser magnetic field because of the relatively great resistance to motion of the particles that is exerted by the water. Occlusion of nonmagnetic particles in magnetic flocs is therefore less serious than in dry separators.

6. Provision of suitable means for disposing of separated products

Disposal of the nonmagnetic product is simple: either free fall onto a belt or conveyance on the feed belt is suitable. The magnetic product must be detached from the magnet. Instead, it is preferable to design the separator that the magnetic product does not stick to the magnet but stick to some pulley or belt. By the motion of pulley or belt, magnetic product is moved beyond the effective field and drops.

7. Provision for production of middling

A middling is readily obtained by using a more intense field after removal of the concentrate. This is facilitated using a keeper to close the magnetic circuit as in Fig. 146. The field at d_2 is denser than at d_1 as well as more strongly convergent. As a result, a concentrate is deflected from the ore path at d_1 and a middling at d_2.

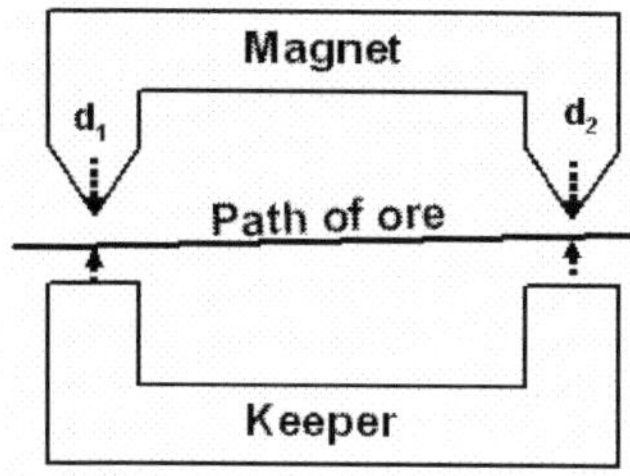

Fig. 146 Keeper to close magnetic circuit

8. Elimination, or reduction to a minimum, of moving parts

15.3.1 Classification of magnetic separators

Magnetic separators can be classified into two groups based on their function:

(a) **Tramp iron magnetic separators** are those used to remove tramp iron and protect the handling and processing equipment

(b) **Concentrators** are those used to separate bulk of magnetic materials from a stream of ore. Magnetic separators used to remove small quantities of deleterious magnetic material from a product (e.g. from china clay) are termed as **purifiers**. The concentrators that remove magnetic material and return to the process are termed as **reclaimers** (e.g. separation of medium particles such as magnetite, ferrosilicon etc. in Heavy Medium Separation process)

15.3.2 Tramp iron magnetic separators

Iron coarser than 3 mm is usually called as tramp iron. Most commonly applied magnetic separators for removing tramp iron are magnetic pulleys, magnetic drum separators, suspended magnets, cross belt magnetic separators, and grate magnets. One of the simplest and most easily installed tramp iron protection devices is the magnetic pulley. The magnetic pulley, installed in the head end of a conveyor, provides continuous removal and automatic discharge of any tramp iron present. Selection of a suitable magnetic pulley is dependent on belt speed, burden depth, type of tramp iron expected to be encountered, and to some extent the size of the particle being handled. Magnetic pulleys are available in both permanent magnet and electromagnetic types and can be supplied in sizes to fit nearly every conveyor belt. The principle of operation is very simple and is shown in Fig. 147.

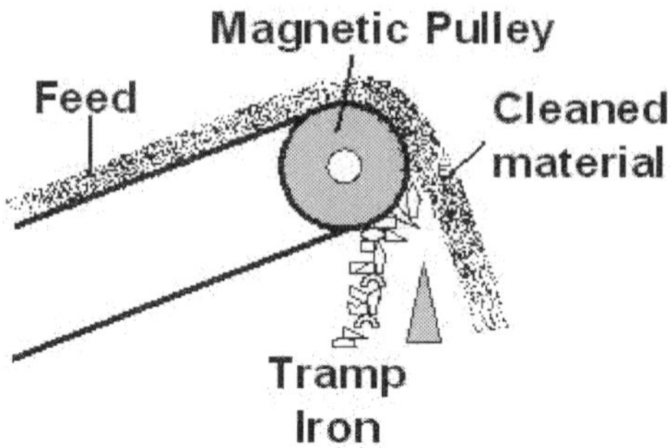

Fig. 147 Magnetic pulley

Permanent magnet pulleys are adaptable to almost all belt conveyors of smaller head pulley diameter. Electromagnetic pulleys are used with large diameter head pulleys handling large volumes for the removal of the larger pieces of tramp iron.

Many types of magnetic drum configurations are available. For removing tramp iron, the feed tonnage and the size of the ore determines the type of magnetic drum applied. Normally, the material is handled in a reasonably thick layer and it is necessary to provide magnetic removal from all portions of the feed bed. Therefore, tramp iron drums are built with radial poles capable of projecting a deep magnetic field and also capable of holding the tramp iron on the magnetic drum face until it can be discharged at an appropriate discharge point. Magnetic drums are also applied in concentration and purification where magnetic particles to be removed are large in size and the magnetic responsiveness of the magnetic particles is high. This operation of picking up strongly magnetic particles at coarser size is called **cobbing** operation. Drum separators consist of a rotating non-magnetic drum containing either permanent or electromagnets. Initially drum separators employed electromagnets. Permanent magnets are used in modern devices utilizing ceramic or rare earth

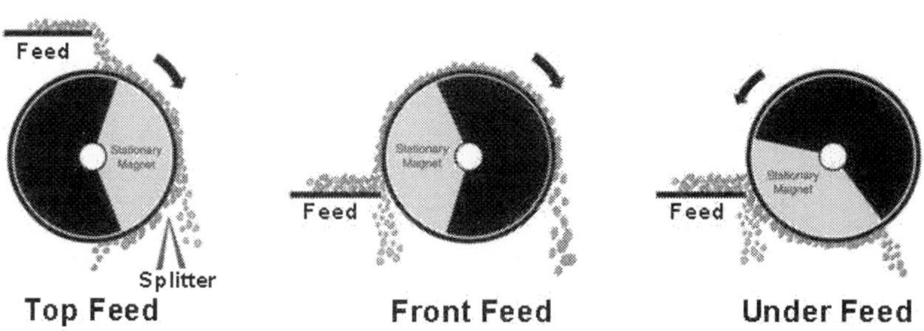

Fig. 148 Low intensity dry magnetic drum separators

magnetic alloys which retain their intensity for an indefinite period. Magnetic particles are attracted by the magnets, pinned to the drum, conveyed out of the field and discharged while the nonmagnetic particles are discharged unaffected. The magnetic drum is normally fed at the top vertical centre, but, with appropriate magnet positioning, it can be fed at any convenient point or even front-fed or underfed (Fig. 148). They are low intensity dry magnetic drums.

Suspended magnets (Fig. 149a) are fixed position electromagnets of rectangular shape and installed over a conveyor belt or head pulley. Tramp iron attracted by the magnet is periodically removed. Cross-belt models (Fig. 149b) are used when continuous removal of tramp iron is required.

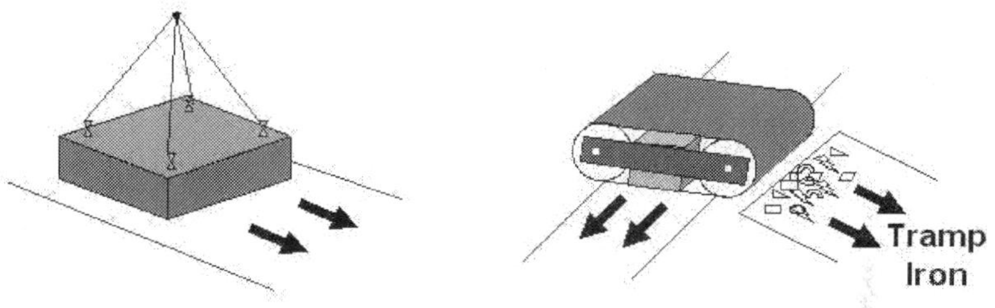

(a) (b)

Fig. 149 (a) Suspended magnet (b) Cross belt magnet

Grate magnet consists of a series of magnetised bars as shown in Fig. 150 and is used to remove tramp iron. Feed is usually vertical through the grate. These magnets must be periodically cleaned.

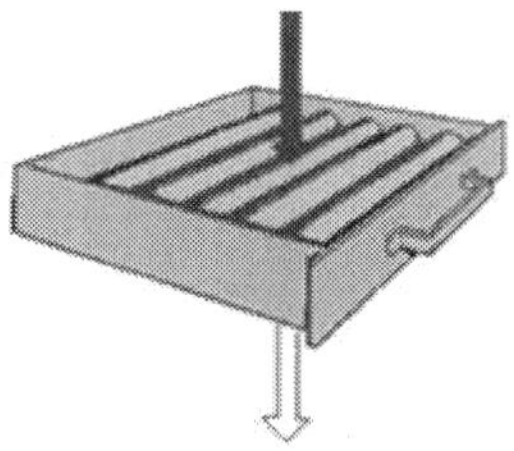

Fig. 150 Grate magnet

Magnetic pulley and grate magnets used for tramp iron removal are also used as concentrators. Grate magnet is especially useful for purification of the product when it contains small amounts of unwanted magnetic particles.

15.3.3 Concentrators

These magnetic separators are classified into dry and wet separators which may further classified into low- and high-intensity separators. Wet separators are generally used for the particles of below 0.5 cm size range.

Dry magnetic separators

Low intensity dry magnetic drum separators of Fig. 148 described in the previous article are the principal magnetic concentrators.

Induced roll separator is a high intensity dry magnetic separator. Induced roll separator (Fig. 151) consists of a horseshoe magnet faced by an iron bridge bar or keeper and of two rolls, one opposite each pole. The magnetic circuit is completely in iron except for the very small clearance between the rolls and the bridge bar and for the gaps between the rolls and primary poles of horseshoe magnet. The rolls consist of alternate laminae of permeable and impermeable material. The magnetic field is strongly convergent toward the roll which is an assemblage of a large number of secondary poles. The field strengths of up to 22,000 gauss are attainable in the gap between primary pole and roll. Non-magnetic particles are thrown off the roll into the tailings compartment, whereas magnetic particles are gripped, carried out of the influence of the field and discharged.

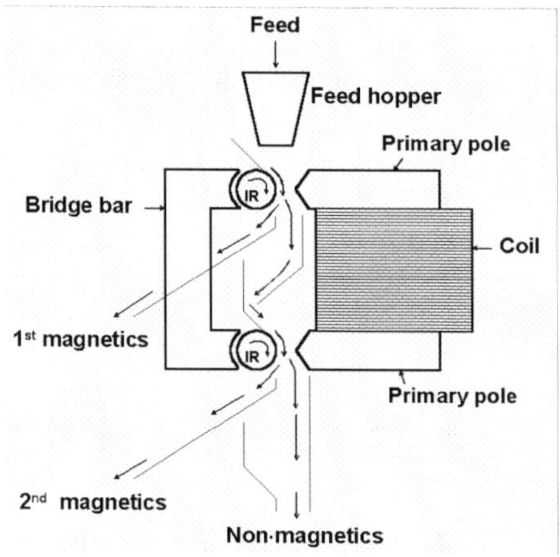

Fig. 151 Induced roll separator

Cross-belt separator and Disc separators are once widely used in the mineral sands industry particularly for recovering ilmenite from heavy mineral concentrates. They are now considered obsolete and are replaced with **rare earth roll magnetic separators** and **rare earth drum magnetic separators.** Rare Earth Drum Separators are made with a high quality rare earth permanent magnetic power source. The rare-earth magnets produce magnetic fields up to 25 times stronger than conventional Ferrite, with no increase in size. The additional strength helps in removing weakly magnetic or very fine iron contaminants from a wide variety of materials.

Rare earth roll magnetic separator (Fig. 152) consists of two small diameter rolls between and around which a belt tracks. Feed material is distributed onto the belt which transports it to the end roll which contains an exotic permanent magnet. This magnet is made up of alternate magnetic and non-magnetic laminations. As the feed is carried onto the magnetic roll by means of a thin belt, there is no bouncing or scattering of particles when they enter the magnetic zone. All the feed particles enter the magnetic zone with the same horizontal velocity. As non-magnetic material is discharged forward of the roll in the normal trajectory, any magnetic particles present are securely held onto the belt by the magnetic force and are discharged down a rear chute as the belt leaves the magnetic field on the underside of the roll. Separation trajectories are set by adjusting the conveyor speed using the inverter control setting on the control panel and adjusting the splitter chutes. Rare earth roll separators are effective at concentrating or removing weakly magnetic minerals from a dry process stream. These minerals include silica, feldspar, hematite and alumina.

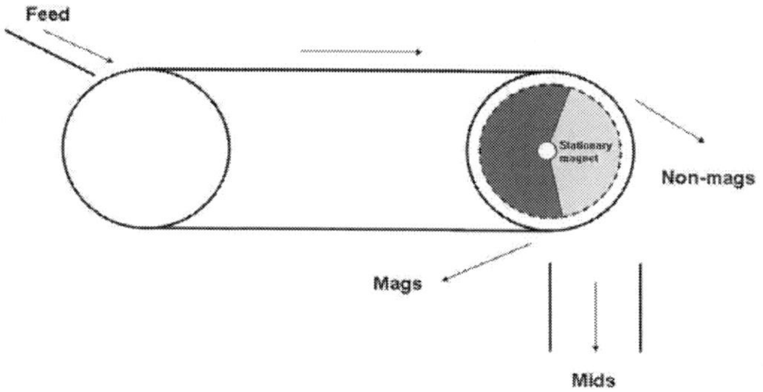

Fig.152 Rare earth roll magnetic separator

Rare earth drum magnetic separators are similar to the Low intensity dry magnetic drum separators except that they employ an exotic permanent magnet capable of producing much higher intensity magnetic field. This magnet is positioned inside a large diameter drum. In some mineral sands application, drum separators have been integrated with one or more rare earth rolls, arranged to treat the middling particles from the drum.

Wet magnetic separators

Drum separators are most common low-intensity wet magnetic separators. They consist of a rotating non-magnetic drum containing three to six stationary magnets, either permanent or electromagnets, of alternating polarity (Fig. 153). Feed in the form of slurry flowing through a bath, is presented to the submerged part of the drum and depending on the magnetic susceptibility, the magnetic particles are attracted by the magnetic field and pinned to the drum surface and the non-magnetic particles are unaffected and discharges from the bath with the slurry. As the drum rotates the magnetic particles leave the magnetic field and are discharged separately usually with the aid of a scraper / brush. Three types of drum separators namely **concurrent** type, **counter-rotation** type and **counter-current** type are in use. In the concurrent type separator, the concentrate is carried forward by the drum and passes through a gap where it is compressed and dewatered before leaving the separator. These separators are most effective for producing an extremely clean magnetic concentrate from relatively coarse materials and are widely used in heavy medium recovery systems. In counter-current type separators, the tailings are forced to travel in the opposite direction to the drum rotation and are discharged into the tailings chute. In counter-rotation type separators, the feed flows in the opposite direction to the rotation. These are used in roughing operations where occasional surges in feed must be handled and magnetic material losses are to be held to a minimum while an extremely clean concentrate is not required and when high solids loading is encountered. These are designed for finishing operation on relatively fine material of size less than about 250 microns.

Drum separators are widely used to treat low grade taconite ores of 40%–50%Fe with mainly magnetite as iron mineral. They are also used in some areas where hematite is finely disseminated in bands in hard siliceous rocks. Very fine grinding is necessary to free the iron minerals that produce a concentrate that requires pelletising before being fed to the blast furnaces.

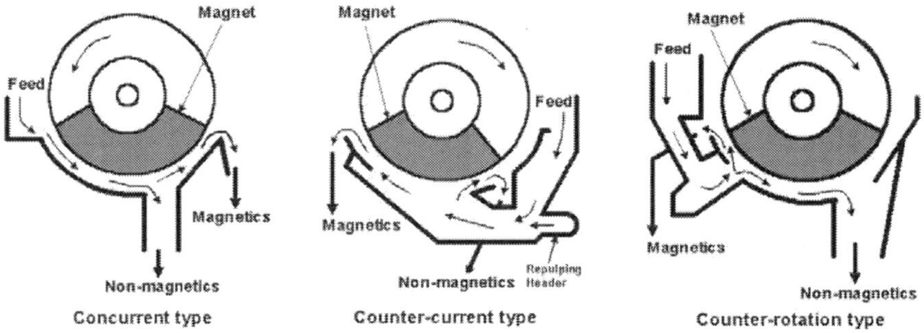

Fig. 153 Low-intensity wet drum magnetic separators

Wet high intensity magnetic separator (WHIMS) (Fig. 154) consists of a rotating carousel, which has vertically inclined salient plates through which feed slurry is passed. As the carousel rotates, it passes through fields of magnetic influence generated by surrounding electromagnets, followed by fields of no magnetic influence. The magnetic grains are initially held up in the plates while the non-magnetic particles are washed through into a launder below. When the plates are in the non-magnetic field, the magnetic particles are washed off into a separate launder below. To operate the WHIMS successfully, the feed must be free of magnetite which would hang up in the plates / matrix and free of +1mm particles which would block the plates / matrix. The feed density and feed rate must be steady. There should be a clean constant supply of high pressure water. Field strength employed in WHIMS is upto15,000 gauss.

To attract and hold paramagnetic particles, both high magnetic field and a high field gradient are required. The **High gradient magnetic separators** (HGMS) produce field strength of about 2 Tesla. In HGMS, the slurry is passed through a container where it is subjected to a high-intensity, high gradient magnetic field. The container is packed with a **capture** matrix made of stainless steel wool. In the presence of magnetic field, paramagnetic particles become magnetised and are trapped in the matrix while the unaffected non-magnetic particles pass through the container. When the matrix is loaded to its magnetic capacity, the slurry feed is stopped and the electric power is cut off. The matrix is then backwashed to remove magnetic particles. Afterwards, the feed and the power are resumed and the entire process is repeated.

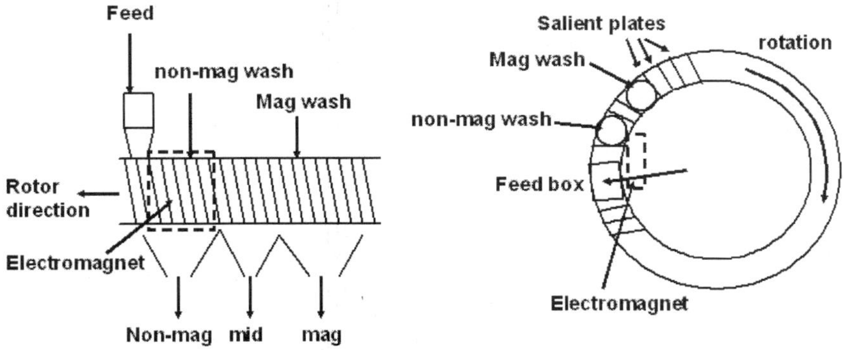

Fig. 154 Wet high-intensity magnetic separator

In continuous HGMS (Fig. 155), the capture matrix is a continuous, segmented metal belt. The slurry is introduced while capture matrix passes through the magnetic section of the separator. The magnetised particles are trapped in the capture matrix. The belt then travels in to the washing section to wash magnetic particles trapped on the belt. The cleaned belt section then returns for the entire process to repeat.

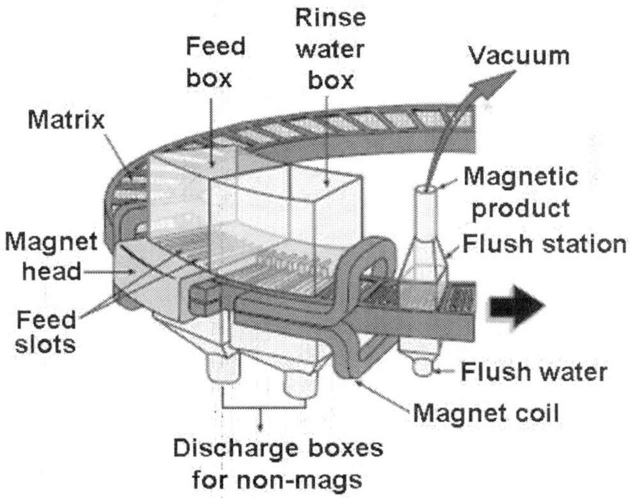

Fig. 155 High gradient magnetic separator

The controlling factors while operating the HGMS are pulp level in the feed boxes, pulp level in passive matrix, pulp level in vacuum receiver, speed of matrix ring, magnetic field adjustment, flow of raw and closed loop water in the cooling system, temperature in magnet coils and water to vacuum system.

High Gradient Magnetic Separation can often be used in separation processes where normally difficulties are experienced with other processes. The paramagnetic minerals at the higher end of the list are normally easy to recover. Many low susceptibility minerals are associated with other minerals or have extra Fe in the crystals, and are hence often possible to separate. HGMS is widely used for recovery of hematite from iron ore. Few of the other applications of HGMS are Ilmenite fines recovery, chromite fines recovery, and recovery of kyanite, apatite and wollastonite.

■■■■

16 Electrical Separation

Electrical separation is a physical separation of particles based on the electrical properties of the mineral particles. Following are the three important electrical separation processes:

1. Electrostatic separation
2. Electrodynamic (High Tension) separation
3. Dielectric separation

Electrostatic and Electrodynamic separation processes utilise the difference in electrical conductivity between various minerals in the ore feed whereas dielectric separation utilises the differences in dielectric constant of mineral particles.

Electrical conductivity is defined as current density (flow of electric charge per unit area of cross section) per unit applied electric field. It is the reciprocal of the resistivity or specific resistance of a conductor. It is measured in $ohm^{-1}cm^{-1}$.

The conductivity of minerals varies enormously from one mineral to another. Native metals such as Au, Ag, and Cu, many metal sulphides like Chalcopyrite, Galena, and Molybdenite as well as some other minerals such as Graphite, Garnet, and Chalcocite are good conductors whereas most gangue and non-metallic minerals such as Calcite, Gypsum, and Monazite are relatively poor conductors of electricity.

Knowledge on charge and charge interactions is helpful in understanding the phenomenon involved.

16.1 CHARGE AND CHARGE INTERACTIONS

The atom consists of **nucleus** surrounded by negatively charged electrons. The proton and neutron are located in the nucleus. The proton is charged positively. The neutron does not possess a charge and is neutral. The amount of charge on a single proton is equal to the amount of charge on a single electron. If an atom contains equal numbers of protons and electrons, the atom is described as **electrically neutral**. Any particle, whether an atom, molecule or ion, which contains less electrons than protons is said to be **positively charged**. Conversely, any particle which contains more electrons than protons is said to be **negatively charged**.

It can be said that positive electrification of a body results from deficiency of the electrons whereas negative electrification of a body results from an excess of electrons.

16.1.1 Conductors and Insulators

Conductors are materials which permit electrons to flow freely from atom to atom and molecule to molecule. An object made of a conducting material will permit charge to be transferred across the entire surface of the object. If charge is transferred to the object at a given location, that charge is quickly distributed across the entire surface of the object. If a charged conductor is touched to another object, the conductor can even transfer its charge to that object. The transfer of charge between objects occurs more readily if the second object is made of a conducting material. Conductors allow for charge transfer through the free movement of electrons.

Insulators are materials which impede the free flow of electrons from atom to atom and molecule to molecule. If charge is transferred to an insulator, the excess charge will remain at the initial location of charging. The particles of the insulator do not permit the free flow of electrons. Conductive objects are often mounted upon insulating objects to prevent transfer of charge from the conductive object to its surroundings.

Examples of conductors include metals, aqueous solutions of salts, graphite, water and the human body. Examples of insulators include plastics, paper, rubber, glass, wood and dry air. Some materials are neither good conductors nor good insulators, since their electrical characteristics fall between those of conductors and insulators. These in-between materials are classified as **Semi-conductors**. Germanium and silicon are two common semiconductors. Different materials may be placed along a continuum, as shown in Fig. 156. The conductivity of a metal might be as much as a million trillion times greater than that of glass.

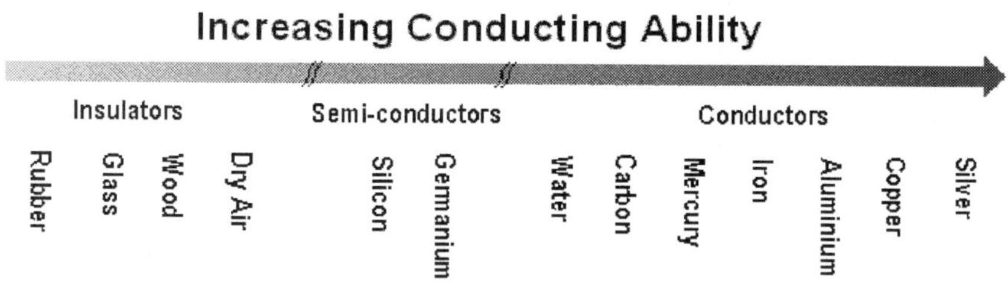

Fig. 156 Classification of Materials based on conducting ability

16.1.2 Polarisation

Polarisation is the process of separating opposite charges within an object. The positive charge becomes separated from the negative charge. By inducing the movement of electrons within an object, one side of the object is left with an excess of positive charge and the other side of the object is left with an excess of negative charge. Charge becomes separated into opposites. The polarisation process always involves the use of a charged object to induce electron movement or electron rearrangement.

Polarisation is not charging. When an object becomes polarised, there is simply a redistribution of the centres of positive and negative charges within the object. Either by the movement of electrons across the surface of the object (as is the case in conductors) or through the distortion of electron clouds (as is the case in insulators), the centres of positive and negative charges become separated from each other. The atoms at one location on the object possess more protons than electrons and the atoms at another location have more electrons than protons.

16.2 METHODS OF CHARGING

1. Charging by Friction: The presence of different atoms in objects provide different objects with different electrical properties. One such property is known as **electron affinity**. The frictional charging process results in a transfer of electrons between the two objects which are rubbed together. Rubber has a much greater attraction for electrons than animal fur. As a result, the atoms of rubber pull electrons from the atoms of animal fur, leaving both objects with an imbalance of charge. The rubber balloon has an excess of electrons and hence is negatively charged. The animal fur has a shortage of electrons and charged positively

As mentioned, different materials have different affinities for electrons. By rubbing a variety of materials against each other and testing their resulting interaction with objects of known charge, the tested materials can be ordered according to their affinity for electrons. Such an ordering of substances is known as a **triboelectric series**.

2. Charging by Induction: The phenomenon of an uncharged body getting charged merely by the nearness of a charged body is known as induction. Induction charging is a method used to charge an object without actually touching the object to any other charged object.

Charging a Two-Sphere System: Two metal spheres are supported by insulating stands so that any charge acquired by the spheres cannot travel to the ground. The spheres are placed side by side (Fig. 157a) to form a two-sphere system. If a rubber balloon is charged negatively and brought near the spheres, electrons within the two-sphere system will be induced to move away from the balloon. Being charged negatively, the electrons are repelled by the negatively charged balloon. They are free to move about the surface of the conductor. Subsequently, there is a mass migration of electrons from sphere A to sphere B. This electron migration causes the two-sphere system to be polarised (Fig. 157b). Sphere A has an overall positive charge and sphere B has an overall negative charge.

Once the two-sphere system is polarised, sphere B is physically separated from sphere A using the insulating stand. Then the negative charge likely redistributes itself uniformly about sphere B (Fig.157c). Meanwhile, the excess positive charge on sphere A remains located near the negatively charged balloon. As the balloon is pulled away, there is a uniform distribution of charge about the surface of both spheres (Fig. 157d). This distribution occurs as the remaining electrons in sphere A move across the surface of the sphere until the excess positive charge is uniformly distributed.

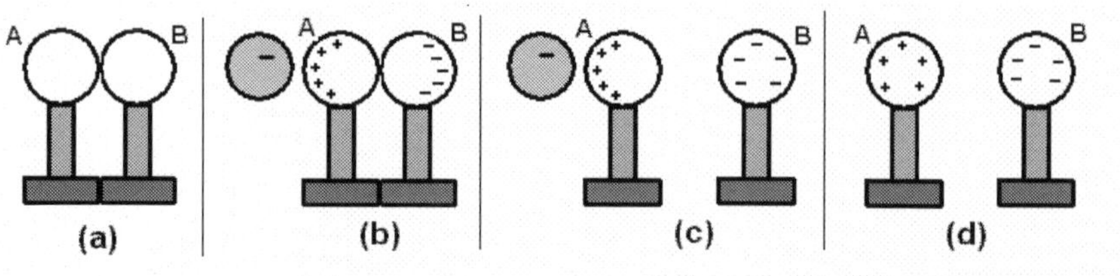

Fig. 157 Charging by induction with negatively charged object

When a positively charged balloon is brought near Sphere A, the positive charge induces a mass migration of electrons from sphere B towards (and into) sphere A. Negatively charged electrons throughout the two-sphere system are attracted to the positively charged balloon. This movement of electrons from sphere B to sphere A leaves sphere B with an overall positive charge and sphere A with an overall negative charge (Fig. 158f). The two-sphere system has been polarised. With the positively charged balloon still held nearby, sphere B is physically separated from sphere A. The excess positive charge is uniformly distributed across the surface of sphere B (Fig.158g). The excess negative charge on sphere A remains crowded towards the left side of the sphere, positioning itself close to the balloon. Once the balloon is removed, electrons redistribute themselves about sphere A until the excess negative charge is evenly distributed across the surface. In the end, sphere A becomes charged negatively and sphere B becomes charged positively (Fig. 158h).

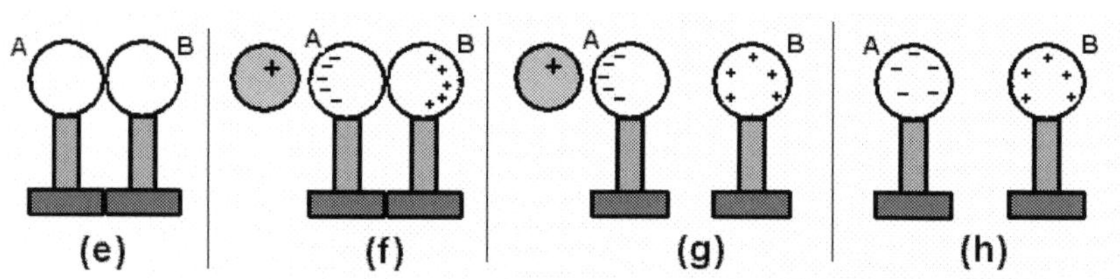

Fig. 158 Charging by induction with positively charged object

Charging a Single sphere: When a negatively charged rubber balloon B is brought near a single sphere, negative electrons within the metal sphere will be repelled by the negatively charged balloon. There will be a mass migration of electrons from the left side of the sphere to the right side of the sphere causing charge within the sphere to become polarised (Fig. 159b). Once charge within the sphere has become polarised, the sphere is grounded.

The grounding of the sphere allows electrons to exit from the sphere and move to the earth (Fig. 159c) and the sphere acquires a positive charge (Fig. 159d). Once the balloon is moved away from the sphere, the excess positive charge redistributes itself (by the movement of remaining electrons) such that the positive charge is uniformly distributed about the sphere's surface (Fig. 159e). Positive and negative charges of A are known as induced charges. The positive charge is called bound charge because it must remain on A so long as negative charge of B remains there. However, the negative charge on the farther end of A is called free charge. If B is removed, then this positive charge will also go to earth, leaving A uncharged.

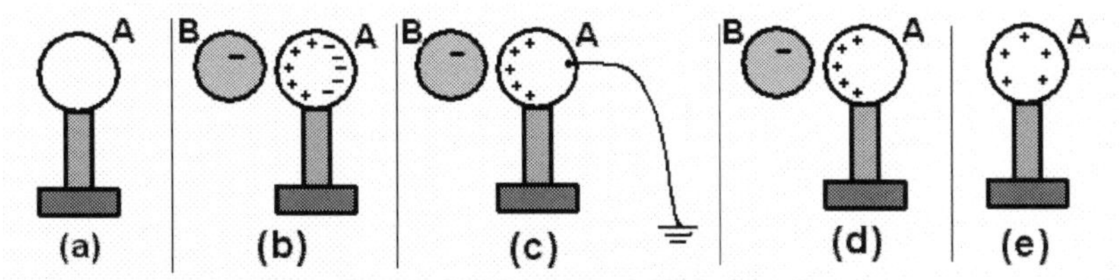

Fig. 159 Charging single sphere by induction

The fundamental principles of induction charging are:

- The charged object is never touched to the object being charged by induction.
- The charged object does not transfer electrons to or receive electrons from the object being charged.
- The charged object serves to polarise the object being charged.
- The object being charged is grounded; electrons are transferred between the ground and the object being charged (either into the object or out of it).
- The object being charged ultimately receives a charge that is opposite that of the charged object which is used to polarise it.

3. Charging by Conduction: Charging by conduction involves the contact of a charged object to a neutral object. If a charged object is touched to a neutral object, the neutral object becomes charged. It is often called **charging by contact**.

When a negatively charged object is made in contact with a neutral object, electrons move from the negatively charged object onto the neutral object. When finished, both objects were negatively charged. In case of charging with positively charged object, electrons move from neutral object to charged object. Having lost electrons to the positively charged body, there is a shortage of electrons on the neutral body and hence it is positively charged. The charged body is still charged positively, only it now has less excess positive charge than it had before the charging process began.

16.3 ELECTROSTATIC SEPARATION

If an ore contains conducting as well as non-conducting minerals/particles, electrostatic separation or high tension separation can be employed. Theoretically, it is not necessary that one of two minerals should be a good conductor and the other be a poor conductor, but the difference in their conductivity will affect the separation.

The basis of any electrostatic separation is the interaction between an external electric field and the electric charges acquired by the various particles. Particles can be charged by

1. contacting dissimilar particles
2. conductive induction
3. ion bombardment

In every separation, two or more charging mechanisms occur.

In charging by contacting dissimilar particles, particles placed on surface are made to repeatedly contact one another as well as the surface; the surface will acquire electrons from one type of particles and give electrons to another type so that two types of particles are charged with opposite charges. When they are passed through an electrostatic separator consisting of two conducting plates across which a high voltage is applied, positively charged particles are attracted towards negative plate and negatively charged particles are attracted towards positive plate. This mechanism is utilised in free fall electrostatic separators. However, this is not the major mechanism in any of the electrostatic separators.

In charging by conductive induction, the particles are placed on a ground conductor in the presence of an electric field. Then the particles will rapidly develop a surface charge by induction. Both conducting and non-conducting particles will become polarised, but conducting particles will have an equipotential surface through its contact with the grounded conductor. The non-conducting particle will remain polarised.

In charging by ion bombardment, corona discharge is obtained by appropriate shaping of the electrodes. Corona discharge is an electrical discharge which occurs when one of the two conducting surfaces (such as electrodes) of differing voltages have a pointed shape. A highly concentrated electric field, set up at the tip of the pointed shape electrode, ionises the air (or other gas) around it. These ions charge the particles by bombardment when the mineral particles are caused to pass within the corona. For the discharge to a large diameter cylindrical surface, a fine wire parallel to the cylinder gives the optimum corona discharge.

The attraction of particles carrying one kind of charge (positive or negative) toward an electrode of the opposite charge is known as **lifting effect** as the particles are lifted toward the charged electrode. Mineral particles tending to become charged with a definite polarity may be separated from each other using lifting effect even though their conductivities may be very similar. For example, quartz is readily negatively charged and can be separated from other non-conductors by using an electrode carrying a positive charge. However, pure electrostatic separation (pure lifting effect) is relatively inefficient.

The other effect which makes the electrostatic separation more effective is the **pinning effect**. Here, non-conducting mineral particles, having received a surface charge from the electrode, retain this charge and are pinned to the earthed surface due to the attraction between the charged non-conducting particle and the grounded surface.

16.4 ELECTROSTATIC SEPARATORS

The earliest electrostatic separators were of the plate electrostatic type. In free fall plate separators, particles fall between two near-vertical plates where one plate is charged positively and another plate is charged negatively with a high voltage gradient between the plates. These separators were used to separate sylvite from halite, feldspar from quartz, and phosphate pebble from quartz. These devices are no longer in use.

In a **rotor-type electrostatic separator** (Fig. 160), a large single electrode produces a strong electric field. When a particle is placed on the grounded rotor in the region of the electrostatic field influence, the particle rapidly develops a surface charge by induction. Whether the particle is a conductor or non-conductor, it will be polarised. However, a conducting particle rapidly becomes an equipotential surface and has same potential as the ground rotor. Therefore, it is attracted toward the electrode, drawn away from the surface, and falls by gravity. The non-conducting particle continues to adhere to the rotor until it falls by gravity.

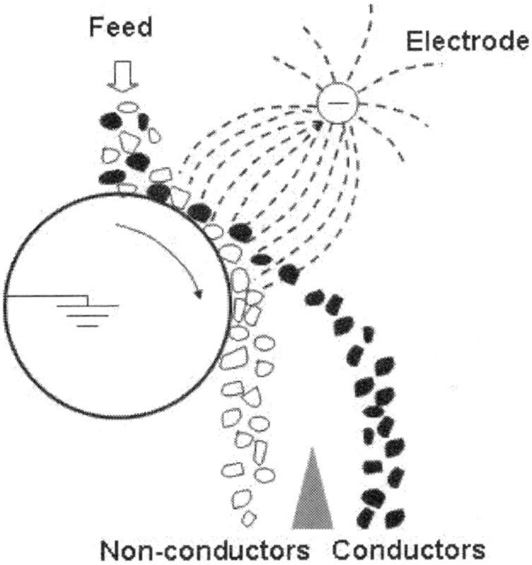

Fig. 160 Electrostatic separator

Since almost all minerals show some difference in conductivity, electrostatic separation is the universal concentrating method. It is widely used in beach sand beneficiation. The feed to electrostatic separation must be perfectly dry. For efficient operation, the feed should be in a layer, one particle deep, which severely restricts the throughput. Electrostatic separation is applicable to a limited size range, normally 20 mesh to 250 mesh. As complete separation is not likely to be obtained in a single pass electrostatic separator, it is usual practice to re-pass separated fractions through other separators as shown in Fig. 161 to get finished products.

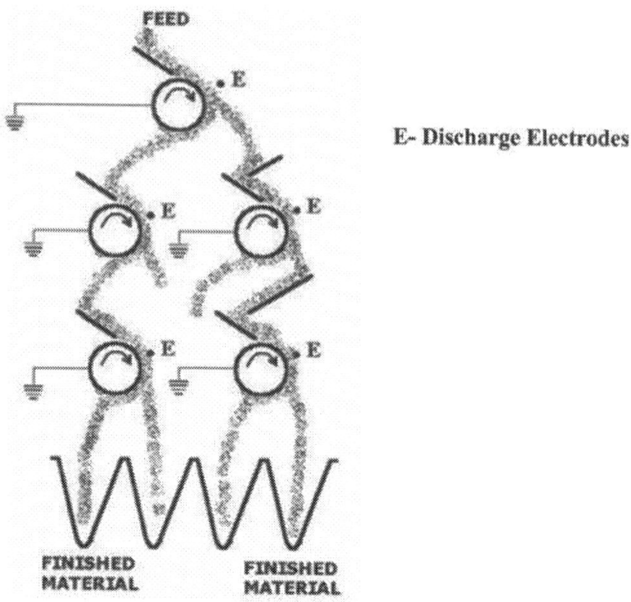

Fig. 161 Multi-pass Electrostatic Separator

The plate electrostatic separator (Fig. 162) and the screen electrostatic separator (Fig. 163) are the two plate type separators. The operating principles are similar. The feed particles slide down a grounded plate into the divergent electric field induced by the large curved electrode. The particles are charged by induction, the conductor particles acquire a charge opposite to the electrode. Thus, the conductor particles are attracted toward the electrode. The non-conductor particles continue down the plate or through the screen.

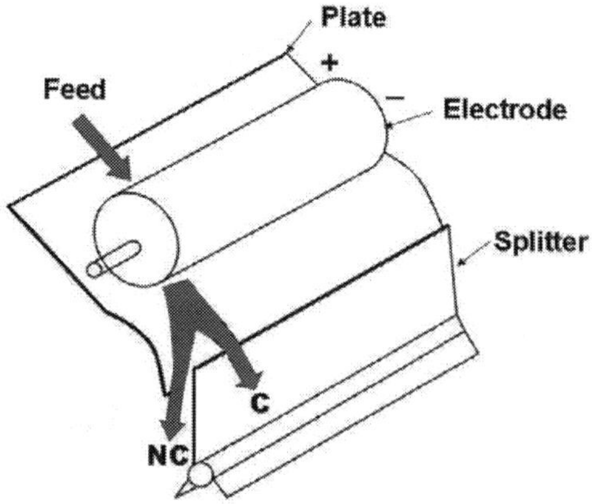

Fig.162 Plate electrostatic separator

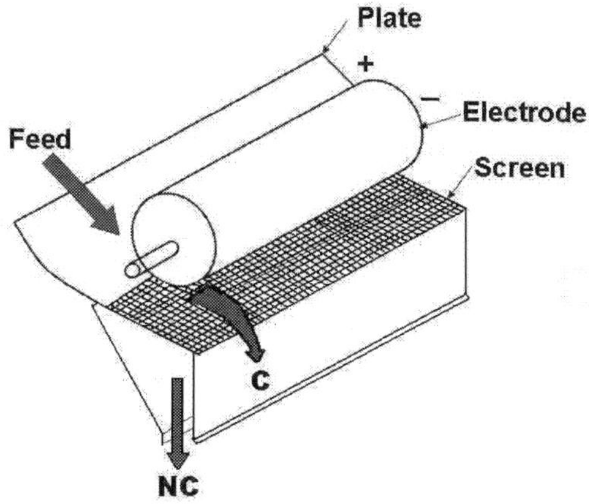

Fig. 163 Screen electrostatic separator

In electro-static separators, it is the fine particles that are most affected by the weak forces involved, and so the conductor product preferentially contains fine conductors. At the same time, there is no tendency for the coarse non-conductors to enter the conductor stream. Because these separators act primarily on the conductor particles, they are mainly used for cleaning a small quantity of conductor particles from a large quantity of non-conductors. Specifically, their main use is in cleaning small amounts of rutile and ilmenite out of zircon concentrates. They are generally used in banks with non-conductors are recleaned through successive separators.

16.5 HIGH TENSION SEPARATOR

In a high-tension separator (Fig. 164), the mixture of ore minerals is fed on to a grounded rotor into the field of a charged ionizing electrode. The electrode assembly, consists of a lengthy fine wire supported by a brass tube, is supplied with DC supply of negative polarity. Then the ionisation of air takes place which can be seen as corona discharge. This gives high surface charge to the poor conducting mineral particles. These particles are attracted and pinned to the rotor. High conducting mineral particles are not charged as the charge is dissipated through the particles to the earthed rotor. Then they come under the influence of the electrostatic field of the non-ionizing electrode and are attracted (by the lifting effect) from the rotor surface. They are discharged away from the rotor. The charge of poor conducting mineral particles is slowly lost as rotor rotates and the particles drop from the rotor as middlings or non-conductor product according to the intensity of the surface charge. Thus, a combined effect of pinning and lifting is experienced in high tension separators. These separators operate on feeds of 60–500 microns.

The size of the particles influences its action in the high-tension separator, since the charge on the surface of a coarse particle is lower in relation to its mass than that on a fine particle. Thus, a coarse particle is more readily thrown from the rotor surface. If the separators are operated with multiple layers of particles on the rotor surface, fine conductor particles are trapped by non-conductors because of their low mass. Thus, the fine particles tend to report preferentially with the non-conductor fraction. As a result, it is normal operating practice to use multiple stages of cleaning. A bank of separators can be arranged for cleaning of the conductor fraction, the non-conductor fraction, or both. Often it is necessary to take a middlings fraction containing fine conductor particles and coarse non-conductors to simultaneously produce clean conductor and non-conductor fractions. If un-liberated particles are present, they will tend to report to the middlings fraction.

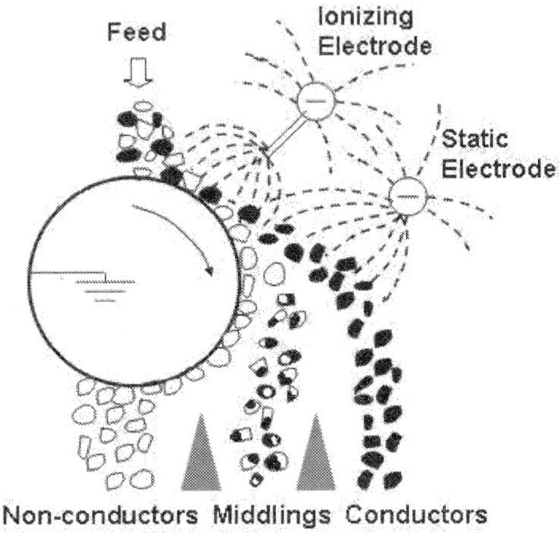

Fig. 164 High Tension Separator

16.6 DIELECTRIC SEPARATION

In Dielectric separation, ore consists of minerals of various dielectric constants is suspended in a non-conducting fluid whose dielectric constant is intermediate between that of two groups of minerals and a converging electric field is set up within the suspension. Then the particles having dielectric constant higher than the fluid travel in the direction of most rapid increase in electric field and the particles having dielectric constant lower than the fluid travel in the opposite direction. The dielectric separation process is applicable to particles finer than 60-mesh and is used as a research tool for study of minerals. It has no industrial application.

■■■■

17 Evaluation of Beneficiation Operations

Evaluation of beneficiation operations is essential for day to day operation of the mineral processing plant. In general, the plant operator collects the samples from feed and products and sent for testing to know their assay value. Using these values weight recovery and metal recovery is calculated. Based on these recoveries, plant is controlled for better performance (recovery) of the plant.

Two material balance equations are usually written for any beneficiation operation. These two equations are the basis for performance evaluation of beneficiation operations. For a two-product beneficiation operation

Let

F = Weight flow rate of the feed

C = Weight flow rate of the concentrate

T = Weight flow rate of the tailing

f = Fraction of the metal (or material) present in the feed

c = Fraction of the metal (or material) present in the concentrate

t = Fraction of the metal (or material) present in the tailing

then the material balance for total material is

$$F = C + T \qquad 17.1$$

i.e. *total material input = total material output*

Similarly, the material balance for valuable metal (or material) they contain is

$$Ff = Cc + Tt \qquad 17.2$$

i.e. *valuable metal (material) input = valuable metal (material) output*

Ratio of concentration, ratio of recovery and ratio of enrichment are three important criteria used for performance evaluation.

Ratio of concentration : It is defined as the ratio of the weight flow rate of the ore fed to the weight flow rate of the concentrate produced

$$\text{Ratio of concentration} = \frac{F}{C} \qquad 17.3$$

From the material balance equations 17.1 and 17.2, ratio of concentration can be computed in terms of assay values as

$$\text{Ratio of concentration} = \frac{F}{C} = \frac{c-t}{f-t} \qquad 17.4$$

Ratio of recovery : There are two types of recovery. One is weight recovery (or concentrate recovery) and another is metal recovery. The weight recovery is defined as the ratio of the weight flow rate of the concentrate to the weight flow rate of the ore fed.

$$\text{Ratio of weight recovery} = \frac{C}{F} \qquad 17.5$$

From the material balance equations 17.1 and 17.2, ratio of weight recovery can be computed in terms of assay values as

$$\text{Ratio of weight recovery} = \frac{C}{F} = \frac{f-t}{c-t} \qquad 17.6$$

The metal recovery is defined as the ratio of the weight flow rate of the valuable metal in the concentrate and the weight flow rate of the valuable metal in the feed. It is the fraction (or percent) of the total amount of valuable metal (material) originally present in the ore which is recovered in the concentrate

$$\text{Ratio of metal recovery} = \frac{Cc}{Ff} \qquad 17.7$$

From the material balance equations 17.1 and 17.2, ratio of recovery can be computed in terms of assay values as

$$\text{Ratio of metal recovery} = \frac{c(f-t)}{f(c-t)} \qquad 17.8$$

Ratio of enrichment: It is defined as the ratio of the assay value of the concentrate to that of the ore.

$$\textbf{Ratio of enrichment} = \frac{c}{f} \qquad 17.9$$

This is equal to the product of the ratio of concentration and the ratio of metal recovery. The ratio of enrichment does not have the direct economic significance of the commonly used ratio of concentration. It gives better idea of the cleanliness of the concentrate.

Calculation of these above ratios are illustrated in examples 17.1 and 17.2.

Example 17.1: *In an iron ore beneficiation operation, samples of feed, concentrate and tailing are analysed, and all stream flowrates are measured. Results are shown in Table 39.*

Table 39: Quantities and assay values

	Flow rates tons/h	Assay value %Fe
Feed	1390	64.77
Concentrate	1112	68.08
Tailing	278	51.53

Calculate ratio of concentration, ratio of enrichment, concentrate recovery and metal recovery. How much iron is lost in tailing?

Solution:

Given

Rate of feed = F = 1390 tons/h

Rate of concentrate = C = 1112 tons/h

Rate of tailing = T = 278 tons/h

%Fe in feed = f = 64.77%

%Fe in concentrate = c = 68.08%

%Fe in tailing = t = 51.53%

Ratio of concentration $= \frac{F}{C} = \frac{1390}{1112} = 1.25$

Ratio of enrichment $= \frac{c}{f} = \frac{68.08}{64.77} = 1.05$

%Concentrate recovery $= \frac{C}{F} \times 100 = \frac{1112}{1390} \times 100 = 80\%$

%Metal recovery $= \frac{Cc}{Ff} \times 100 = \frac{1112 \times 68.08}{1390 \times 64.77} \times 100 = 84\%$

Iron lost in tailing $= \frac{Tt}{100} = \frac{278 \times 51.53}{100}$ = 143.3 tons/h

Example 17.2: *Tin ore assaying 10% Sn is fed to a concentration plant at the rate of 100 tons/h. If the grades of concentrate and tailing are 40% Sn and 1% Sn, calculate the recovery of tin and tin lost in tailing in tons/h.*

Solution:

Given

Flow rate of the feed = F = 100 tons/h

Assay value of feed = f = 10% Sn

Assay value of concentrate = c = 40% Sn

Assay value of tailing $= t = 1\%$ Sn

Total material balance $F = C + T \Rightarrow 100 = C + T$

Tin balance $Ff = Cc + Tt \Rightarrow 100 \times 10 = 40\,C + T$

On solving $C = 23.077$ tons/h; $T = 76.923$ tons/h

$$\%\text{ Recovery of tin} = \frac{Cc}{Ff} \times 100 = \frac{23.077 \times 40}{100 \times 10} \times 100 = 92.31\%$$

$$\text{Tin lost in tailing} = \frac{Tt}{100} = \frac{76.923 \times 1}{100} = 0.76923 \text{ ton/h}$$

To achieve the recovery to the maximum possible extent, the ore is repeatedly treated by using another unit of the same device. Following is the terminology in use in such cases:

Roughing : It is the operation of obtaining valuable mineral particles as concentrate at the earliest stage of beneficiation of the ground ore. The unit used is called Rougher. This concentrate is called rough concentrate and usually contains considerable quantity of gangue mineral particles.

Scavenging : It is the operation of obtaining the last recoverable fraction of valuable mineral particles from the tailings of rougher before discarding as final tailing from the beneficiation plant. The unit used for this purpose is called Scavenger.

Cleaning : It is the operation of re-beneficiating the rough concentrate to improve its quality by removing the gangue particles present in the rougher concentrate. The unit used for this cleaning stage is called Cleaner.

The arrangement of these three beneficiation units is called a circuit or flowsheet. Fig. 165 is a typical flowsheet showing three-stage beneficiation.

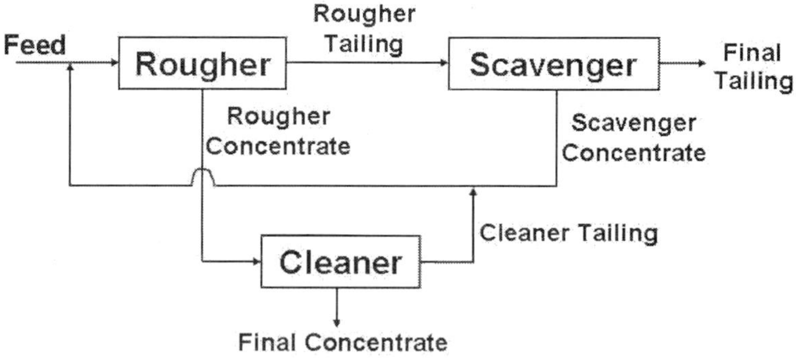

Fig. 165 Flowsheet of three-stage beneficiation

17.1. PROBLEMS FOR PRACTICE

17.1.1: *A beneficiation plant treats 200 tons/day of lead ore. Assay analyses of samples of feed, concentrate and tailing determined as 4.4% Pb, 55% Pb* *and 0.05%Pb. Calculate the amount of concentrate recovered, lead recovery and loss of lead in tailing.* *[15.83 tons/day, 98.94%, 0.0921 ton/day]*

17.1.2: *Estimate the recovery of chromite from the following data:*

The ore contains 18.4% Cr_2O_3. The assay of product and rejects are 52.5% and 10.8% Cr_2O_3, respectively. *[52%]*

17.1.3: *Copper ore treated by a flotation plant produces concentrate of 28% Cu and tailing of 0.1% Cu. The ore assays 0.9% Cu.*

Calculate a) Cu recovery in concentrate

b) Enrichment ratio *[89.3%; 31.1]*

■■■■

18 Dewatering

Dewatering in mineral processing plants is simply the removal of water from the pulp. With respect to original pulp, it separates the pulp into two parts: one is relatively solid-free and the other is relatively liquid-free. As most mineral-processing operations are conducted using substantial quantities of water, the products are obtained in the form of pulp. Hence, water must be removed from the pulp to get final products in dry form. If the solid particles in the pulp are relatively coarser, screening the pulp results in removal of water by passing the water through apertures and retaining the solid particles on the screen. Draining is a fairly effective method to remove water from coarse sands; however, if fine sands and slimes are present, they tend to run off with the water.

Pulps containing fine sands and slimes, e.g. flotation pulps , are dewatered usually in the following three stages:

1. Sedimentation
2. Filtration
3. Drying

Sedimentation means gravity settling or subsidence of solids suspended in liquid. This operation, coupled with continuous overflow of water and withdrawal of partially dewatered solids (thick pulp) from the bottom, is called thickening. Much of the water is first removed by the thickening operation and the thick pulp so obtained is then filtered to produce moist filter cake. The filter cake, in many cases, is subjected to further processing directly. When it is required to eliminate the moisture, the filter cake is thermally dried to about 95% solids by weight. Removal of water by drying is the most expensive operation. Hence, it is usual practice to eliminate as much water as possible by filtration, leaving only the moisture contained in the filter cake for removal by thermal drying.

18.1 THICKENING

Thickening is an operation of concentrating a relatively dilute slimy pulp into a thick pulp by allowing solid particles to settle under the influence of gravity force. Thickening is also considered as a classification with a difference that in thickening all the solid particles are allowed to settle, whereas only certain solid particles are allowed to settle in classification.

Thickening is a complex phenomenon to analyse and its rate is difficult to predict as a large number of factors are involved. As the process of sedimentation (or thickening) proceeds, the slurry becomes more and more concentrated and, as a result, the rate of settling of particles decreases. Sedimentation is thus a time-dependent phenomenon and its rate decreases with time. A preliminary study of the phenomenon—known as batch sedimentation test (Fig. 166)—is usually performed in the laboratory on a sample of the slurry. A sample of the slurry is taken in a graduated cylinder and is kept under observation.

At the outset, the particles settle at their maximum hindered settling velocity and the rate of sedimentation will be essentially constant. Soon, a zone of clear liquid (A) develops at the top, below it appears the thickening zone (B) of essentially uniform concentration of solids more or less similar to that of the original slurry and the heavy sludge or sediment that accumulates at the bottom constitutes the compression zone (D). A transition zone (C) may also appear between the thickening zone and compression zone. As settling proceeds, the interface between the thickening zone and the clear liquid zone descends down, decreasing the height of the thickening zone and increasing that of the clear liquid and compression zones. As the process is in progress, the density and viscosity of the suspension increases and the sedimentation rate gradually decreases. It reaches a critical point where the thickening zone (B) and the transition zone (C) disappears and only two zones, thickening and compression zones, exists. Further sedimentation results in compaction of the compression zone D by oozing out some liquid into clear liquid zone due to the weight of the clear liquid in zone A. This process of removal of some liquid from compression zone due to weight of clear liquid is known as exudation. It continues till the ultimate height of the compression zone is reached.

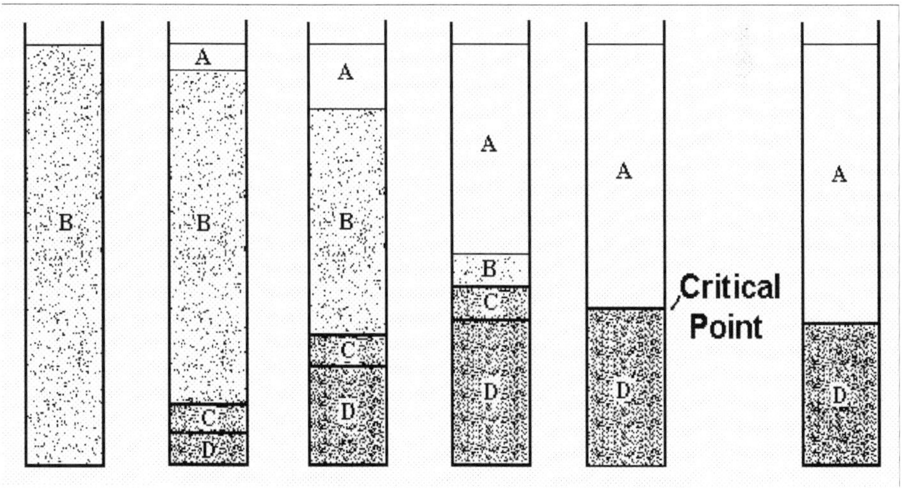

Fig. 166 Batch Sedimentation Test

As very fine particles of few microns size settle extremely slowly, the thickening operation is very slow and cannot produce clarified liquid. To increase the settling rate, these fine solid particles are agglomerated or flocculated.

18.1.1 Flocculation

Flocculation is defined as clustering, coagulation or coalescence of fine particles to form big particles known as floccules or flocs in a liquid medium. Flocculation involves the formation of much more open agglomerates than those resulting from coagulation and relies upon molecules of reagent acting as bridges between the suspended particles. Flocculation is achieved by the addition of certain reagents known as flocculating agents or flocculants

The reagents used to form the bridges are long-chain organic polymers. Formerly natural materials such as lime, alum, starch, gypsum, sulphuric acid, copper sulphate, sodium aluminate, glue, gelatine and guar gum, which are surface active, were used. Synthetic materials, loosely termed as polyelectrolytes such as separan and aerofloc, are now increasingly used. Inorganic salts cannot perform this bridging function, but they are sometimes used in conjunction with an organic reagent as a cheaper means of charge neutralisation; however, an ionic polyelectrolyte can and often perform both functions.

Flocculation is to be performed before the separating action of sedimentation process can be effectively utilised. Some pulps are naturally in a separable condition and for them flocculation is of little importance. Others require special treatment to destabilise the suspension and induce aggregation so that the separation can be obtained. Of these pulps, some need only gentle agitation to promote flocculation while others require treatment with flocculants. Flocculation promotes the coalescence of separate particles into flocs with sufficient diameter to settle at an economic rate.

18.1.2 Thickener

Thickening is usually performed in large tanks called **thickeners**. A thickener consists of a large cylindrical tank with a very short conical bottom as shown in Fig. 167.

The thickener shown in the Fig. 167 is a sketch of a bridge thickener in which the bridge runs across the thickener tank. The bridge support the rakes and the motor rotating the rakes sit on a platform in the centre of the tank. The rakes are bolted to the central column which is rotated by the motor. The bridge thickeners have a maximum diameter of about 30 metres.

Following are the elements of a thickener:

- A **feed well** is a hallow cylinder centrally located at the top of the thickener. The function of the feed well is to dissipate energy of movement in the feed stream to cause the feed to enter the tank in a relatively quiescent condition and to provide a means of introducing the slurry at an appropriate depth in the thickener.

- A **tank** provides holding time to produce settled solids and clarified liquor. The sloped bottom assists movement of the concentrated solids toward the discharge point.
- **Rake arms** are rotating radial arms. A series of blades are suspended from the rake arms and these blades are shaped to rake the settled solids towards the central outlet. Rake arms serve the following three functions:
  1. Move the settled solids towards the discharge point
  2. Maintain a degree of fluidity in the thickener to ensure hydraulic removal
  3. Increase underflow solids concentration by providing channels for water to escape from the thickening solids in the compression zone
- **Cone, or Trench, Scrapers** perform an action similar to the Rake Arms so that underflow solids can be discharged.
- An **overflow launder** collects clarified liquor for transport to an outlet at a uniform rate.
- A **rake drive** provides driving force (torque) to move the Rake Arms and Blades against resistance of the thickened solids.
- A **rake lift** provides a means of lifting the Rake Arms out of contact if the torque exceeds a certain value thus preventing damage due to overloading.

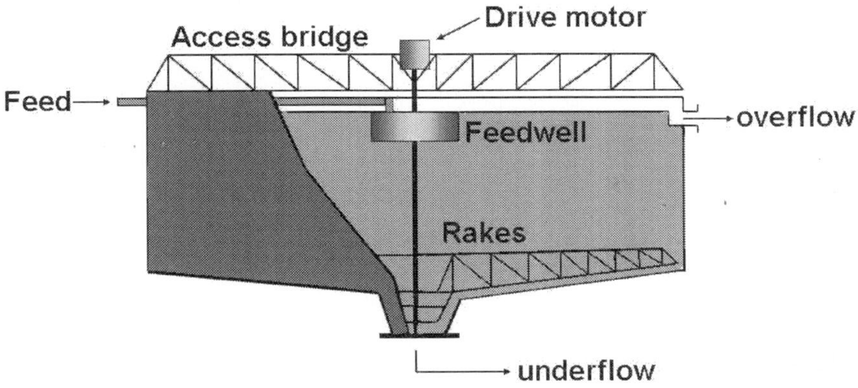

Fig. 167 Thickener

The methods of supporting the drive mechanism primarily depend on the tank diameter. In relatively small thickeners of less than 45 m diameter, the drive head is usually supported on a superstructure spanning the tank with the arms being attached to the drive shaft. Such thickeners are referred to as **bridge** or **beam thickeners** (Fig. 167). For larger thickeners of about 180 m in diameter, drive mechanism is supported on a stationary steel or concrete centre column. In the **traction thickener**, a single long arm is mounted with one end on the central support column while to the other are fixed traction wheels that run on a rail on top of the tank wall. **Cable thickeners** have a hinged rake arm fastened to the bottom of the drive cage or centre shaft. The rake arm is pulled by cables connected to a torque or drive-arm structure.

The feed is introduced continuously through a central semi-submerged feed well at a rate that allows the upper surface of the solid matter to settle at a safe distance below the overflow level. Thus, clear water overflows into the top peripheral launder. The settled solids are swept by a slowly revolving raking mechanism attached to the central rotating shaft and positioned slightly above the tank bottom. Slow agitation of the slurry helps in reducing the apparent viscosity of the suspension. The sludge swept by the rakes is directed towards the centre to facilitate easy removal by a suitable pump such as diaphragm pump.

The solids concentration in a thickener varies from clear overflow to that of the thickened underflow being discharged. Although the variation in concentration is continuous, the concentrations at various depths may be grouped into four zones, as shown in Fig. 168. The lines of demarcation between these zones are not well defined

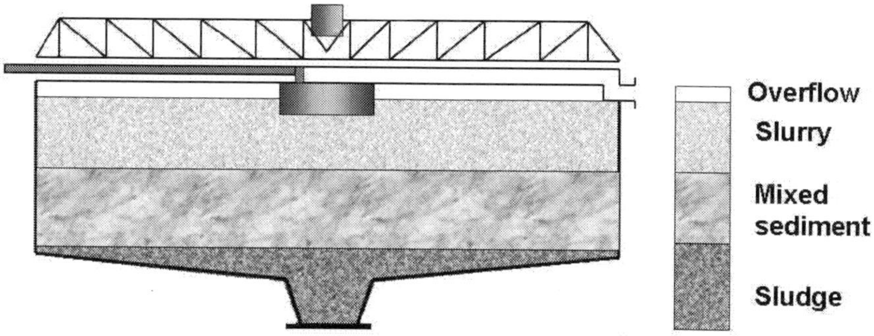

Fig. 168 Zones in a thickener

Thickener operation starts with a system that introduces the feed slurry. The feed velocity must be sufficiently high to prevent solid material sand out in the feed launder, but not so high as to cause excess turbulence in the thickener feed well and tank. As a general rule, the feed to a thickener should contain very little material larger than 250 microns. Coarse solids will usually overload a mechanism. Coarse solids of 250–840 microns should generally be kept out of a thickener.

Flocculation of the feed must be controlled and must be complete. The flocculant chemicals are added in the feed well and they must be uniformly dispersed if they are to function appropriately.

The operator must rely upon representative measurements of density, size distribution and quantity of solids in feed and underflow. Flocculant dosage and pH of the liquor is also to be measured for appropriate operation. Increase in pulp level is indicative of accumulation, possibly due to flocculation or insufficient underflow withdrawal. Increase in torque indicates overload may be occurring. Interpretation of the meaning of these measurements provides the operator with actions required to stay out of trouble. There is no universal set of rules as to what changes in the above measurements mean. There are only generalities for similar installations.

The thickeners often incorporate substantial storage capacity so that if the filtration section is shut down for maintenance, the concentrator can continue to feed material to the dewatering section. During such periods the thickened underflow should be recirculated into the thickener feed well. At no time should the underflow cease to be pumped, as chokage of the discharge cone rapidly occurs. Since the capital cost is the major cost of thickening, selection of the correct size of thickener for a particular application is important.

The area of a thickener must provide sufficient detention time to allow the slowest settling particle to reach the bottom of the unit. Thickener size is often expressed in area per unit weight of dry solids per day. (Metre squared/ton of dry solids/day).

During thickening, the feed pulp containing about 15–30% solids is thickened to a pulp of 55%–65% solids. The main aim of thickening is to obtain thick pulp. If it is desired to obtain clear liquid rather than the thick pulp, the thickening operation is called as **clarification**. This operation involves a dilute pulp of 1%–5% solids. The unit used for both the thickening and clarification may be the same but called as thickener and clarifier, respectively. However, the clarifier cannot make 100% removal of solids in a reasonable length of time; although some operations in chemical plants produce overflows containing 3–4 ppm of solids. The treatment of tailing is done to obtain clarified water for recirculation.

The operation of thickeners involves a delicate balance of the feed rate, the overflow rate and the underflow withdrawal rate and is dependant on the concentration of the feed, overflow and underflow streams. The feed stream generally enters the feed well at a speed of about 15 m/min; however, this would depend on its characteristics such as concentration (liquid/solid ratio), particle size, particle shape and viscosity. The characteristics of the overflow and underflow streams depend on the sedimentation time and particle properties like, specific gravity, shape, size and wettability.

Rakes help to increase the sedimentation rate and also break up large agglomerates. The rakes are operated between 8 and 18 m/min. To prevent damage to the rakes and torque meters the recommended operation is to discharge the sludge at regular intervals at predetermined set conditions. It is necessary for the operator to detect the build-up on the rakes and operate to avoid the jamming and seizure of the rakes. Usually the built-up mud tends to form islands which grows and develops a moment that could easily damage the rake mechanism. During normal operation, the rise rate varies from about 0.01–0.03 m/min/m of cross-sectional area and the detention time is between 2 and 5 hours.

18.1.3 Hi-capacity thickener

Conventional thickeners suffer from the disadvantage that large floor areas are required, since the throughput depends above all on the area, while depth is of minor importance. The availability of new and powerful flocculants made possible thickeners with a very small area per unit weight of solids per day. Very thorough mixing and dispersing of flocculant and slurry are required to substantially reduce thickening area required. **High-capacity thickener** (Fig. 169) is typified by a reduction in unit area requirement from conventional installations.

The feed well of a Hi-Capacity Thickener contains a special flocculation chamber. The feed enters a flocculation chamber via a hollow drive shaft. A series of mixing impellers of required shear and pumping capacity are placed on a vertical shaft driven by a variable speed drive. Flocculant solution is added at each impeller to improve homogenous mixing. This staged mixing action is said to improve thickening since it makes most effective use of the flocculant. The flocculated feed leaves the mixing chambers and is injected into a blanket of slurry where the feed solids are further flocculated by contacting previously flocculated material. Direct contact between rising fluid and settling solids, which is common to most thickeners, is averted with slurry blanket injection.

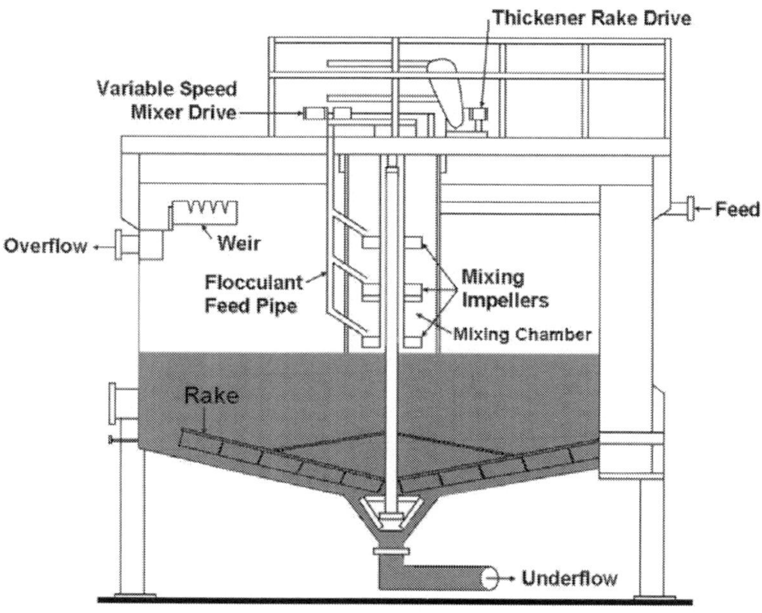

Fig. 169 High Capacity Thickeners

To increase the stability of the compression zone, radially mounted inclined plates are partially submerged in the slurry blanket to dampen action which would tend to cause the bed to erupt. The settling solids in the slurry blanket slide downwards along the inclined plates producing faster and more effective thickening than vertical descent. The height of the slurry blanket is automated using a level sensor.

Three control systems for Hi-Capacity Thickener operation which aid in reducing operating costs and insure consistent equipment performance are

- The simplest system is a Pulp Blanket Level Control System which automatically monitors and maintains the pulp blanket at a set level by means of sonic type level controls. The controls maintain the Pulp Blanket near a maximum by timed on-off pump operation.
- A more precise control system is available in which the underflow pumps operate continuously but with pumping volume controlled according to the need to maintain the pulp blanket level within a maximum and minimum range. Variable speed pumps are used for removal of underflow, and sensing of pulp blanket level is done through sonic level controls.
- An automatic Hi-Capacity Thickener control system is available to control flocculant consumption, underflow solids concentration and pulp blanket level. The system measures feed flow rate, feed density, flocculant flow rate, underflow density and pulp blanket level. The measured values are signalled to a microprocessor which computes these signals to a mass feed rate, and adjusts the flocculant dosage accordingly. A Sonic sensing system signals pulp blanket level and the underflow pump rate is automatically adjusted to maintain the pulp blanket at the appropriate level.

Hi-Capacity Thickeners have a wide field of application, but are not intended to replace all conventional thickeners. Some slurries are only slightly responsive to flocculation, or simply require such a large quantity of reagent as to make this approach uneconomical. Others thicken so slowly that a large compression volume is required to obtain maximum underflow concentration.

High density thickeners (or high compression thickeners) are an extension of high capacity thickening utilising a deeper mud bed to increase capacity and underflow density. High rate rakeless thickeners use a deep tank and a steep bottom cone to maximise underflow density while eliminating the rake and rake drive. In some applications underflow with the consistency of paste can be produced from high density and rakeless thickeners.

18.2 FILTRATION

Filtration is the separation of finely divided solid particles from a fluid by driving the pulp to a **membrane** or **septum** (commonly called as the **filter medium**), porous to the fluid but impervious to the solid, through which the fluid called **filtrate** passes. The volume of filtrate collected per unit time is termed as the **rate of filtration**. As the filtration process proceeds, solid particles accumulate on the filter medium forming a packed bed of solids called **filter cake**. Pressure difference between two sides of filter medium, more pressure at the upstream side and less pressure at the downstream side, is the driving force.

The simplest type of filter may be assumed to consists of a tube of small bore (Fig. 170A) through which the fluid is sucked while the solid particles accumulate at the entrance. As the device is operated, solids at first pass through the tube, but they quickly arch or bridge across the opening, allowing only clear liquid to pass afterward.

In practice, the slurry is passed through a semi-permeable membrane which is designed to hold the solids and permit the liquid to pass through. In the early stages of separation across this membrane, the solids deposit forming a second semi-permeable medium or cake (Fig. 170B). These two layers then form the filtering medium for the remainder of the slurry. The structure of the filtering cake changes continuously as more particles deposits with time.

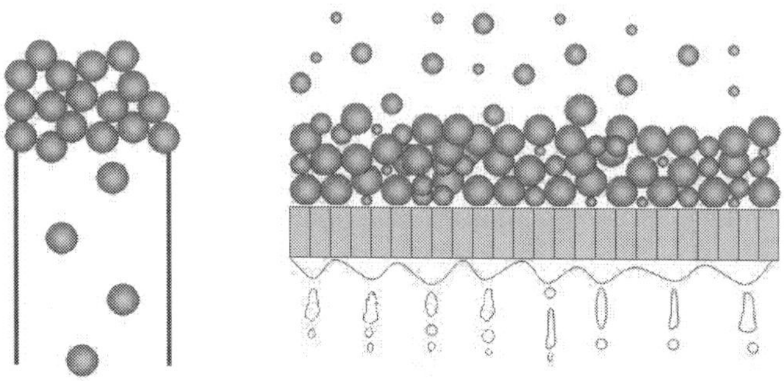

A B

Fig. 170 Mechanism of filtration

The main changes relate to permeability and porosity of the filtering zone. The permeability of the cake depends on the particle size, shape, thickness (depth) of solids and on the liquid properties, such as viscosity. The filtration rate is affected by differential pressure that is applied on the membrane to improve performance. Once a thick cake is formed the permeability decreases to the extent that the process is stopped. Filtration can continue by removing the cake of the deposited solids.

The process of filtration is predominantly carried out at conditions of either constant pressure or constant volume flow rate. When pressure is applied for filtration, it takes some time to build up to the required level. During this period, some filtration takes place. Once the required pressure is attained it can be held constant for filtration to proceed at that pressure. This type of filtration is called Constant Pressure Filtration. In industrial situations, a constant volume rate of flow of filtrate is often required to meet the demands of downstream operations, like flotation circuits. To maintain constant volume flow rate, i.e. productivity, the filtering pressure must be increased.

The filtering process is completed when nearly all the liquid has been removed from the pulp and the filter cake is removed from the filtering medium. Before removing the cake, it can be washed to remove the adhering fluid, the fluid that is retained in the pore spaces in the cake and any solute in the feed that is entrapped within the cake. A good filter medium should be able to bridge solid particles across the pores, have a low resistance to filtrate flow, resist chemical attack, have sufficient strength to withstand the filtration pressure and mechanical wear and should allow efficient discharge of the cake.

Following are the factors that effects the rate of filtration:

- Available filtering area (direct)
- Pressure difference across the cake and filter medium (direct)
- Number of pores per unit area of the filter medium (direct)
- Average cross section of pores within the filter cake (direct)
- Thickness of the filter cake (inverse)
- Size range of the particles (inverse)
- Degree of flocculation (direct)
- Pulp temperature (direct)

18.2.1 Types of filters

The entire filtration process consists of a cycle of four main steps:

1. Filtering
2. Washing of cake
3. Drying of cake and
4. Removal of cake

The designs of filters, therefore, depend on the filtering process and the cycle adopted. Cake filters are most frequently used in mineral processing plants where the recovery of large amounts of solids is the requirement. Cake filters are of two types—Pressure filters and Vacuum filters. Pressure filters are used where higher flow rates and better washing and drying are required. Vacuum filters are most widely used filters in mineral processing plants.

18.2.2 Pressure filters

Because of the virtual incompressibility of solids, filtration under pressure has certain advantages over vacuum. Higher flow rates and better washing and drying may result from the higher pressures that can be used. However, the continuous removal of solids from the pressure-filter chamber can be extremely difficult and consequently, although continuous pressure filters do exist, the vast majority operate as batch units.

Filter presses are the most frequently used type of pressure filter. They are made in two forms: the plate and frame press and the recessed plate or chamber press.

The **plate and frame press** is one of the oldest types of batch pressure filter employed in the mineral industry. It consists of vertical plates clamped together alternately by an externally operated screw system or hydraulic ram. A series of hollow frames which are placed side by side and hung from two parallel rails on either side of the plates. The filtering medium is placed against the sides of the plates and the slurry is pumped between them. The slurry pressure presses the pulp against the medium forcing the liquid through the filter medium and leaving the solids as a cake, on both surfaces of the frame. The plates are usually square shaped with ribbed or studded surfaces. Circular plates are also available in industry. Plates may number as many as 100 in a single filter yielding a very large filtration area. The usual number of plates in commercial practice varies between 25 and 50. The size of plates varies from about 450 × 450 mm to 2000 × 2000 mm and frames from 10 mm to 202 mm in thickness. They are usually made of steel to withstand pressures in excess of 1800 kPa.

In operation, feed in the form of slurry is pumped in through a common channel entering the filters through individual ports (Fig. 171). This ensures uniform distribution of feed in each chamber. The feed can be charged either through top or bottom ports in the frame. The filtrate passes through the filter medium and runs down the grooved surfaces of the plates and removed through ports to a common discharge channel. The cake remains in the frame and, when the frame is full, the filter press is opened and each plate and frame is moved manually or with automatic plate shifters so the cake can be discharged through the floor to a conveyor belt or hopper. The filter press is closed again, the feed is admitted and pressure is gradually increases as the cake deposits.

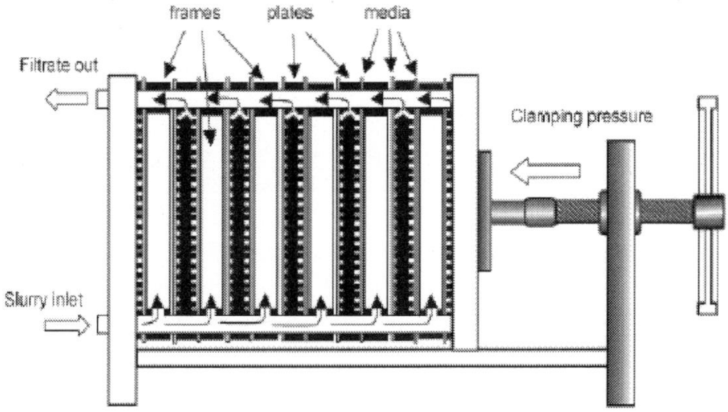

Fig. 171 Plate and frame filter press

In some installations, the discharge from individual filters can be controlled. This introduces the flexibility of accepting or rejecting the product from a particular plate. Due to tear or bursting of the filter medium in a particular plate, it discharges dirty or turbid filtrate. In such cases, it can be shut off while the other plates are in operation. The normal and often used medium is woven cotton or plastics which permit filtering rates ranging between 0.1 and 0.6 $m^3/h/m^2$. Non-woven plastics with various apertures are also used. For washing the cake the same hook-up for feed and discharge pipes are used to supply the wash water and to discharge the effluent. The cake is recovered by dismantling the entire unit. Steam is used in some cases to assist in drying the filter cake. Some filters have provision for multistage washing, vibration and hot air drying of cakes.

Feed pump selection is important to maximise filtration rate. Generally, this will be a relatively constant rate until the maximum operating pressure is approached at which time it becomes constant pressure filtration. General operating pressure range is 60–250 psi.

The **Chamber filters** (Fig. 172) are improved plate and frame filters. These filters use recessed plates which when clamped together form chambers. The recess can be up to 25 mm. By recessing the plate, it forms its own frame and permits a thicker cake than the plate-and frame filters. Cake thickness is usually a maximum of 2 inches (50 mm). All the chambers are connected by means of a comparatively large port in the centre of each plate. The feed usually enters through this central port. The filtrate escapes through a manifold at the top, the cake gradually depositing in the chambers. The other features of the filter plate are essentially similar to that of the plate and frame filters.

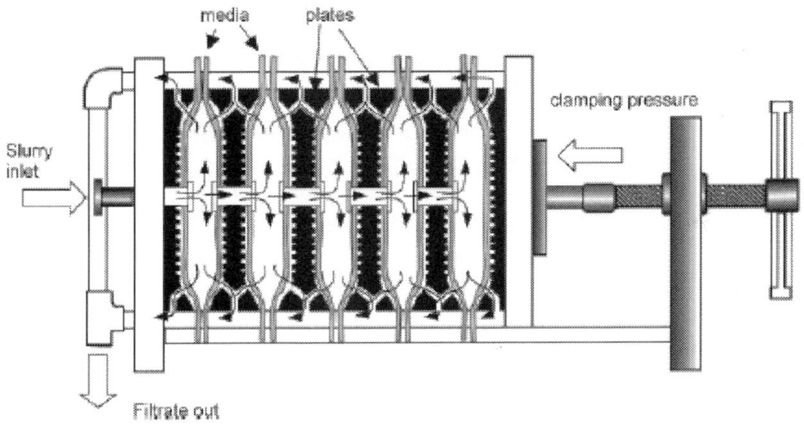

Fig. 172 Recessed plate filter press

Chamber filters are usually designed to operate with a maximum of 153 plates with surface areas varying between 0.2 and 2.6 m per chamber. The plates are connected to water lines for washing and to steam line for hot drying of the deposited cake. The cake is released from the medium by reversing the clamping device which is hydraulic or mechanical. Recessed plate filters are preferred where the cake is not very permeable, e.g. cakes produced on filtering slurries with excessive fine clays, or metallurgical slurries like in iron and alumina industries where the hydroxides must be filtered.

While designing plate filters, it is important to remember that the entire filter surface is not effective. For example, the available filtering surface of a filter of size 1450 × 1450 mm will be about 12% less than the maximum 2.10 m^2. For smaller filter sizes of say 250 × 250 mm, the filtering surface could be less by about 30%.

Automatic pressure filters are now widely used in most new flotation plants. Opening of the plate pack, starting of pump and ancillary equipment, operation of the valve and discharge of the cake are all automatically controlled. Modern pressure filters can process up to 150 tons/h dry solids per unit of concentrate in filters with filtration areas of up to 144 m^2. Residual cake moisture depends on the material being filtered but values in the range of 7.5%–12.5% are typical.

18.2.3 Continuous vacuum filters

There are many different types of continuous vacuum filters, and they all incorporate filter media suitably supported on a drainage system. Pressure is reduced by connecting to a vacuum system. The three different types of vacuum filters are drums, discs and horizontal filters.

The **Rotary Vacuum Drum Filter** consists of a cast cylinder mounted horizontally and rotated at a very low rpm. The cylinder has perforated surface over which the filter medium is wrapped tightly. The periphery of the drum is divided into compartments. Each compartment is provided with a number of drain lines passing through the inside of the drum and connected to a rotary valve assembly on the central drum shaft. The drum is immersed to the required depth in the slurry which is kept agitated to prevent settling of solids.

By the action of automatic rotary valve on the drum shaft, vacuum is applied to the immersed compartments and results in cake build-up on the filter medium surface. As the drum rotates, the cake is raised above the slurry level and wash water is sprayed on the surface if required. On further rotation, the cake is dried. The vacuum is continuously applied till the end of drying stage. Air is admitted in to the compartment before it re-enter the slurry, thus permitting the cake to be blown away from the filtering surface where it is removed by scraper on to a belt conveyor. Fig. 173A shows cycle of operation such as cake formation, washing, drying and blowing. Fig. 173B shows the cake discharge from the drum filter. The heart of the filter is the valve head (Fig. 174) which at a predetermined position causes a change from vacuum to pressure (or blowback).

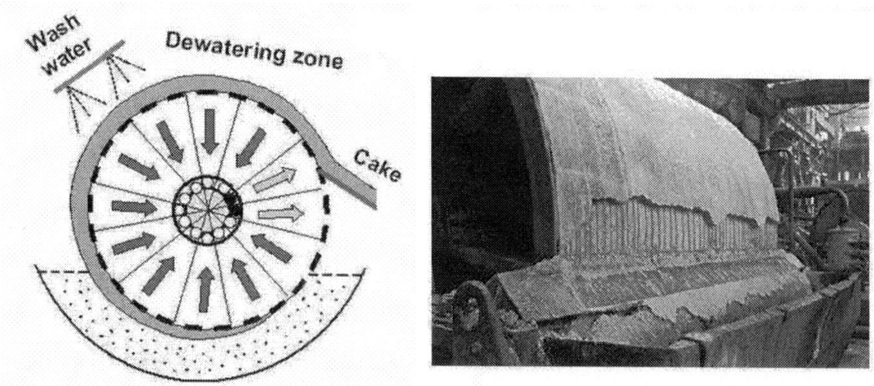

A. Cycle of operation B. Cake discharge

Fig. 173 Rotary vacuum drum filter

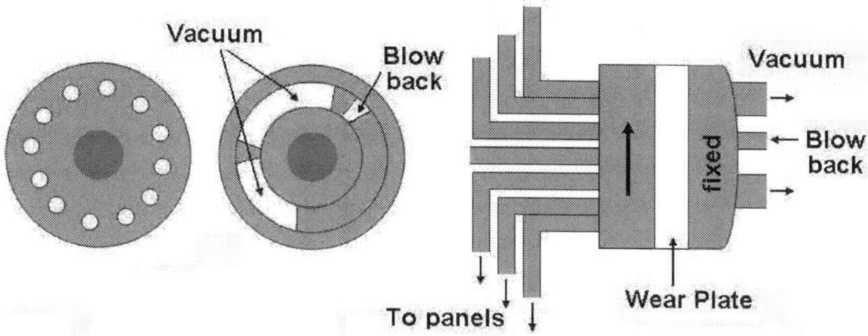

Fig. 174 Rotary valve head

Periodically the filter cloth of the drum must be changed. This takes several hours, during which feed must accumulate in the thickener.

The **Rotary Vacuum Disc Filter** (Fig. 175A) is similar to the drum filter except that instead of a cylindrical drum it consists of a number of circular discs mounted on a horizontal tubular shaft and spaced by means of hubs. A line of holes is drilled from outside of the shaft into each of the tubes for the insertion of the sector (Fig. 175B) nipple. Each disc consists of ten sectors. Filter medium consists of cloth bags are covered over the sector and fastened to the nipple. The completed sector is joined to the central shaft by screwing the nipple into the holes of the shaft. The cycle of operation is similar to that of drum filter (Fig. 173A). Fig. 175C shows the cake discharge from the disc filter.

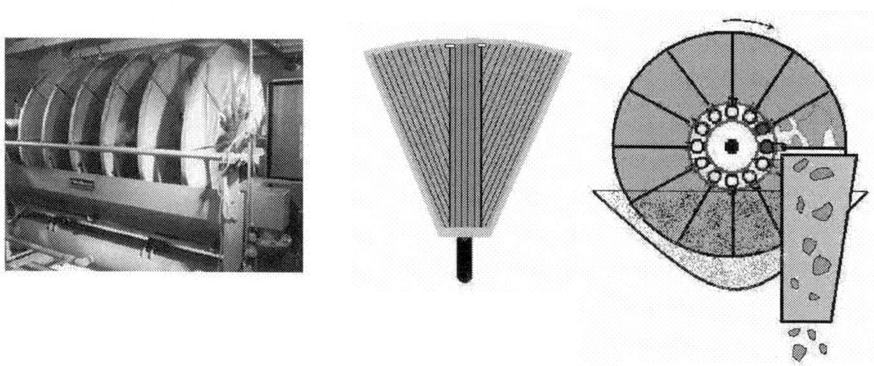

A. Rotary vacuum disc filter B. Sector C. Cake discharge

Fig. 175 Rotary Vacuum Disc Filter
(Courtesy of Westech Process Equipment India Pvt. Ltd)

Among the drum and disc filters, the disc filter has the following advantages:

1. It has more filtering area when compared to drum filter
2. Head room required per unit filter area is less

3. It is possible to filter several products in single unit
4. Worn cloth can be replaced without interrupting the process for long time

Correct selection of the filter medium is an essential part of efficient filter operation. A good filter medium should have the ability to bridge solids across the pores, have a low resistance to filtrate flow, have sufficient strength to withstand the filtration pressure and mechanical wear, avoid clogging of pores with particles, resist chemical attack and allow efficient discharge of the cake.

Canvas, cotton cloth, nylon and felt are mostly used as filter cloths. Cotton cloths are widely used as they are cheap, have good mechanical strength, resistance to wear and readily discharge cakes. Tight weaves have high flow resistance, plug quickly and have poor chemical resistance. Felt overcomes some of these disadvantages and has proved to be suitable for disc filters. Nylon is also widely employed as it has superior resistance to abrasion and clogging, and the cake moves better from its smooth surface. Its longer working life makes it cheaper than cotton. The most suitable cloth for a given job is usually arrived at by experimentation.

Ceramic Filters: Ceramic filter discs utilises microporous ceramic sectors instead of conventional filter cloth. Suction is created through capillary action. The microporous filter medium allows only liquid to flow through and no air penetrates the filter medium. Consequently, ceramic filters require only a small vacuum pump to transfer filtrate from the discs to the filtrate receiver. Filter cake is removed from the ceramic discs by scraper, eliminating the need for compressed air for blow off.

The capillary action dewatering system is exceptionally versatile for many applications. The disc material is inert, resistant to almost all chemicals and slurry temperatures and has a long operational life. Ceramic filters are best suited to filter feed slurries with consistent, high solids concentration and to solids with a d80 coarser than 40 microns.

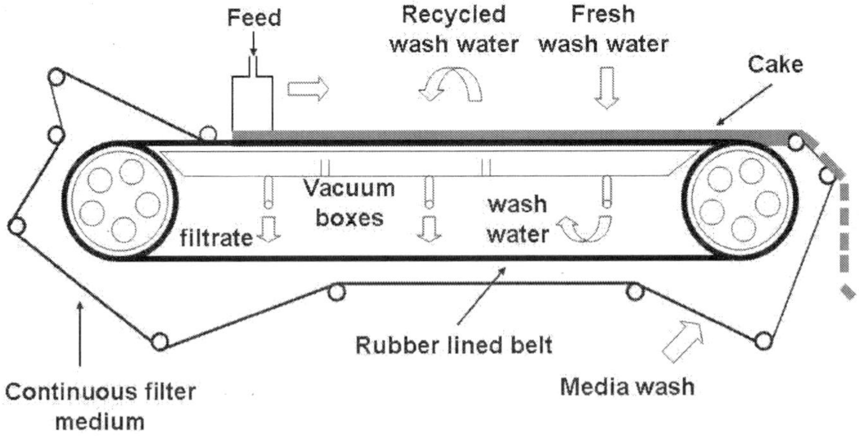

Fig. 176 Horizontal belt filter

The **horizontal belt filter** (Fig. 176) consists of an endless perforated rubber drainage deck supporting a separate belt made from a suitable filter cloth. At the start of the horizontal travel, slurry flows by gravity on to the belt. Filtration immediately commences due to gravity and the vacuum applied to the suction boxes at the underside of the drainage deck. The cake formed is dewatered and dried by drawing air through it and then discharged as the belt reverses over a small-diameter roller. If required, one or more washes can be incorporated.

The advantages of the horizontal belt filters are independent choice of cake formation and washing and drying times, effective filtration of heavy dense solids, effective filtration of sludges, the ability to flood wash and adaptability to counter-current washing or leaching. However, they are more expensive to install and operate than drum filters, and they require relatively large floor areas for a given filter area. This restricts these filters to hydrometallurgical operations where their washing ability justifies their higher costs.

Fig. 177 shows the general arrangement of the vacuum system employed in vacuum filters. It consists of vacuum pump for producing vacuum which is connected to a filtrate tank through a moisture trap for collecting any moisture entered from the filtrate tank. The filtrate is sucked into filtrate tank by the vacuum produced by the pump. The filtrate pump continuously pumps out the filtrate collected. The capacity of the vacuum pump is determined mainly by the amount of air sucked through the cake during washing and drying periods when there is a simultaneous flow of both water and air. The vacuum pump should have adequate capacity and the entire system must be free of leaks and provide unobstructed passage from filter to pump. Narrow pipes, awkward bends, rusty and rough conducting channels reduce the vacuum available at the filter, however good the pump may be. Mere increase of pump speed cannot increase the vacuum if air flow is obstructed by bad piping.

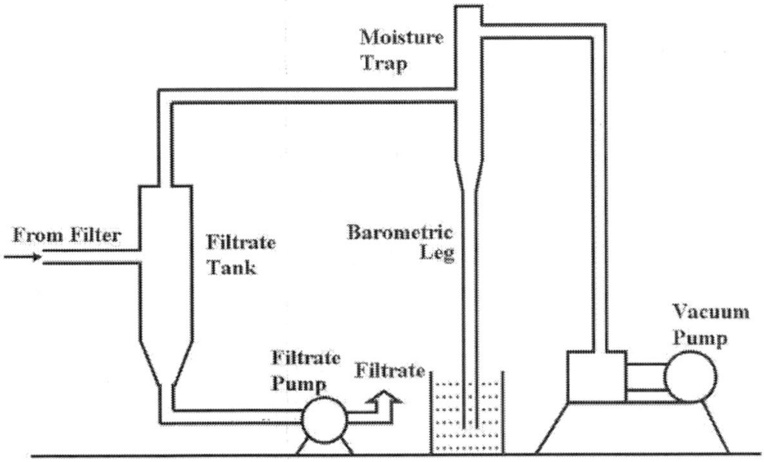

Fig. 177 General arrangement of the vacuum system in vacuum filters

The filter layout requires protection of the pump against a carry-over of filtrate, which is being drawn from the filtrate tank. This may be provided by means of a barometric leg if the vertical distance between moisture trap and seal pit exceeds 30 feet. This prevents water sucking into the vacuum pump.

■■■■

19 Coal Characterisation

Coal is a solid stratified rock and natural fossil fuel which occurs in layers in the earth's crust and was formed many millions of years ago from the remains of decaying trees and vegetation. Large trees, i.e. many resembling giant ferns, grew in dense forests on the low-lying land or in the shallow waters of the lake and succeeding generations of trees as they died and accumulated on the floor of the lake to form a vegetable sludge.

From time-to-time, sinking of the floor of the lake or swamp drowned the forests and sediments (sand and mud) carried by rivers were deposited in layers above the vegetable sludge. Eventually, with a halt in the sinking, the water became shallow again, trees re-established themselves and the entire cycle was repeated. This occurred many times over, accumulating 20 or 30 or even more layers of vegetable sludge separated by layers of sand and mud. The last two materials, with little more change than hardening, eventually formed sandstone and shale, but the conversion of the plant material into coal involved a considerable change in composition. Each layer of vegetable sludge which forms further as coal bed is called **Coal Seam**. Formations of series of coal seams separated by layers of sand and mud (sedimentary rock) are known as **Coal Measure**.

Fallen trees and plants are attacked by oxygen from the surrounding air and slowly converted to carbon dioxide and water. The vegetation is water logged, however, air cannot penetrate to the cellulose and decay takes place anaerobically (i.e. in absence of air) by the action of bacteria. The products of decomposition have been buried to a considerable depth by the time the bacterial action ceases. The resulting product thus formed is called as **PEAT**. It is further altered by the agencies of heat and pressure by the reactions named as coalification reactions to the varying degrees of completeness leading to formation of lignite, bituminous and anthracite coals.

There are two modes of origin of coal seams called **in situ** and **drift**. If coal seams occupy more or less the site on which the original plants grew and where their remains accumulated, it is called in situ origin. The coal of this origin is not polluted much with the extraneous dirt and the seams thus formed are not very thick. In drift origin, plants were drifted from one place—where it actually grew and died—by flood or river transportation and accumulated in a lake or estuary. Under some earthquake or other geological sequence, everything had gone underground and covered up with earth strata. The coal of drift origin is naturally contaminated by extraneous dirt.

Since coal is derived chiefly from vegetal matter, it consists mainly of the elements that compose plants; however, it differs from plants in composition in as much as certain proportions changed during the formation of coal. As the plants accumulated under water, some silt also settled along with them. Thus, at a later stage, some mineral matter got mixed with the coal as it formed. Hence, the coal as obtained from the earth's crust is not strictly a coal but consists of **pure coal** (or **coal substance**), **mineral matter** and **moisture**.

Coal substance is that part of the plant's organic material (or cellulose) which has been later converted during the coalification reactions to form coal beds. It contains the elements carbon, hydrogen and oxygen with small amounts of nitrogen and sulphur and only traces of inorganic material which are not chemically combined with the organic material forming most of the coal.

Mineral matter refers to such impurities as they exist in the coal. The mineral matter is non-combustible. The residue from this mineral matter after coal has been burned is called **ash**. Ash forming mineral matter is of two types: **inherent mineral matter** and **extraneous** or **adventitious mineral matter**. The inherent mineral matter represents inorganic elements present in plants giving raise to coal beds. It is very small in amount, about 2% or less of the total ash, different in composition and not possible to separate by usual beneficiation methods. The extraneous mineral matter is due to the substances which got associated with the decaying vegetable material during its conversion into coal and also due to the rocks and dirt getting mixed up during mining and handling of coal. The former of extraneous mineral matter is in a fine state and intimately associated with the organic mass of coal. Hence, difficulties are experienced in removing this from coal by mechanical methods. The second type of extraneous mineral matter is more amenable to coal cleaning methods. The major portion of mineral matter of commercial coal is extraneous.

The moisture in coal can be considered to occur in two parts. One part called the **free moisture** occurs on the surface of the coal and in its cracks and joints. The other part called **inherent moisture** is the amount of moisture in three forms as **Hygroscopic moisture** which is the water held inside the capillaries of the coal substance, **Decomposition moisture** which is the water incorporated in some of the coal's organic compounds and **Mineral moisture** which is the water that forms part of the crystal structure of clays and other minerals present in the coal.

There are two kinds of coal analysis. One is proximate analysis where the percentage of moisture, volatile matter, ash and fixed carbon are determined. The other is ultimate analysis where the elements Carbon, Hydrogen, Nitrogen, Sulphur and Oxygen are determined.

19.1 PROXIMATE ANALYSIS

Analysis of coal is essential from the viewpoint of the selection of coal for different purposes such as combustion, carbonisation, gasification and liquefaction. Proximate analysis is the most often used type of analysis for characterizing coals in connection with their utilisation. The proximate analysis determines the percentage of moisture, volatile matter, ash and fixed carbon. Proximate analysis is much more readily made and gives a preliminary indication of quality and suitability for various uses.

Determination of moisture

About 1 g of −72 mesh B.S. Sieve coal sample is kept in a silica crucible and heated in a hot air oven at 105–110 °C for one hour. Thereafter, the crucible is taken out, cooled in a desiccator and weighed. The process of heating, cooling and weighing is repeated number of times till the constant weight of coal is achieved. The loss in weight is the weight of moisture.

$$\% \text{ Moisture} = \frac{\text{Weight of moisture}}{\text{Weight of original sample}} \times 100 \qquad \ldots 19.1$$

Determination of volatile matter

About 1 g of air-dried coal of –72 mesh B.S. sieve is weighed in a standard cylindrical silica crucible with lid. The crucible is placed in a muffle furnace of specified dimensions at 925°C for a period of exactly seven minutes. Remove the crucible from the furnace and place on a cold iron plate to cool it rapidly. This prevents any oxidation of the contents in the crucible. Transfer the crucible to a desiccator while still warm. Allow it to cool and weigh. The loss of weight is taken as weight of volatile matter and air-dried moisture.

% Volatile matter and air-dried moisture =

$$\frac{\text{Weight of volatile matter and air - dried moisture}}{\text{Weight of original sample}} \times 100 \qquad 19.2$$

% Volatile matter =

$$\% \text{ Volatile matter and air-dried moisture} - \% \text{ air-dried moisture} \qquad 19.3$$

Determination of ash

About 1 g of air-dried coal of −72 mesh B.S. sieve is taken in a clean dry silica crucible. The coal is distributed so that the thickness of the layer does not exceed 0.14 g per square centimetre. The crucible is put into a muffle furnace at 500°C for 30 minutes and raised the temperature to 815±10°C in another 30–60 minutes. The crucible is kept for one hour or more at this temperature until there is no loss in weight. The crucible is then taken out, cooled in a desiccator and weighed. The weight of the material left in the crucible is the weight of the ash.

$$\% \text{ Ash} = \frac{\text{Weight of ash}}{\text{Weight of original sample}} \times 100 \qquad 19.4$$

It is also practical to heat the coal sample directly at 750°C till a constant weight is obtained.

Determination of fixed carbon

Fixed carbon is the pure carbon present in the coal. It is the carbon available in the coal for combustion. Higher the fixed carbon content, higher will be its calorific value. **Total carbon** is the fixed carbon plus the carbon present in the volatile matters, e.g. carbon monoxide, carbon dioxide, methane and hydrocarbons. Total carbon is always more than fixed carbon in any coal. High total carbon containing coal will have higher calorific value.

Fixed carbon is not determined directly. It is simply the difference between the sum of the other components and 100.

$$\% \text{ Fixed carbon} = 100 - (\%\text{Moisture} + \%\text{Volatile matter} + \%\text{Ash}) \qquad 19.5$$

In example 19.1, necessary calculations are made to determine proximate analysis of coal from the laboratory observed values.

Example 19.1: *Three samples from medium coking coal were collected for the determination of proximate analysis. Following is the laboratory observed data obtained:*

Sample 1:

Weight of the empty crucible	*= 18.765 g*
Weight of the crucible with coal sample	*= 19.649 g*
Weight of the crucible with coal sample after heating at 105°C till constant weight	*= 19.570 g*

Sample 2:

Weight of the empty crucible	*= 19.109 g*
Weight of the crucible with coal sample	*= 20.012 g*
Weight of the crucible with coal sample after heating at 925± 10°C for 7 min	*= 19.664 g*

Sample 3:

Weight of the empty crucible	*= 18.567 g*
Weight of the crucible with coal sample	*= 19.490 g*
Weight of the crucible with coal sample after heating at 750°C till constant weight	*= 18.945 g*

Calculate the proximate analysis of the coal.

Solution:

Moisture determination (Sample.1)

Weight of the coal	= 19.649 − 18.765 = 0.884 g
Weight of moisture	= 19.649 − 19.570 = 0.079 g

$$\% \text{ Moisture} = M = \frac{0.079}{0.884} \times 100 = 8.94\ \%$$

Volatile matter determination (Sample.2)

Weight of the coal = 20.012 − 19.109 = 0.903 g

Weight of volatile matter + moisture = 20.012 − 19.664 = 0.348 g

Since fresh sample is used

$$\% \text{ (Volatile matter + Moisture)} = \frac{0.348}{0.903} \times 100 = 38.54\%$$

% Volatile matter = V = 38.54 − 8.94 = 29.60%

Ash determination (Sample.3)

Weight of the coal = 19.490 − 18.567 = 0.923 g

Weight of ash = 18.945 − 18.567 = 0.378 g

$$\% \text{ Ash} = A = \frac{0.378}{0.923} \times 100 = 40.95\%$$

% Fixed carbon = FC = 100 − 8.94 − 29.60 − 40.95 = 20.51%

Proximate analysis of the coal taken is

Moisture = M = 08.94 %

Volatile matter = V = 29.60 %

Ash = A = 40.95 %

Fixed carbon = FC = 20.51 %

19.1.1 Expression of analytical results on different bases

The analytical results of proximate analysis can be modified by appropriate corrections to allow expressions on different bases. The most commonly used bases for reporting of analytical results are as follows:

As received or As sampled

When the coal is received by a consumer from the mine, samples for analysis are collected, then its analysis is reported on **As received basis** or **As sampled basis**. Many physical and chemical changes occur during the transportation of coal from mine to the consumer and during the processing such as size reduction, washing etc. Hence, analysis of coal at the consumer end is reported on **As received basis**.

Air-dried

Freshly mined coal loses its moisture due to exposure to atmospheric air, during transportation and storage. The data obtained by analysing the coal at this stage is on an **air-dried basis**. The data on **As received** and **Air dried** bases may be the same because in both cases coal loses its moisture similarly, depending on humidity and temperature of atmospheric air.

Dry or moisture free

When it is required to completely eliminate the effect of moisture on analytical data, the coal analysis is reported on **Dry** or **moisture free basis**. It is the data expressed as the percentages of the coal after all the moisture has been removed.

Dry, ash free (d.a.f)

If the effect of moisture and ash is to be eliminated, then the data is reported on **dry ash free basis**. The coal is considered to consist of volatile matter and fixed carbon on the basis of recalculation with moisture and ash removed. This does not allow for the volatile matter derived from minerals present in the air-dried coal. This is the simplest way to compare the organic fractions of coals without diluting the effects of inorganic components. This data is suitable for comparing low ash coals (ash <10%).

Dry, mineral matter free (d.m.m.f)

Here, it is necessary that the total amount of mineral matter rather than ash is determined so that the volatile matter content in the mineral matter can be removed. In case of high ash coals, the mineral matter content is about 10% more than its ash whereas mineral matter is almost equal to its ash in case of low-ash coals. Hence, the data expressed on **dry mineral matter free basis** is most suitable for comparing high ash coals (ash >10%).

19.1.2 Calculations on different bases

The details of calculations to express the proximate analysis on different bases are given as follows:

As received or As sampled basis

The results of proximate analysis are expressed as percentages of the coal including the total moisture content.

$$\% \text{ Moisture} = \frac{\text{Weight of moisture}}{\text{Weight of coal sample as received}} \times 100 \qquad 19.6$$

$$\% \text{ Volatile matter} = \frac{\text{Weight of volatile matter}}{\text{Weight of coal sample as received}} \times 100 \qquad 19.7$$

$$\% \text{ Ash} = \frac{\text{Weight of ash}}{\text{Weight of coal sample as received}} \times 100 \qquad 19.8$$

$$\% \text{ Fixed carbon} = 100 - (\text{Moisture\%} + \text{Volatile matter\%} + \text{Ash\%}) \qquad 19.9$$

Air-dried basis

The results are expressed as percentages of the air-dried coal, including inherent but not surface or free moisture.

$$\% \text{ Moisture } = \text{M} = \frac{\text{Weight of moisture}}{\text{Weight of air - dried coal sample}} \times 100 \quad 19.10$$

$$\% \text{ Volatile matter } = \text{V} = \frac{\text{Weight of volatile matter}}{\text{Weight of air - dried coal sample}} \times 100 \quad 19.11$$

$$\% \text{ Ash } = \text{A} = \frac{\text{Weight of ash}}{\text{Weight of air - dried coal sample}} \times 100 \quad 19.12$$

$$\% \text{ Fixed carbon } = \text{FC} = 100 - (\text{M} + \text{V} + \text{A}) \quad 19.13$$

Dry or Moisture free basis

The results are expressed as percentages of the coal after the inherent moisture has been removed. Using the percentages on air-dried basis, the results of proximate analysis on dry basis can be calculated as follows

On Dry basis

$$\% \text{ Ash} = \frac{\text{A}}{100 - \text{M}} \times 100 \quad 19.14$$

$$\% \text{ Volatile matter} = \frac{\text{V}}{100 - \text{M}} \times 100 \quad 19.15$$

$$\% \text{ Fixed carbon} = \frac{\text{FC}}{100 - \text{M}} \times 100 \quad 19.16$$

Dry, ash free (d.a.f) basis

The coal is considered to consist of only volatile matter and fixed carbon, on the basis of recalculation with ash and moisture removed. The calculations are as follows.

On d.a.f basis

$$\% \text{ Volatile matter} = \frac{\text{V}}{100 - \text{M} - \text{A}} \times 100 \quad 19.17$$

$$\% \text{ Fixed carbon} = \frac{\text{FC}}{100 - \text{M} - \text{A}} \times 100 \quad 19.18$$

Dry, mineral matter free (d.m.m.f) basis

The coal is also considered to consist of solely volatile matter and fixed carbon, but it is necessary that the total amount of mineral matter rather than ash be determined.

Allowance is also made in dry, mineral-matter-free data for the contribution to the air-dried volatile matter that comes from the mineral components. This may be done directly or indirectly. As it is difficult to determine volatile matter that comes from the mineral matter, it is suggested and agreed that the mineral matter of coal contributes to the volatile matter by an amount approximately equal to 10% of the ash.

Accordingly

$$\% \text{ Mineral Matter} = MM = A + 0.1\ A = 1.1\ A \qquad 19.19$$

The calculations are as follows

On d.m.m.f basis

$$\% \text{ Volatile matter} = \frac{V - 0.1A}{100 - M - 1.1A} \times 100 \qquad 19.20$$

$$\% \text{ Fixed carbon} = \frac{FC}{100 - M - 1.1A} \times 100 \qquad .19.21$$

All the above bases are indicated pictorially in Fig. 178.

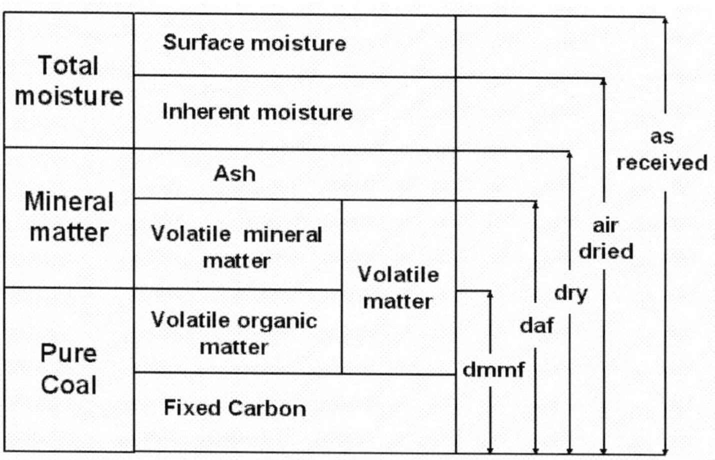

Fig. 178 Pictorial Presentation of proximate analysis on different bases

Detailed calculations on different bases are illustrated in example 19.2.

Example 19.2: *A coal has 2.34% moisture, 23.45% volatile matter and 45.67% ash. Calculate ash% on dry basis, volatile matter on d.a.f basis and fixed carbon on d.m.m.f basis.*

Solution:

$$\% \text{ Fixed carbon} = 100 - 2.34 - 23.45 - 45.67 = 28.54\%$$

Proximate Analysis of Coal

M = 02.34 %
V = 23.45 %
A = 45.67 %
FC = 28.54 %

$$\% \text{ Ash on dry basis} = \frac{A}{100 - M} \times 100 = \frac{45.67}{100 - 02.34} \times 100 = 46.76\%$$

% Volatile matter on d.a.f basis

$$= \frac{V}{100 - M - A} \times 100 = \frac{23.45}{100 - 02.34 - 45.67} \times 100 = 45.10\%$$

% Fixed carbon on d.m.m.f basis

$$= \frac{FC}{100 - M - 1.1A} \times 100 = \frac{28.54}{100 - 02.34 - 1.1 \times 45.67} \times 100 = 60.18\%$$

Alternately, fixed carbon can also be calculated through volatile matter

% Volatile matter on d.m.m.f basis

$$= \frac{V - 0.1A}{100 - M - 1.1A} \times 100 = \frac{23.45 - 0.1 \times 45.67}{100 - 02.34 - 1.1 \times 45.67} \times 100 = 39.82\%$$

% Fixed carbon on d.m.m.f basis = 100 – 39.82 = 60.18%

19.2 VARIETIES OF COAL

The various classifications of coal have been given by different geologists. Most classifications are based on some property of coal which varies with increasing maturity or rank of coal. The maturity or stage of coalification reached by a coal is referred to as its **rank**. Rank is not measured in percentages or other units, though it is judged by composition particularly the carbon or the oxygen content of the coal. The term is used only in a comparative sense, i.e. in expressions such as high, medium and low ranks. It does not imply any superiority or inferiority of quality; low rank, for example, does not imply poor quality.

For the classification to be used and developed for commercial purposes, the properties used are the proximate analysis, sulphur in coal and calorific value of coal in cal/g. **Calorific value** of coal is the quantity of heat generated by complete combustion of unit weight of coal.

Units for the calorific value are

Calories/g	(In C.G.S system)
Kilo Calories/kg	(In M.K.S system)
Btu/lb	(In F.P.S system)

Various classifications proposed are less familiar. Certain varieties are recognised almost universally in science and commerce. These varieties are classified by ranks, i.e. according to the degree of metamorphism in the series from lignite to anthracite. Those varieties are **Peat**, **Lignite** or **Brown coal**, **Sub-**

bituminous coal or **Black lignite**, **Bituminous coal**, **Semi-anthracite** and **Anthracite**. They are not sharply separated and they grade into one another. Two coals with a certain percentage of fixed carbon may have very different calorific value.

From a commercial viewpoint coal may be broadly classified as **coking coals** and **non-coking coals**. Coking coals are the coals when heated out of contact with air, volatile matter gets removed and carbon particles join each other to form a porous cellular mass with sufficient strength called **coke**. These coals are also called as **metallurgical coals**. Coking coals form a part of the bituminous group. Peat, Lignite, Sub-bituminous, Semi-anthracite and Anthracite are all non-coking coals. Some bituminous coals are also non-coking coals. Non-coking coals may resemble coking coal in all outward appearance. When they are heated in absence of air, they leave a powdery residue. Many non-coking coals also leave a solid coherent residue but not possess the physical and chemical properties of the coke and not suitable for manufacture of coke. Non-coking coals are mainly used for power generation.

19.3 FLOAT AND SINK

Both coking and non-coking coals are processed to obtain the coal product called **clean coal** of required ash. The other product obtained by processing is called **refuse** or **rejects**. As mineral matter is intimately associated with the coal substance, ordinary mechanical cleaning processes can only separate coal particles of lower ash content from those of higher ash content. The degree of cleaning possible with any coal, therefore, depends upon the distribution of its ash forming constituents. The ash in the coal can be divided as **Free ash** and **Fixed ash**. Impurities which exist as individual discrete particles when the coal has been broken to the size at which it is cleaned contribute to the free ash content of the coal. The term fixed or inherent ash is used to designate the impurities that are structurally a part of the coal. The former type of impurities can be removed by mechanical means while the latter are difficult to remove.

The coal particles of lower ash are separated from coal particles of higher ash. As the density of coal particle depends on its ash: low-ash coal particles have low density and high-ash coal particles have high density. It can be said that the particles of different densities are separated during the processing of coal. Separation of coal particles of different densities in laboratory is called **float and sink**. By determining the ash% of thus fractionated coal particles, their separability characteristics are assessed. In case of coal, separability is termed as washability.

Washability of coal means amenability of coal to improvement in quality by beneficiation techniques. Washability characteristics of coal vary from locality to locality and from seam to seam in the same locality due to variance in the extent and nature of impurities associated with coal. It is, therefore, necessary to assess the cleaning (beneficiation) potentiality of a coal before sending it to a processing plant.

The normal procedure to assess the cleaning potentialities of a coal is to perform float and sink analysis on the representative sample of desired sizes. In the case of coking coals containing a large proportion of middlings or sinks, separate float and sink analyses are performed on middlings/sinks crushed to various sizes to find out the possible recovery of additional clean coal. For the evaluation of efficiency of washing baths, float and sink analyses are separately carried out on the individual washed products. Thus, the specific objectives of float and sink analyses of raw and/or washed products are as follows:

1. To know the cleaning possibilities of a coal including recovery of products, ash content, gravity of cut etc.
2. To study the feasibility of a washery project including selection of washing units and flow scheme for treatment
3. To assess the efficiency of different washing baths in operation
4. To predict the yield of products achievable in practice

19.4 WASHABILITY CHARACTERISTICS OF COAL

Washability characteristics of coal are determined by subjecting a coal sample for float and sink analysis. Float and sink, commercially called as Heavy Liquid Separation (HLS), is an operation where particles of different specific gravities are separated using a suitable heavy liquid. The principle of float and sink is

> *"When two particles of different specific gravities are immersed in a liquid having specific gravity intermediate between that of two particles, lighter particle would float and heavier particle would sink"*

Float and sink test is based on the difference in specific gravity of coal particles. Specific gravity of clean coal (free from shale, clay, sandstone, etc.) is about 1.20 (1.12–1.35) while specific gravity of impure coal varies from 1.60 and 2.60.

The usual organic liquids employed for float and sink test are Benzene (sp.gr. 0.86), Tetra Chloro Ethylene (sp.gr 1.62) and Bromoform (sp.gr. 2.80). By mixing benzene with tetra chloro ethylene in appropriate proportions, it is possible to prepare a range of liquid baths with specific gravities lying between 1.25 and 1.62. For testing of samples at specific gravity more than 1.62, mixtures of bromoform and tetra chloro ethylene are used in different proportions. Since tetra chloro ethylene and bromoform are both toxic, all tests using these liquids are necessarily conducted in well ventilated places and where possible, under an open shed out of doors.

Chlorinated salt solutions, specially zinc chloride, are often recommended for float and sink testing of coal samples above 6 mm in size. Zinc chloride can be conveniently used to prepare test solutions ranging in specific gravity from 1.25 to 1.70.

Laboratory float and sink test is performed on coal to know the washability characteristics of coal thereby to determine the economic separating density. Suitability of heavy medium separation can also be assessed from this test.

Liquids covering a range of specific gravities in incremental steps are prepared and the representative sample of coal is introduced into the liquid of lowest specific gravity. The sink product is removed, washed and placed in a liquid of next higher specific gravity, whose sink product is then transferred to the next higher specific gravity and so on. All the float products are collected, drained, washed, dried and then weighed together with the final sink product to give the specific gravity distribution of the sample by weight.

A sample of less than 1 g is prepared separately from each float fraction as well as final sink fraction and ash percentages are determined as detailed in article 19.1.

It is desirable to size the coal and make separate float and sink tests on the separate sizes. Conducting float and sink tests on separate sizes not only assists separation but also gives information regarding the distribution of the impurities according to the size which is essential to select suitable beneficiation process.

The laboratory observations of float and sink analysis on an individual size fraction are to be tabulated. Table 40 shows such tabulated values for a hypothetical coal. Hypothetical data is considered for the convenience of discussion.

Table 40: Laboratory observed values of float and sink analysis

Specific gravity of liquid used	Weight of watch glass (g)	Weight of watch glass with coal floated (g)	Weight of empty crucible (g)	Weight of crucible with coal (g)	Weight of crucible with coal heated at 750°C till constant weight (g)
1.30	18.62	31.62	16.843	17.792	16.881
1.35	18.75	38.25	17.346	18.262	17.401
1.40	18.56	44.56	19.982	20.956	20.060
1.45	18.46	50.96	16.827	17.784	16.961
1.50	17.32	30.32	17.924	18.907	18.160
1.55	18.48	24.98	17.465	18.430	17.774
1.60	18.32	24.82	17.638	18.619	18.070
1.75	18.62	25.12	17.239	18.196	17.775
2.00	18.35	21.60	16.347	17.333	16.939
2.00 (sink)	17.52	20.77	16.466	17.439	17.060

From the laboratory observed values of Table 40, weight of the coal floated at each specific gravity, its weight percentages, cumulative weight percentages and ash percentages in floated coal at each specific gravity are calculated and shown in Table 41.

In Table 41, first four columns are the results of float and sink test. Column 5 is the calculated ash% in floated coal at each specific gravity. Column 6 represents the coal floated on percentage basis, also known as differential weight percentage or fractional yield. Column 7 represents cumulative yield percentages of cleans (floats) for each gravity fraction and obtained by simply adding fractional yields of column 6 from topmost value to the value at respective specific gravity.

Table 41: Calculated values of float and sink analysis

Sp. gr. of liquid used	Wt of coal floated (g)	Wt of coal taken for ash determination (g)	Wt of ash (g)	Ash% in floated coal	Wt% of floated coal	Cum wt% of floated coal
1	2	3	4	5	6	7
1.30	13.00	0.949	0.038	4.0	10.0	10.0
1.35	19.50	0.916	0.055	6.0	15.0	25.0
1.40	26.00	0.974	0.078	8.0	20.0	45.0
1.45	32.50	0.957	0.134	14.0	25.0	70.0
1.50	13.00	0.983	0.236	24.0	10.0	80.0
1.55	6.50	0.965	0.309	32.0	5.0	85.0
1.60	6.50	0.981	0.432	44.0	5.0	90.0
1.75	6.50	0.957	0.536	56.0	5.0	95.0
2.00	3.25	0.986	0.592	60.0	2.5	97.5
2.00(sink)	3.25	0.973	0.594	61.0	2.5	100.0
	130.00				**100.0**	

From the values of Table 41, it is evident that every coal particle in the sample tested has its own ash and their ash vary from 4% to 61% ash.

From the above results ash percentages corresponding to each cumulative yield percent of floats as well as sinks are calculated. Table 42 shows values after such calculations.

Table 42: Results of cumulative yields of floats and sinks

Sp. gr	Wt%	Ash%	Ash product	Cumulative floats			Cumulative sinks		
				Wt%	Ash product	Ash%	Wt%	Ash product	Ash%
1	2	3	4	5	6	7	8	9	10
1.30	10.0	4.0	40.0	10.0	40.0	4.0	90.0	1802.5	20.0
1.35	15.0	6.0	90.0	25.0	130.0	5.2	75.0	1712.5	22.8
1.40	20.0	8.0	160.0	45.0	290.0	6.4	55.0	1552.5	28.2
1.45	25.0	14.0	350.0	70.0	640.0	9.1	30.0	1202.5	40.1
1.50	10.0	24.0	240.0	80.0	880.0	11.0	20.0	962.5	48.1
1.55	5.0	32.0	160.0	85.0	1040.0	12.2	15.0	802.5	53.5
1.60	5.0	44.0	220.0	90.0	1260.0	14.0	10.0	582.5	58.3
1.75	5.0	56.0	280.0	95.0	1540.0	16.2	5.0	302.5	60.5
2.00	2.5	60.0	150.0	97.5	1690.0	17.3	2.5	152.5	61.0
2.00 (sink)	2.5	61.0	152.5	100.0	1842.5	18.4	0.0	0.0	0.0

In Table 42, the values in column 4 are the product of values in columns 2 and 3, and are called as Ash product or Ash point. Columns 5 and 6 are the simple addition of values in respective columns 2 and 4. The values in column 7 (cumulative ash percentages or ash percentages in total yields at respective specific gravities) is obtained by dividing values of column 6 by values in column 5. The last value 18.4% of column 7 is the ash percent of the coal sample taken for float and sink test. Ash percentages in cumulative sinks is obtained essentially the same manner but the values are cumulated from bottom to top to get cumulative sinks percent, cumulative ash product and cumulative ash percentages.

All the above results are confined to one particular size fraction. Similar tables are to be prepared for all size fractions of coal separately by similar calculations.

Finally, the combined float and sink results of the overall coal can be obtained by adding weights of floated coal in all size fractions at respective specific gravities and computing ash percentages in each floated fraction and then carrying out similar calculations to get cumulative ash percentages.

In addition to expressing the float and sink results of a coal in the form of table as shown in Table 42, they are represented in the form of curves, familiarly known as washability curves developed in 1902 (Coe 1938).

The graphical representation of float and sink results helps to understand more clearly the washing or cleaning possibilities of coal at any desired ash level of floats or at any desired specific gravity of cut and is, therefore, more suitable for studying the implications of these results for practical use. The four different kinds of standard washability curves which are conveniently drawn for this purpose are as follows:

1. Characteristic curve, also known as elementary ash curve or fractional yield ash curve
2. Total floats-ash curve or cumulative floats curve
3. Total sinks-ash curve or cumulative sinks curve
4. The yield gravity curve

The characteristic curve is a primary washability curve and is constructed by plotting the cumulative yield of floats expressed in percentages of the total coal against the ash percentages of the individual fractions. At any yield level of the total floats, it shows directly the ash contents of those particles of the floats which have the highest ash content.

The total floats-ash curve is obtained by plotting the cumulative yield of floats expressed in percentage at each specific gravity against the cumulative ash percentage of floats at that specific gravity. As the total floats ash curve directly tells us the recovery of clean or washed coal obtainable for any desired ash content of the floated coal, this curve finds more frequent use in practice than the characteristic curve.

Total sinks-ash curve is drawn between cumulative yield of sinks expressed in percentage at each specific gravity and the cumulative ash percentage of sinks at that specific gravity. This curve is complementary to the total floats-ash curve and shows directly the ash content of the total sinks at any yield level of sinks or floats.

The yield gravity curve is drawn between cumulative yield of floats expressed in percentage at each specific gravity and the specific gravity. The desired specific gravity of separation corresponding to any yield level of floated coal can be obtained by this curve.

All these four curves are usually drawn on a common diagram as shown in Fig.179.

A. Characteristic Curve

B. Total floats-ash Curve

C. Total sinks-ash Curve

D. Yield gravity Curve

E. Near-gravity material Curve

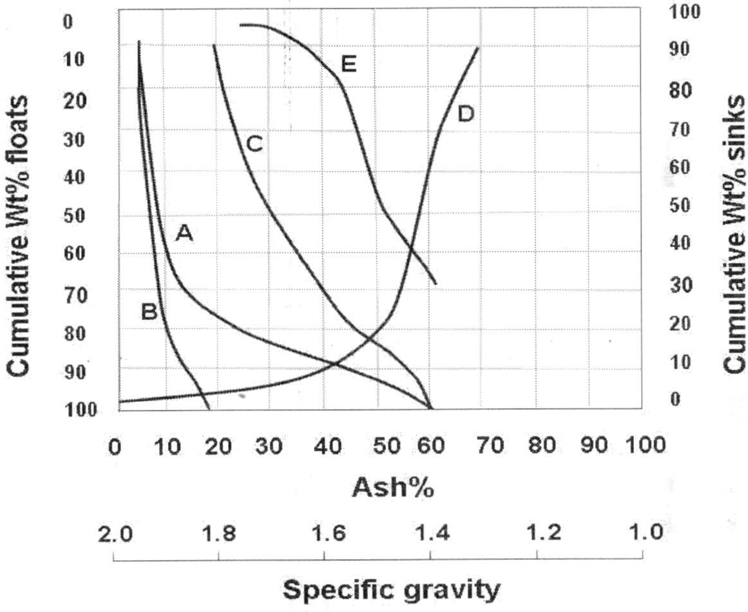

Fig. 179 Washability Curves

The advantages of drawing all the curves on a common diagram are that all the essential information required for studying the cleaning possibilities of a coal can be readily obtained by cross projections. For example, if one is interested to recover 10% ash clean coal, he/she must read first from the total floats ash curve the percent yield of cleans corresponding to 10% ash. Then from this yield point, a horizontal line is drawn to cut the total sinks ash curve and the yield gravity curve. At the cut point of the total sinks ash curve, the ash content of the sinks is read from the ash-axis and at the cut point of the yield-gravity curve, the required specific gravity of separation is read from the gravity axis.

Of these families of curves, the characteristic curve is rightly regarded as the parent curve from which other two curves (total floats-ash and total sinks-ash curves) may be derived either by direct calculation or by graphical plotting.

Both the shape and position of the characteristic curve are of much significance for assessing whether the coal is easy or difficult to wash. If the characteristic curve approaches straight line, the coal becomes more difficult to wash. This is more true when the curve approaches vertically, i.e. becomes nearly parallel to vertical axis. This nearly vertical shape with steep slope represents relatively small ash and correspondingly small specific gravity differences, hence more difficult to wash.

Flat slope represents the reverse physical condition and correspondingly easy separation. If the characteristic curve lies mostly near the vertical axis and then takes a sudden turn toward horizontality, the coal becomes very easy to wash.

The chief limitation of the standard washability curves is that the yield of middlings of any stipulated ash content cannot be directly read from them.

These standard washability curves do not provide any common and accurate basis for the comparison and correlation of the washability characteristics of different coals. The need for such a basis is felt more when there is wide divergence between the ash contents of raw coals and then the yield levels of clean coal cannot be maintained at any uniform ash or specific gravity level. Hence, more standard criteria have been proposed from time to time to compare and correlate the washability characteristics of different coals according to the ease or difficulty of washing.

These are

1. Near-Gravity Materials (N.G.M)
2. Yield Reduction Factor (Y.R.F)
3. Washability Index (W.I)
4. Optimum degree of Washability (O.D.W)
5. Washability Number (W.N)

19.5 NEAR-GRAVITY MATERIALS

In common usage, the near-gravity material is defined as the percent yield of total coal, which is within ±0.10 specific-gravity range of the effective density range of separation. Sometimes, an additional curve known as a ±0.10 specific-gravity distribution curve or Near-Gravity Material Curve can be constructed along with the standard washability curves to directly measure the ease or difficulty of washing at different densities of cut. The curve can be easily derived from the yield gravity curve.

The near-gravity material curves may be of variable nature. Generally, coal contains more of fixed ash show high percentages of near-gravity materials and those containing more of free dirt show lower percentages of near-gravity materials. However, precise amount of near-gravity materials always depends on the quantum of crowding of the specific gravity fractions around the desired density of separation.

In 1931, Bird (1931) proposed the following classification based on near-gravity materials to indicate roughly the degree of difficulty faced in the washing of a coal at any selected density of separation. The basis of Bird's classification is shown in Table 43.

Table 43: Bird's Classification

±0.10 sp.gr distribution present	Degree of difficulty	Gravity process recommended	Type
0–7	Simple	Almost any process High tonnages	Jigs, Tables, Spirals
7–10	Moderately difficult	Efficient process High tonnages	Sluices, Cones, HMS
10–15	Difficult	Efficient process Medium tonnages Good operation	HMS
15–20	Very difficult	Efficient process Low tonnages Expert operation	HMS
20–25	Exceedingly difficult	Very efficient process Low tonnages Expert operation	HMS
above 25	Formidable	Limited to a few exceptionally efficient processes Expert operation	HMS with close control

As per Bird's classification, most Indian coals can be considered as formidable to washing, whereas most American or continental coals fall under simple to moderately difficult washing groups. In Indian coals (except Assam coals), the near-gravity materials hardly come down to below 25% to 30% at the 15% to 17% ash level. Most inferior coal seams of India exhibit much worse cleaning characteristics because they often contain 60% to 80% near-gravity materials at the desired density of separation.

Table 44 shows the values of ±0.10 near-gravity material at respective specific gravities deduced from the yield gravity curve for the coal under consideration. The Near-Gravity Material curve is also drawn along with standard washability curves in Fig.179.

19.6 YIELD REDUCTION FACTOR

Yield Reduction Factor is expressed as the percent reduction or sacrifice in the yield for each percent reduction or decrease in ash content at any selected ash level of the clean coal. In the present case, the ash content of the raw coal is 18.4%. Let the coal is washed at 1.50 specific gravity. The percent yield and its ash are 80.0% and 11.0%, respectively. A reduction of 7.4% ash is effected by sacrificing 20.0% yield; hence, the Yield Reduction Factor works out to be 20.0 divided by 7.4, i.e. 2.7. It has been observed that the yield reduction factor for any coal seam remains reasonably constant so long as the size grading of the coal and clean coal ash level do not undergo any appreciable change.

Table 44: Values of ±0.10 near-gravity material

Sp.gr	Wt% of ±0.10 near-gravity material
1.35	70.0
1.40	70.0
1.45	60.0
1.50	45.0
1.55	22.0
1.60	14.0
1.65	10.0
1.70	6.0
1.75	5.0

The lower the Yield Reduction Factor, the better is the washability characteristics of the coal. The comparatively easy-washing coals of the Jharia field of India often show a 3–3.5 yield-reduction factor when the raw coal is crushed to below 75 mm and separated at 16.0% ash level. Under similar conditions of treatment, the yield reduction factors of the difficult washing coals are as high as 7–10. Except Assam coals, the average yield reduction factor values for Indian coals lie between 4 and 7 as against 1.2 to 2.5 obtained for American and continental coals.

On the basis of yield reduction factors at 15.0% ash level of cleans from raw coals crushed to below 15 mm, coal seams of different washability characteristics may be subdivided as easy coals if yield reduction factor is less than 4, medium coals if the yield reduction factor ranges from 4 to 6, and difficult coals if yield reduction factor is more than 6.

19.7 WASHABILITY INDEX

The most important criterion in defining the washability characteristics of coals for comparing and correlating the results is the Washability Index proposed for the first time by CIMFR (Sarkar et al 1962) at the Fourth International Coal Preparation Congress held at Harrogate (U.K) in 1962. This Index is expressed by a number ranging between zero and 100, and is independent of the overall ash content of the raw coal or the level of clean coal ash. The coal is difficult to wash when this index is low and easy to wash when the index is high..

A graph is to be drawn between cumulative weight percent of floats and cumulative ash as percentage of the total ash in the raw coal at different yield levels. For any coal, this yield ash distribution curve originates at zero yield (corresponding to zero ash) and terminates at 100 yield (corresponding to 100 ash). For any hypothetical coal which is absolutely non-washable, this curve has to take the form of a straight line running from zero yield (with zero ash) to 100 yield (with 100 ash). On the other hand, the greater the concavity of this curve, the better is the cleaning characteristics of the coal. When the area bounded by the curve and the diagonal through zero yield and 100% ash is expressed as the percent of the total area bounded by the two axes and the diagonal, Washability Index in terms of number is obtained, which has necessarily to lie between zero and 100.

The washability indices of a large number of coals from different coalfields of India, particularly from the Jharia coalfield, have been evaluated. It has been observed that the Indian coals, except the coals from the Assam coalfield which is of tertiary origin, gave indices between 15 and 43. For Assam coals, these indices varied between 47 and 56. The indices for British, German and American coals range between 45 and 76, indicating that they are comparatively much easier to wash. On the other hand, the indices of Japanese, Australian and South African coals range between 20 and 49, which is more or less similar to those obtained for Indian coals.

Since the reserves of Indian coals having washability indices above 40 are limited, it may be convenient to adopt a suitable classification for defining the Washability Index. Thus, coals having washability indices above 40 may be termed as easy coals, those between 21 and 40 as medium coals and those below 20 as difficult coals.

Table 45 shows the values calculated for drawing the curve to determine washability index for the hypothetical coal under consideration. The last column is the cumulative ash percentages with respect to total ash present in the coal.

Table 45: Calculated values to determine W.I

Sp.gr	Wt%	Ash%	Ash product	Cumulative floats		
				Wt%	Ash product	Ash%
1.30	10.0	4.0	40.0	10.0	40.0	2.2
1.35	15.0	6.0	90.0	25.0	130.0	7.1
1.40	20.0	8.0	160.0	45.0	290.0	15.7
1.45	25.0	14.0	350.0	70.0	640.0	34.7
1.50	10.0	24.0	240.0	80.0	880.0	47.8
1.55	5.0	32.0	160.0	85.0	1040.0	56.4
1.60	5.0	44.0	220.0	90.0	1260.0	68.4
1.75	5.0	56.0	280.0	95.0	1540.0	83.6
2.00	2.5	60.0	150.0	97.5	1690.0	91.7
2.00 (sink)	2.5	61.0	152.5	100.0	1842.5	100.0

Fig. 180 shows the curve drawn with the above values. The washability index for this coal is calculated as 45.5.

$$W.I = \frac{\text{Area of ACBA}}{\text{Area of AOBA}} = 45.5$$

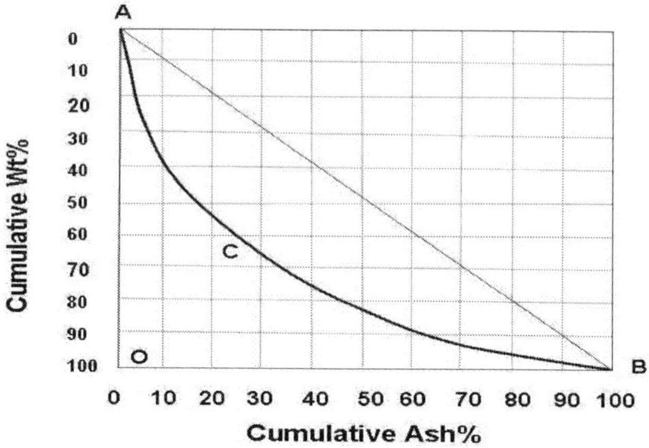

Fig. 180 Curve to determine Washability Index

19.8 OPTIMUM DEGREE OF WASHABILITY (ODW)

For every coal, there is an optimal cut point, where one can expect the maximum advantage in coal-dirt separation.

While the washability index gives a general impression about the overall washability characteristics of a coal, it fails to locate this optimal cut point. With respect to a coal of some specified top size, the degree of washability at any level of cut can be represented by the following expression:

Degree of washability at any particular level of cut

$$= \frac{\text{Recovery percent of clean coal} \times (\text{Ash\% in raw coal - Ash\% in clean coal})}{\text{Ash\% in raw coal}} \quad 19.22$$

The degree of washability of any coal at zero recovery level must be zero because one of the factors in the numerator becomes zero. Hence, also at the recovery level of 100%, where there is no difference between the ash level of raw coal and that of clean coal, the product of the two factors representing the numerator becomes zero, signifying that the degree of washability at that level is again zero. Between these two recovery levels there is a value that represents the optimum degree of washability (ODW).

By plotting the degree of washability against recovery percent of cleans at different specific gravity levels, a peak point is observed for each coal. The degree of washability corresponding to this peak point is designated as the optimum degree of washability. This optimum degree of washability has its values lying between 0 and 100 and generally bears a rectilinear relationship with the washability index. A coal can be considered to have better washability characteristics, when it has higher ODW value and the lower corresponding clean coal ash.

Table 46 shows the values calculated for drawing the curve to determine optimum degree of washability for the coal under consideration.

Table 46: Calculated values to determine O.D.W

Sp.gr	Wt%	Ash%	Cum Wt%	Cum Ash%	Degree of Washability
1.30	10.0	4.0	10.0	4.0	7.8
1.35	15.0	6.0	25.0	5.2	17.9
1.40	20.0	8.0	45.0	6.4	29.4
1.45	25.0	14.0	70.0	9.1	35.4
1.50	10.0	24.0	80.0	11.0	32.2
1.55	5.0	32.0	85.0	12.2	28.6
1.60	5.0	44.0	90.0	14.0	21.5
1.75	5.0	56.0	95.0	16.2	11.4
2.00	2.5	60.0	97.5	17.3	5.8
2.00 (sink)	2.5	61.0	100.0	18.4	0.0

The curve drawn between degree of washability and cumulative weight percent is shown in Fig. 181 and the value of optimum degree of washability is obtained as 36.

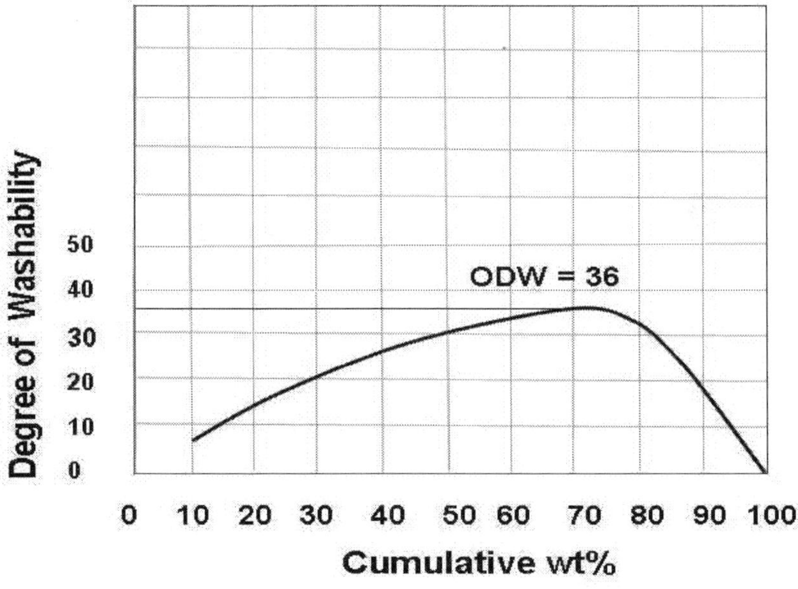

Fig. 181 Curve to determine optimum degree of washability

19.9 WASHABILITY NUMBER

To classify coals in accordance with their washability characteristics, it may be more convenient to express the final value as the ratio of the Optimum Degree of Washability to the clean coal ash at this optimum level.

Since in coals having extremely intractable cleaning characteristics—the value of this ratio is often less than 1—a magnified scale is used by multiplying the resultant value by 10 and rounding up the decimal (if any) to the nearest whole number. This number calculated as

$$\frac{\text{O.D.W}}{\text{Clean coal ash}} \times 10 \qquad 19.23$$

has been designated as Washability Number of a coal.

For the coal under consideration, the clean coal ash corresponding to optimum degree of washability is 8.8%.

Hence, the washability number is $\frac{36}{8.8} \times 10 = 41$

In examples 19.3–19.5, different types of calculations are shown.

Example 19.3: *When a sample of coal is subjected to float and sink analysis, the values recorded are shown in Table 47.*

Table 47: Float and sink analysis data for example 19.3

Sp.gr of liquid used	Weight of watch glass (g)	Weight of watch glass with coal floated (g)	Weight of empty crucible (g)	Weight of crucible with coal (g)	Weight of crucible with coal heated at 750° constant weight (g)
1.30	18.48	27.07	16.842	17.707	16.885
1.40	18.25	38.96	17.468	18.405	17.554
1.50	17.87	49.37	17.354	18.199	17.520
1.60	17.54	26.03	17.637	18.457	17.877
1.70	18.36	21.37	16.987	17.767	17.299
1.80	17.65	20.13	17.123	17.918	17.492
1.80 (sink)	18.53	33.75	17.234	18.069	17.865

Calculate the ash percentage of the coal sample taken.

Solution:

Values of wt%, Ash% and Ash product are calculated and tabulated in Table 48.

Table 48: Calculated values for example 19.3

Sp. gr	Wt of coal	Wt%	Weight of Coal taken	Weight of Ash	Ash%	Ash product
1.30	8.59	9.55	0.865	0.043	4.97	47.4635
1.40	20.71	23.01	0.937	0.086	9.18	211.2318
1.50	31.50	35.00	0.845	0.166	19.64	687.4000
1.60	8.49	9.43	0.820	0.240	29.27	276.0161
1.70	3.01	3.34	0.780	0.312	40.00	133.6000
1.80	2.48	2.76	0.795	0.369	46.42	128.1192
1.80 (sink)	15.22	16.91	0.835	0.631	75.57	1277.8887
	90.00	100.00				2761.7193

Ash % of coal sample = 2761.7193/100 = 27.62%

Example 19.4: *A student who wants to study the washability characteristics of coal has collected a sample of coal and performed float and sink analysis. The observations he recorded are shown in Table 49.*

Table 49: Float and sink analysis for example 19.4

Specific gravity of liquid used	Weight of watch glass (g)	Weight of watch glass with coal floated (g)	Weight of empty crucible (g)	Weight of crucible with coal (g)	Weight of crucible with coal heated at 750°C till constant weight (g)
1.30	18.48	27.07	16.842	17.707	16.885
1.40	18.25	38.96	17.468	18.405	17.554
1.50	17.87	49.37	17.354	18.199	17.523
1.60	17.54	26.03	17.637	18.457	17.877
1.70	18.36	21.37	16.987	17.767	17.299
1.80	17.65	20.13	17.123	17.918	17.492
1.80(sink)	18.53	33.75	17.234	18.069	17.865

Using the above data, draw the washability curves.

If this coal is to be washed to give the clean coal of 15% Ash, determine the percent yield and the specific gravity of separation.

Determine percent rejects and its Ash percent.

Determine percent ±0.1 near-gravity material at the separation density.

Calculate yield reduction factor.

Solution:

Calculations of weight% and ash% from float and sink analysis data are done and shown in the Table 50.

Table 50: Calculation of Weight% and Ash%

Specific gravity of liquid used	Weight of coal floated (g)	Weight% of coal floated (g)	Weight of coal in crucible (g)	Weight of ash in crucible (g)	Ash%
1.30	8.59	9.55	0.865	0.043	4.97
1.40	20.71	23.01	0.937	0.086	9.18
1.50	31.50	35.00	0.845	0.169	20.00
1.60	8.49	9.43	0.820	0.240	29.27
1.70	3.01	3.34	0.780	0.312	40.00
1.80	2.48	2.76	0.795	0.369	46.42
1.80(sink)	15.22	16.91	0.835	0.631	75.57
Total	90.00	100.00			

Calculation of cumulative weight percentages, cumulative ash percentages and weight percent ±0.1 near-gravity material are done and shown in the Table 51.

Table 51: Calculation of Cumulative Weight%, Ash% and NGM

Sp. gr	Wt%	Ash %	Ash product	Cumulative floats			Cumulative sinks			NGM
				Wt%	Ash product	Ash%	Wt%	Ash product	Ash%	
							100.00	2774.32		---
1.30	9.55	4.97	47.46	9.55	47.46	4.97	90.45	2726.86	30.15	---
1.40	23.01	9.18	211.23	32.56	258.69	7.95	67.44	2515.63	37.30	58.01
1.50	35.00	20.00	700.00	67.56	958.69	14.19	32.44	1815.63	55.97	44.43
1.60	9.43	29.27	276.02	76.99	1234.71	16.04	23.01	1539.61	66.91	12.77
1.70	3.34	40.00	133.60	80.33	1368.31	17.03	19.67	1406.01	71.48	6.10
1.80	2.76	46.42	128.12	83.09	1496.43	19.01	16.91	1277.89	75.57	---
1.80 sink	16.91	75.57	1277.89	100.00	2774.32	27.74	---	---	---	---

Ash% of the sample of coal = 27.74 (Last column of cumulative floats)

With the calculated data, five washability curves are drawn and shown in Fig.182.

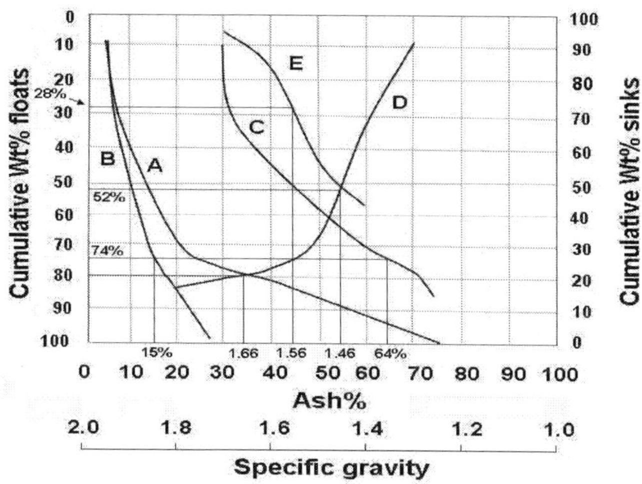

Fig. 182 Washability curves for example 19.4

From the Fig. 182,

Yield percent corresponding to 15% ash	= 74%
Specific gravity of separation	= 1.56
% Rejects	= 100 – 74
	= 26%
Ash% in rejects	= 64%
±0.1 near-gravity material at the separation specific gravity	= 80 – 52
	= 28%

$$\text{Yield reduction factor} = \frac{100-74}{27.74-15.0} = 2.04$$

Example 19.5: *Washability test data of a coal sample is shown in Table 52.*

Table 52: Washability test data for example 19.5

Sp.gr.	Differential% of floats	Ash% in differential floats
1.30	7.6	5.9
1.40	21.3	12.7
1.50	36.7	21.8
1.60	13.8	30.5
1.70	11.6	39.6
1.80	4.8	48.3
1.80 (sink)	4.2	63.4

Determine the specific gravity at which the separation should be performed for the beneficiation of this coal so that the ash content of the clean coal is not more than 17%. What is the expected yield of clean coal corresponding to this separation? Determine the ash percent in refuse and maximum ash of a particle in the yield. Calculate the yield reduction factor.

Solution:

The required values are calculated and tabulated as shown in Table 53.

Table 53: Calculated values for example 19.5

Sp. gr	Wt%	Ash%	Ash product	Cumulative floats Wt%	Cumulative floats Ash product	Cumulative floats Ash%	Cumulative sinks Wt%	Cumulative sinks Ash product	Cumulative sinks Ash%
1	2	3	4	5	6	7	8	9	10
1.30	7.6	5.9	44.84	7.6	44.84	5.90	92.4	2448.95	26.50
1.40	21.3	12.7	270.51	28.9	315.35	10.91	71.1	2178.44	30.64
1.50	36.7	21.8	800.06	65.6	1115.41	17.00	34.4	1378.38	40.07
1.60	13.8	30.5	420.90	79.4	1536.31	19.35	20.6	957.48	46.48
1.70	11.6	39.6	459.36	91.0	1995.67	21.93	9.0	498.12	55.35
1.80	4.8	48.3	231.84	95.8	2227.51	23.25	4.2	266.28	63.40
1.80 sink	4.2	63.4	266.28	100.0	2493.79	24.94	-----	-------	------

From the 7th column of Table 53, for floats of 17% ash the following are read:

(a) Recovery of the clean coal = 65.6% (Fifth column)

(b) Ash percent in rejects = 40.07% (10th column)

(c) Maximum ash in any particle in the clean coal product = 21.8% (Third column)

(d) Specific gravity of separation = 1.50 (First column)

Reduction in yield = 100 – 65.6 = 34.4 %

Reduction in ash = 24.94 – 17.0 = 7.94 %

$$\therefore \text{Yield Reduction Factor} = \frac{34.4}{7.94} = 4.33$$

19.10 COAL PROCESSING

Coal Beneficiation means the separation of coal particles of low ash from the run-of-mine coal by physical and/or physico-chemical treatments. As it comes from the mine, coal is known as **run-of-mine coal** and consists of a range of sizes from chunks to small particles mixed with some dirt and rocks. In most cases, this ROM coal is subjected to various operations principally size reduction and screening, beneficiation (separation) and dewatering (if wet beneficiation operations are used) to meet certain market requirements as to sizes, ash, sulphur, moisture and heating values. Coal Processing is a term that is used to designate the various operations performed on the ROM coal to prepare it for specific end uses such as feed to a coke oven or a coal-fired boiler or to a coal conversion process without destroying the physical identity of the coal. The terms coal preparation, coal dressing, coal cleaning, coal washing, coal upgradation and coal concentration are also applied to the same field. All of them are synonymous terms; different terms are used in different countries and different locations. When wet gravity methods are used for beneficiation of coal, it is called **coal washing** and such plants are called as **Coal Washeries.** In India, almost all coal beneficiation plants use wet methods; hence, they are called coal washeries. The low ash coal product is called **washed coal**.

The mineral matter, which gives ash after burning the coal, present in the coal is intimately associated with the coal substance in ROM coal. When ROM coal is reduced in size, all the coal particles produced will never contain the same quantity of mineral matter. Every coal particle may differ from other coal particles in its mineral matter (or ash) content. These coal particles are separated or beneficiated to obtain low-and high-ash coal particles as two products. The product of low-ash coal particles is called as **clean coal** or **combustibles**. The other product of high-ash coal particles is called as **refuse** or **rejects**. Sometimes, another product—medium ash-coal particles called as **middling**—is also separated depending on the requirement.

The methods used in the beneficiation of coal are analogous in many ways to those used in the beneficiation of ores. The principles involved in coal beneficiation are similar to that of ore beneficiation. In some instances, the same machines could be used in both fields perhaps with slight modifications. An essential feature of coal beneficiation processes is that the treatment must be rapid and inexpensive, because coal is a cheap product. Accordingly, simple plants consisting of a few units, each capable of handling a large tonnage, are required.

Another factor important in coal processing is that the coal—during all stages of its handling, i.e. from mining, through beneficiation, transportation and finally to its delivery to the consumer—must undergo a minimum of degradation because the value of coal depends on its particle size. In general, the larger the particle size the better price the coal commands on the market.

Gravity separation principles form the basis of most of the coal beneficiation processes. In case of Coal Beneficiation, the valuable part, the low-ash coal, is light and the impurities to be removed by washing or cleaning are heavier whereas in beneficiation of ores, the valuable mineral is heavy and impurities (gangue) is light.

Coal Processing forms a link between coal mining and coal utilisation. Coal Processing literally means increasing the commercial value of coal by suitable preparation. The function of the modern coal processing plant is to produce the maximum yield of clean coal of suitable quality for the consumer at an economic cost.

Coal processing methods and the flowsheets of selected coal washeries are discussed in chapter 20.

19.11 PROBLEMS FOR PRACTICE

19.11.1: *From the observations as shown in Table 54 obtained from a float and sink experiment, calculate the ash percentage of the coal sample taken.*

Table 54: Float and sink analysis for problem 19.11.1

Sp.gr of the liquid	Weight of the coal floated g	Weight of the coal taken in crucible g	Weight of the ash in crucible g
1.30	15.89	0.921	0.047
1.40	34.23	0.831	0.068
1.50	7.91	0.760	0.149
1.60	3.15	0.625	0.199
1.80	3.36	0.634	0.280
1.80 (sink)	5.46	0.740	0.487

If the coal is washed at 1.5 sp.gr., what is the degree of difficulty based on near-gravity material.

[very difficult to wash]

19.11.2: *Using the float and sink data of a coal sample shown in Table 55, calculate ash % in coal sample, clean coal and rejects when the coal is washed at 1.5 specific gravity.*

Table 55: Float and sink data for problem 19.11.2

Specific gravity	Weight of the coal floated (g)	Weight of the ash in floated coal (g)
1.30	11.4	0.5
1.40	24.4	2.0
1.50	5.6	1.1
1.60	2.3	0.7
1.80	2.4	1.0
2.00	3.9	2.5

[15.6, 8.7%, 48.8%]

19.11.3: *The washability test data of a sample of coal is given in Table 56*

Table 56: Washability test data for problem 19.11.3

Sp.gr	Wt%	Ash%
– 1.30	36.0	2.6
+ 1.30 – 1.35	25.7	7.5
+ 1.35 – 1.40	11.0	14.1
+ 1.40 – 1.45	5.3	19.2
+ 1.45 – 1.50	3.5	23.9
+ 1.50 – 1.55	2.0	28.1
+ 1.55 – 1.60	1.2	31.7
+ 1.60 – 1.65	1.1	34.3
+ 1.65 – 1.70	0.7	37.5
+ 1.70	13.5	67.4
		36.28

When this coal is washed, a clean coal of 8.8% ash is obtained. Find

(a) Recovery of clean coal

(b) Ash percent in rejects

(c) Maximum ash in any particle in the clean coal product

(d) Specific gravity of separation

Calculate yield reduction factor and assess the washability character.

[85.8%, 65.9%, 34.3%,1.65, 1.75, easy to wash]

■■■■

20 Processing of Ores and Coal

The ores are processed to obtain high-grade concentrate. The operations employed to process or beneficiate the ores are chiefly depend on

Liberation size

Physical properties such as Specific gravity,
Surface property
Magnetic susceptibility
Electrical conductivity
Required grade of the concentrate
Operating characteristics of the equipment used
Financial feasibility

Detailed laboratory investigations are to be performed to assess the various properties of the minerals present in the ore before planning for its processing.

It is to be noted that practically it is not possible to achieve 100% liberation, i.e. all the valuables will not detach during comminution. Hence, concentrate contains small amounts of gangue, and tailing contains small amounts of valuable. In some cases, a middling product, containing a considerable amount of valuables, is obtained. In such cases, the middling product is an indicative of more locked particles. Hence, this product is again comminuted for more liberation and re-beneficiated for additional valuables.

To achieve 100% liberation, all the ore is to be comminuted to a size far less than the grain size. This leads to consumption of more energy and time. Comminution operation becomes more expensive. If beneficiation is done after this kind of comminution, the cost of the entire process will be high and may exceed the cost of the product. Hence, such kind of beneficiation is not practiced.

Various operations of a processing plant can well be presented in the form of a flowsheet. Flowsheet is a diagrammatic representation of the flow of the material in the plant, starting from raw material to final products and the equipment used for the process adopted. Flowsheets are varied greatly in form according to the information to be conveyed. The simplest form of representing a flowsheet is to simply write the names of the equipment/operations in the sequence as employed in the plant and indicate the direction of flow of material with arrows from one equipment to the other. The second form is to represent the equipment/operations with block diagrams and material flow path with arrows. Both forms fail to convey how the equipment works without adding an explanation.

The third form is to represent the equipment with standard drawings. The input of the feed and output of the products of an equipment/operation are shown through which working of equipment is known without explanation. The flowsheet is self-explanatory. The same flowsheet can also be represented in an abbreviated segregated form showing only principal equipment, and the flow path is represented by the arrows as usual.

The flowsheet is devised with the various operations employed for the process, the material fed to each operation and the point of feeding, the products obtained and the points at which they are taken out. The accessory apparatus such as storage bins, pumps, pipelines, conveyors, feeders, weighing equipment, measuring instruments, sampling apparatus, etc. and the reagents used are not usually shown but denoted by numbered arrows which indicate their location and refer to details in the accompanying list. Simplified flowsheets are useful for study and comparison of practices followed in different processing plants.

20.1 PROCESSING OF ORES OF METALLIC MINERALS

1. Iron Ore

Iron is the second most common metal in the earth's crust. The chief minerals of iron are Hematite (Fe_2O_3) and Magnetite (Fe_3O_4). Other iron minerals are Goethite (FeO(OH), Siderite ($FeCO_3$), Limonite ($2Fe_2O_3.3H_2O$), Pyrite (FeS_2) and Pyrrohotite. The gangue minerals generally associate with iron ores are Silicates, Alumina, Clay, Lime, Ilmenite, Magnesia, Quartz, Gibbsite, etc. Iron ores are classified on the basis of the quality of associated gangue as siliceous, aluminous, argillaceous, calcarious, titaniferous, etc. The presence of elements such as Phosphorous, Sulphur, Arsenic, Titanium, Zinc, Lead, Nickel, Vanadium, Copper, Chromium, etc. in the ore has a considerable bearing on the operation of iron making process and the quality of the product. These are termed as **impurities**. If the ore is associated with substantial amount of these impurities, the ores are often referred to as phosphoric, sulphurous, arsenical, etc. Different types of Iron ores are Oolitic, lateritic, banded, massive, soft and powdery ores. Banded Hematite Quartz, Banded Hematite Jasper, Banded Magnetite Quartz and Banded Magnetite Jasper are the different banded formations of iron ores. It is relatively easy to beneficiate banded ores. **Blue dust** is the naturally occurring extremely friable high-grade Hematite Iron ore powder.

Hematite and magnetite ores are the most prominent iron ores found in India. Of these, hematite is considered to be the most important iron ore owing to its high-grade quality and lumpy nature, which is consumed by a large number of pig- and sponge-iron industries in India. Almost all present-day production comes from hematite reserves. The blast furnace route of iron making is predominant in India.

The raw materials used in the blast furnace for pig iron are lumpy iron ore, agglomerates (sinters and pellets), metallurgical coke and fluxes (limestone and dolomite).

Indian hematite, though rich in iron content have higher percentage of alumina (1%–7% Al_2O_3) and low silica content, i.e. high alumina silica ratio (1.5–3.0 for lumpy ore and 3–4 for fines). This adverse alumina to silica ratio is detrimental to blast furnace as well as sinter plant productivity and should be less than 1.5 and preferably below 1. High alumina content generates highly viscous slag. This needs higher amount of fluxes and coke, thereby producing large slag volumes and consequently decreasing the productivity of the blast furnace. In Indian iron ores, alumina is contributed by clay (kaolinite), gibbsite, lateritic material as well as solid solutions in hydrated iron oxides.

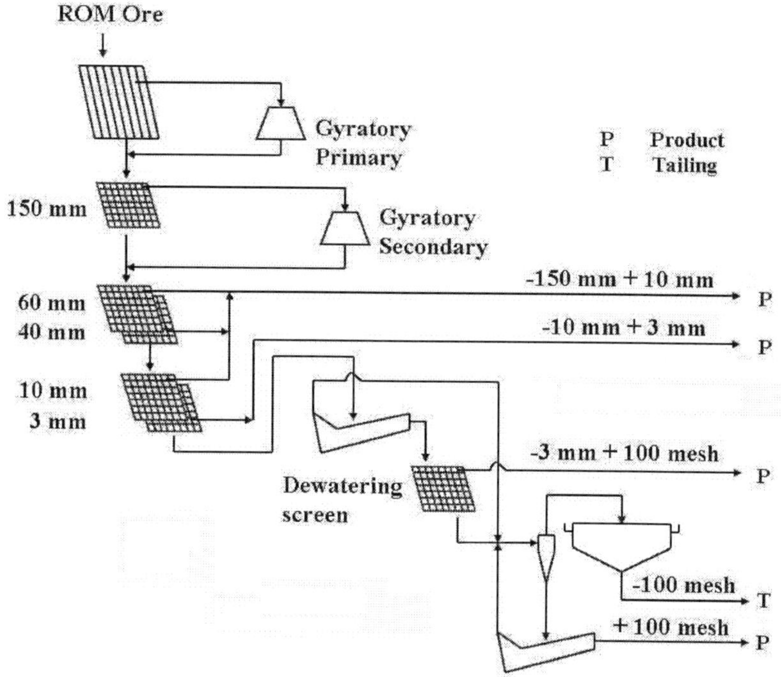

Fig. 183 Process flowsheet of Bailadila iron ore washing (Source: Mineral resources & beneficiation plant practices in India, IIME Bhubaneswar chapter)

The beneficiation practices in vogue for improvement of iron content and reduction of alumina and silica, in the country, comprise dry and wet processes. Most of the high-grade (>65%Fe) hematite iron ore are subjected to dry process to meet the size requirement involving multistage crushing and screening to obtain calibrated lumps, i.e. –40+10 mm for blast furnace, –18+6 mm for sponge iron and fines of –10/6 mm stacked at mine site as waste. In some iron ore mines, fine fraction of –10 mm generated after dry processing is further processed in mechanical classifiers and hydrocyclones to obtain –10+0.15 mm size products that constitute the feed for sintering. The classifier/ hydrocyclone overflow, i.e. –015 mm (100 mesh) size product constitutes the slime and is dumped into tailing pond.

The Indian iron ores, being soft in nature on account of high-alumina-bearing laterite and the friable portion breaking up preferentially into a finer size during crushing and sizing, generate large quantities of fines. This leads to concentration of alumina in iron ore fines. The slimes (−100 mesh) are also rich in alumina of 6%–10%.

Hematite deposit at Bailadila is the richest ore available in India with 66%–67%Fe. Average grade of the hematite ores at other places is 57%–64%Fe. Iron ore of −10 mm size is considered as fines as this size of ore is not suitable for extraction of iron in blast furnace. Processing of Iron ores in India (called as iron ore washing) consists of stage crushing and screening followed by classification and dewatering of fines for many years. The sized products from various screens are combined to produce final products. The process flowsheet of iron ore washing of Bailadila deposits is shown in Fig. 183.

For the last two and a half decades, iron ore fines dumped by washing plants are beneficiated and utilised for pelletisation. Beneficiation process consists of grinding, classification, gravity and magnetic concentration and one such flowsheet is shown in Fig 184. Iron ore fines of –10 mm are ground in the ball mill and classified in hydrocyclone. Underflow from the hydrocyclone is concentrated in spiral concentrators in two or three stages—rougher, scavenger and cleaner. The tailings from the scavenger along with the slimes from hydrocyclone are processed through High Gradient Magnetic Separator (HGMS) for reclaiming an additional Fe before rejecting the tailings through the tailing thickener. The concentrate from the spiral cleaner as well as the high-gradient magnetic separator is finally ground in the ball mills to 60% of –45 microns size and utilised for pelletisation. The general process flowsheet is shown in Fig. 184.

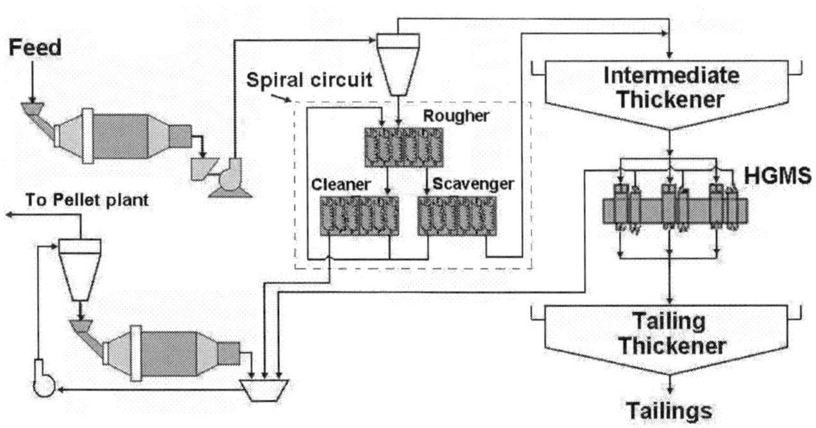

Fig. 184 General process flowsheet of beneficiation of iron ore fines

The Kudremukh iron ore beneficiation plant is the only plant treating magnetite ore, and the ore is of low grade of about 38%Fe. Owing to environmental concerns and consequent to the Supreme Court's order, the plant was stopped at Kudremukh in December 2005.

2. Manganese Ore

Manganese is an essential constituent of majority of steels. Manganese combines the twin benefits of relatively low price with outstanding technical usage. It imparts numerous beneficial mechanical properties such as harness, toughness and workability to steel in addition to improving its strength.

The most important manganese ore minerals are Pyrolusite (MnO_2), Psilomelane (Mn oxides), Hausmannite (Mn_3O_4), Braunite ($3Mn_2O_3.MnSiO_3$) and Wad (Mn oxides). Other minerals are Manganite ($Mn_2O_3.H_2O$) and Rhodochrosite ($MnCO_3$). Normally, the impurities are silica, alumina, magnesia, phosphorus and lime. Their presence results in an excess of slag and a greater consumption of fuel during metallurgical treatment. In general, two types of manganese ore—float ore and bedded deposits—are mined in India. Float ore is screened and then subjected to hand jigging to remove the light gangue minerals. In this case, no size reduction is adopted; therefore, the jig concentrates are not of good quality. In case of bedded deposits, the ore mined is crushed and then selectively handpicked.

For the purpose of beneficiation, manganese ores are classified as follows:

1. **Simple Ores:** These are further classified as high silica-high iron type.

   Usual gangue minerals: Quartz, Feldspar, Clay, Micaceous minerals, Amphiboles, Pyroxenes, Barites etc. and differing appreciably in specific gravity from manganese minerals.

   Beneficiation methods: Crushing, washing, handpicking, gravity separation (HMS, jigging and shaking tables), magnetic separation and flotation

2. **Ferruginous Ores:** Manganese present as Braunite, Pyrolusite and Vredenburgite.

   Usual gangue minerals:

   a) Iron as hematite or hydrated oxides.
   b) Feldspar, Mica, Magnetite, little or no garnet, Jaspery silica, Iron stained quartz or schist
   c) Ferruginous clay, Kaolinite, Dolomite traces of silliminite and tourmaline

   Beneficiation methods: Straight Magnetic separation or after-reduction roast

3. **Garnetiferous Ores:** Manganese present as Braunite, sometimes highly replaced by Psilomelane.

   Usual gangue minerals: a) Garnet as high as 30%

   b) Quartz, Hematite, Rhodonite, Feldspar, Apatite and Hydrated iron oxides

Beneficiation methods: As Garnet has similar density and magnetic susceptibility as manganese minerals, it cannot be separated by simple gravity or magnetic separation methods and requires costly electrostatic separation or froth flotation for its removal. Both processes also eliminate quartz if present in the ore.

4. **Complex Ores:** Beneficiation methods adopted for complex ores varies with the type of gangue minerals and impurities present.
   1. Complex Ores having varied types of gangue minerals or extremely fine size liberation of values require elaborate treatments for their upgradation by beneficiation techniques involving fine grinding. When they contain undesirable impurities such as phosphorus, sulphur, high silica, alumina and sometimes arsenic, antimony and copper, they do not yield satisfactory results with beneficiation techniques. They are better treated by thermal or chemical methods.
   2. High calcite ores associated with free silicon requiring careful beneficiation by fine grinding and flotation results in considerable tailing losses at times. Removal of apatite if present associated with manganese values becomes difficult by beneficiation methods. When most of the phosphorus is intimately associated with the metallic minerals, gravity and similar methods of concentration tend to raise the phosphorus content.

Generally, techniques employed for beneficiating manganese ores are mainly depending on the nature of the ore and on associated gangue minerals. The beneficiation flow sheet includes crushing, screening, jigging and heavy media separation for producing coarse concentrates, tabling and magnetic separation for producing fine ore concentrates are generally adopted. Wet scrubbing of the lumpy ore followed by washing and screening is used to remove clay minerals or coating (if any). Reduction roasting followed by magnetic separation is used to reduce the iron content and to maintain Mn/Fe ratio in the concentrate.

Following is the general flowsheet for beneficiation of manganese ore.

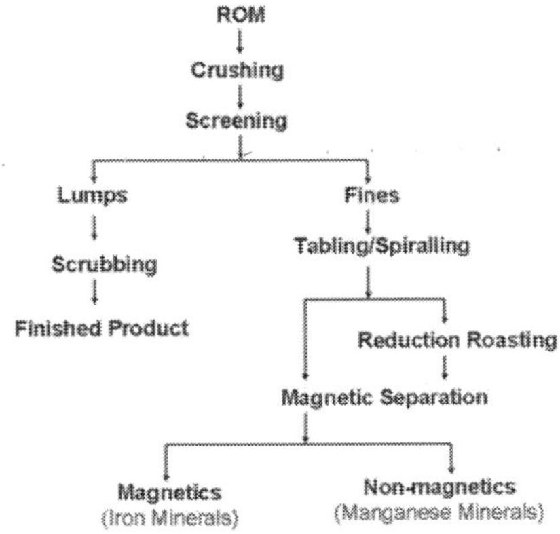

Fig. 185 General process flowsheet for beneficiation of manganese ore

In the manganese ore beneficiation plant, flowsheet as shown in Fig. 186, the crushed ore of −37 mm is dried and crushed in a hammer mill in closed circuit with an 8-mm screen. Undersize is screened in a double deck screen of 4 and 2 mm. Oversize is crushed in a rolls crusher in closed circuit with a 2-mm screen. Undersize is subjected to magnetic separation to obtain the concentrate.

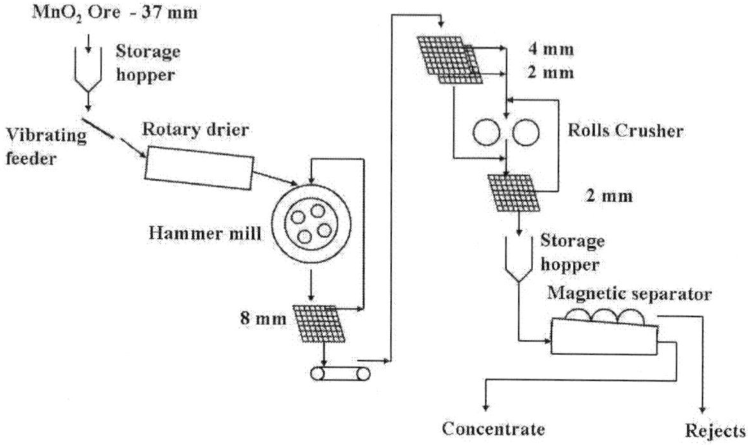

Fig. 186 Process flowsheet of manganese ore beneficiation

3. Chrome Ore

Chromite, the only ore of chromium, has a chemical composition $FeO.Cr_2O_3$ with 68% Cr_2O_3 and 32% FeO. Iron minerals such as Hematite, Magnetite, Geothite and Limonite are associated with chromite ore in variable proportions. The main gangue minerals are quartz, gibbsite, clay, mica, serpentine, calcite, garnet etc.

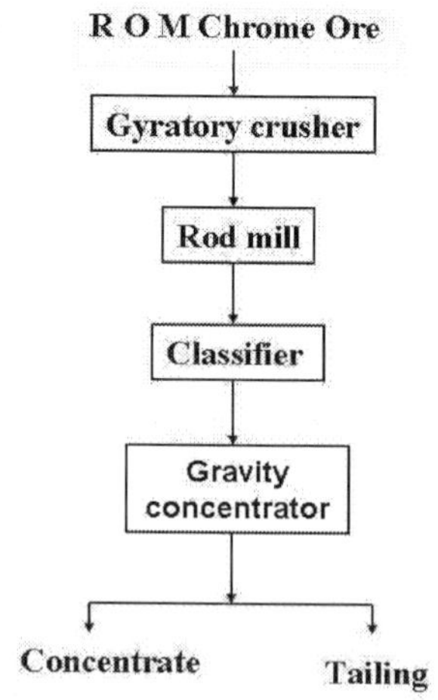

Fig. 187 Suggested flowsheet for beneficiation of chrome ore

In most cases, good-quality chrome ore as mined is of sufficient purity and hardly needs any beneficiation except hand picking. In case of disseminated deposits, concentration is achieved using vibrating screens, jigs and tables.

The general flowsheet for beneficiation of chrome ore is shown in Fig. 187.

Depending upon the grade of the chromite ore and the associated minerals present in the ore and liberation size, the grade of the concentrate varies when the chrome is processed by beneficiation methods under similar conditions. Therefore, separate study is necessary for beneficiation of chrome ore available at various places. Studies on various ores concluded that Indian chrome ore can be processed by beneficiation methods to yield good-grade concentrates with high recoveries. Use of a hydraulic classifier and spiral concentrator is very promising for the beneficiation of chrome ore to obtain the required grade of the concentrate. The process flowsheet shown in Fig. 188 employs a jaw crusher for primary crushing and gyratory crusher in closed circuit with a screen for secondary crushing. The crushed product is grounded in a ball mill in closed circuit with a high-frequency vibrating screen.

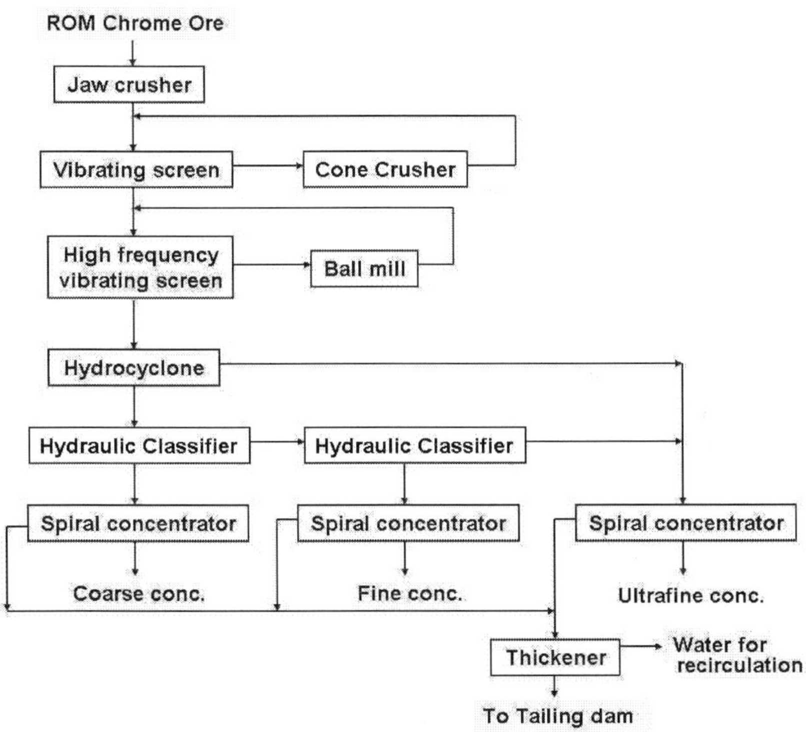

Fig. 188 Process flowsheet for beneficiation of chrome ore

The grounded product is classified in hydrocyclone. Underflow is subjected to hydraulic classification. Underflows are treated in two spiral concentrators to obtain two separate concentrates of different grades. Overflow is treated in separate spiral concentrator to obtain third concentrate. Tailings are dewatered by thickener and sent to the tailing dam.

The chromium iron ratio of concentrates obtained by beneficiation methods is not more than 2 and is at the maximum 2.5 for some type of ores, which is suitable for chemical and refractory industries but not for ferro-chrome manufacture. The process of reduction roast followed by magnetic separation gives the chromium iron ratio of 3.0. The method consists of converting the iron oxide to magnetite using pure hydrogen or carbon monoxide or industrial gases such as blast-furnace gas, coke-oven gas or producer gas. The reduced ore is crushed to a size at which the iron minerals are liberated and passed through a magnetic separator. The intensity of the field can be adjusted according to the grade of chromite desired. Non-magnetic product is the chromite concentrate.

4. Lead, zinc and copper Ores

Lead, Zinc and Copper metals are extracted from their respective sulphide minerals Galena (PbS), Sphalerite (ZnS) and Chalcopyrite ($CuFeS_2$). The ores of these minerals are of four distinct types in India:

1. Those containing lead alone as an economic metal
2. Those containing copper alone as an economic metal
3. Lead–Zinc ores
4. Lead–Zinc–Copper ores

Lead–Zinc beneficiation plants are situated at Zawar Mines, Rampura-Agucha Rajpura-Dariba and Sindesar-Khurd in Rajasthan. Lead beneficiation plants are situated at Sargipalli in Orissa and Agnigundala in Andhra Pradesh.

In Lead–Zinc ore of zawar mines, Rajasthan, galena (PbS) and sphalerite (ZnS) are the predominant lead and zinc minerals. Associated minerals are cerrussite, anglesite, Smithsonite, Hemimorphite, Pyrite, Pyrrhotite, Arsenopyrite, Chalcopyrite, Native silver, Gold and Cadmium. The chief gangue minerals are quartz, dolomite, feldspar, muscovite, biotite, sericite, tourmaline and barite. The ore is relatively clean and requires liberation in the range of 60%–65% of −74 μm. The assay value of the ore typically varies from 0.75% to 2.00% Pb and 2.53% to 4.50%Zn.

ROM Ore is crushed and screened in two stages to −15 mm and grounded in a ball mill with hydrocyclone in a closed circuit. The product is 60%–70% −74 μm. Zinc sulphate and Sodium cyanide are added into the ball mill for effective depression of sphalerite and pyrite, respectively. Cyclone overflow is conditioned with Potassium ethyl xanthate as a collector and sent to a lead-flotation circuit consisting of rougher, scavenger and cleaner cell banks. Cresylic acid as a frother is added in the rougher. Froth of the rougher is cleaned in cleaner cells in stages with the addition of Sodium cyanide for further depression of pyrite. The sink from rougher is treated in scavenger cells where froth is re-fed to the rougher. The froth of cleaner cells is thickened and filtered in disc filter to obtain lead concentrate. The sinks from the lead scavenger are grounded in a ball mill to 90% −45 μm for further liberation. The ground product is conditioned with Copper sulphate and Sodium isopropyl xanthate as an activator and collector, respectively. The conditioned pulp is treated in three bank cells called rougher, scavenger and cleaner. Cresylic acid is added in zinc rougher. The froth of rougher is cleaned in three stages by adding Lime and Sodium cyanide to maintain the required pH and depression of pyrite. The sinks of rougher are treated in scavenger, and scavenger sinks is the final tailing which is thickened for recirculated water. The scavenger floats and cleaner sinks are fed to a re-grind ball mill. The froth of cleaner cells is thickened and filtered to obtain zinc concentrate. In the process flowsheet shown in Fig. 189, flotation reagents and their places of addition are mentioned.

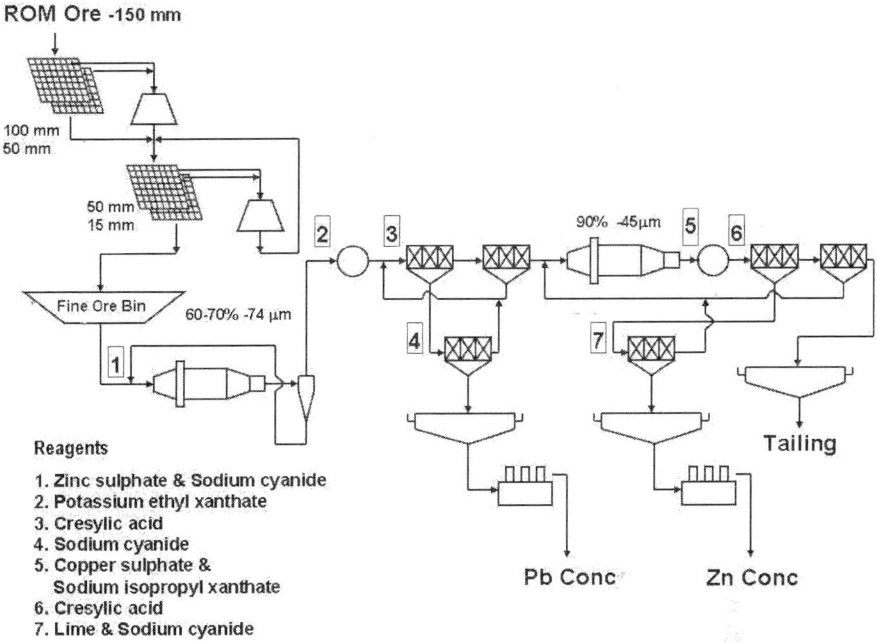

Fig. 189 Process flowsheet for beneficiation of lead-zinc ore at zawar mines (Source: Mineral resources & beneficiation plant practices in India, IIME Bhubaneswar chapter)

The principal copper mineral in copper ore is Chalcopyrite ($CuFeS_2$). Associated minerals are Chalcocite, Bornite, Cuprite, Malachite, Azurite, Pyrite, Pyrrhotite, Arsenopyrite, Sphalerite, Magnetite, Bismuthenite and Native bismuth. Quartz, carbonates and silicates are the usual gangue minerals.

Two stage crushing followed by grinding with classifier in a closed circuit is the ore preparation process for copper ore. Grounded copper ore is subjected to froth flotation process in two stages—rougher and cleaner—with sodium isopropyl xanthate as a collector, pine oil as a frother, sodium cyanide as a depressant, soda ash as a dispersant and lime as a pH regulator. The process flowsheet of copper beneficiation plant at Mosabani, Bihar is shown in Fig. 190.

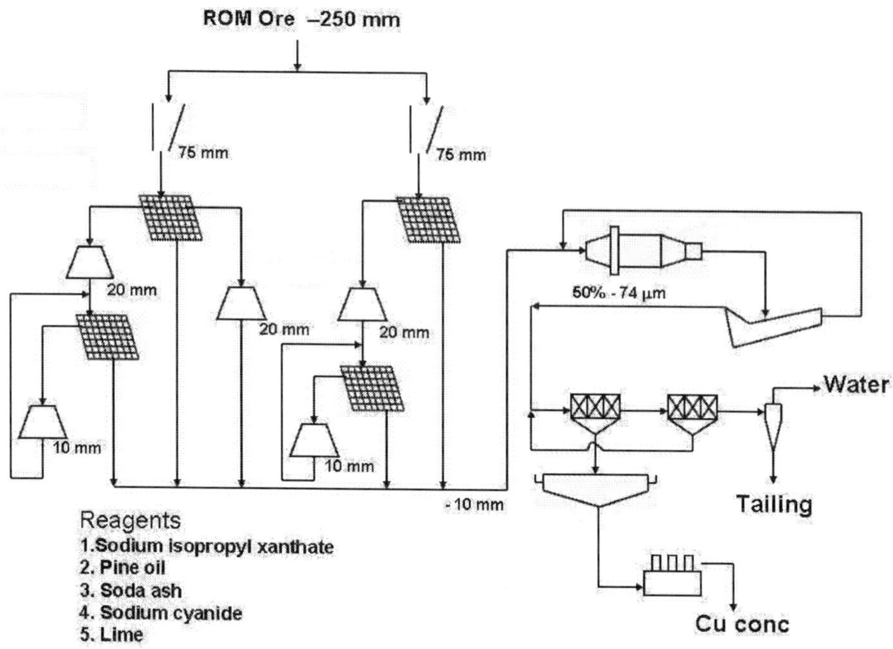

Fig. 190 Process flowsheet for beneficiation of copper ore at Mosabani (Source: Mineral resources & beneficiation plant practices in India, IIME Bhubaneswar chapter)

The Lead–Zinc–Copper beneficiation plant at Rangpo, Sikkim is the only one in India which processes a complex Lead–Zinc–Copper sulphide ore. The ore consists of galena, sphalerite, chalcopyrite as principal ore minerals and associated minerals are pyrite, pyrrhotite, with small amounts of bornite, arsenopyrite and sulphides of nickel, cobalt and bismuth. Quartz, feldspar, chlorite, garnet, biotite and zircon occur as gangue minerals are closely interlocked. The ore assays about 1.1%Pb, 3.0%Zn and 1.2%Cu.

ROM Ore is crushed in two stages and grounded in ball mill with spiral classifier in closed circuit with a ball mill. Galena and chalcopyrite are floated by bulk flotation using potassium ethyl xanthate as a collector and pine oil as a frother in a circuit of rougher, scavenger and cleaner cells. Zinc sulphate, sodium bi-sulphate, sodium cyanide and sodium silicate are used to depress sphalerite and pyrite. The tailings of this circuit is further treated and the sphalerite is floated by using copper sulphate as activator. Galena and chalcopyrite are treated separately in a circuit of another bank of flotation cells. By selective flotation, chalcopyrite is floated as concentrate and the tailing of this circuit is Lead concentrate. Fig. 191 shows the process flowsheet.

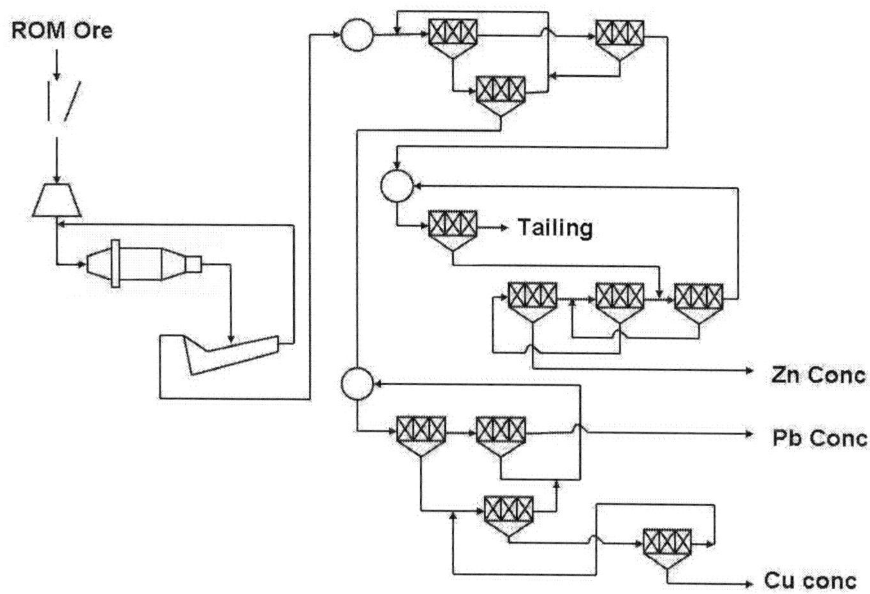

Fig. 191 Process flowsheet for beneficiation of lead-zinc-copper ore at Rangpo

5. Uranium Ore

In India, uranium mines are at Jaduguda in Singbhum district of Bihar. The average grade of the ore is between 0.05% and 0.1% U_3O_8, the principal mineral being uraninite (50%–80% U_3O_8) associated with autunite and torbernite. Quartz form up to 60% of the gangue and is associated with chlorite, magnetite, tourmaline, apatite, ilmenite and others. The ore contains about 0.08% Cu, 0.02% Mo and 0.1% Ni.

The processing of uranium ore consists of two-stage crushing of ROM ore followed by grinding with the addition of reagents to the grinding circuit. The conditioned pulp is subjected to bulk sulphide flotation for recovery of by-products of molybdenum and nickel concentrate. The tailings are collected in a thickener and sent to a uranium processing unit. The bulk sulphide float is collected in a thickener and underflow is cleaned three times with the addition of sodium silicate in the second and third stages. The float from the third cleaning stage is sent to washing tank and washed to free it from reagents. The washed product is cleaned three times with the addition of sodium sulphide at each stage. The float is thickened, washed and re-cleaned two times with the addition of sodium silicate. The float product from the re-cleaning step is the final molybdenum concentrate. It is filtered using a drum filter and dried. The major impurities in the concentrate are mica and silica. The tailings from the re-cleaning step are washed and re-circulated to the first stage of flotation with sodium sulphide.

The tailings from the cleaner circuit (with addition of sodium sulphide) contain the sulphides of copper and nickel along with some uranium and molybdenum and are collected in a thickener. The underflow of the thickener, after conditioning with lime, is floated to recover part of the nickel as tailings. The float contains copper above 20% and goes to uranium processing unit. The process flowsheet for recovery of by-products is shown in Fig. 192.

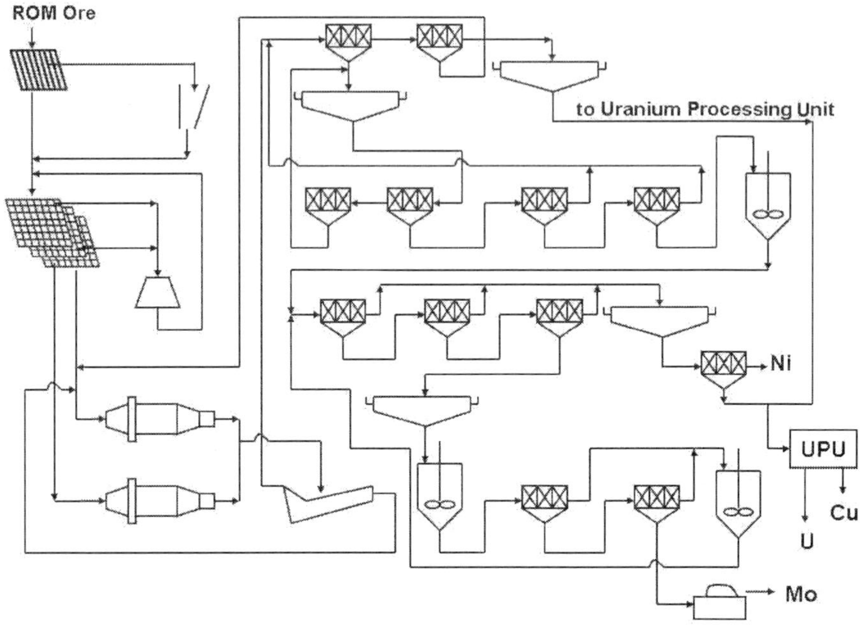

Fig. 192 Process flowsheet for by-product recovery from uranium ore (Source: Mineral resources & beneficiation plant practices in India, IIME Bhubaneswar chapter)

The uranium processing unit (Fig. 193) consists of eight pachuka tanks in series with the pulp overflowing from one to the other. Agitation of the pulp in each of these tanks is achieved by blowing compressed air. The acid-leaching in the first four tanks consists of treating the pulp with 98% sulphuric acid. Oxidising agent such as MnO_2 is added to oxidise ferrous iron of the iron-bearing minerals. About 90% of the uranium in the ore is leached out at 40°C in about 12 hours. The leached uranium-bearing solution contains U_3O_8 (0.4 g/l) and is removed by filtration.

The filtrate is subjected to ion-exchange using columns with De Acidit FF 530 resin. In ion exchange, uranyl sulphate anions formed in sulphate solution are removed by adsorption on resin. The adsorbed ions are subsequently removed by treating the loaded resins with a solution containing high concentrations of either nitrate or chloride ions. By treatment with magnesia, the product is finally stored as a complex magnesium diuranate compound. The uranium concentrate produced at Jaduguda plant is processed at Trombay to produce uranium metal of nuclear purity.

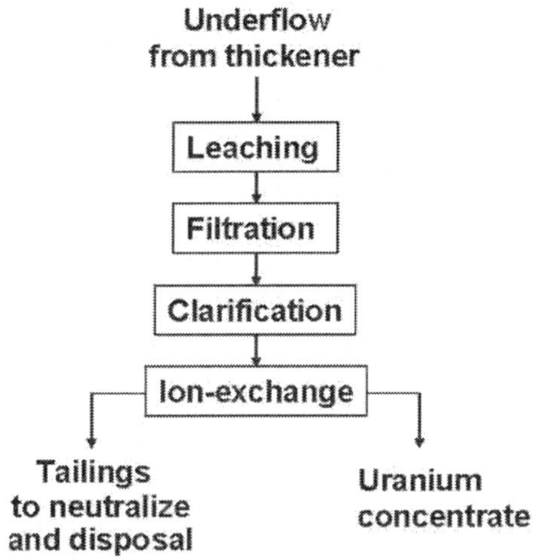

Fig. 193 Uranium processing flowsheet

6. Beach sand

India is one of the leading producers of heavy minerals from beach-sand deposits. Economic minerals in beach sand deposits are ilmenite, rutile, zircon, garnet, monazite, sillimanite and leucoxene. Quartz is the dominant gangue mineral with leucoxene, calcareous spar and mica as subordinate associates. The properties of minerals present in beach sand are given in Table 57.

Table 57: Properties of minerals present in beach sand

Mineral	Chemical formula	Composition	Grain size Crystal structure Colour	Sp.gr	Conductivity Susceptibility
Ilmenite	FeO TiO_2	54% TiO_2 Remaining is ferrous and ferric oxides traces of P_2O_5, V_2O_5, MgO and Cr_2O_3	Medium to fine Hexagonal Metallic black	4.54	Highly cond Highly mag
Rutile	TiO_2	95% TiO_2 FeO, SiO_2 and P_2O_5	Medium to fine Tetragonal Slight greenish	4.26	Cond Non mag
Zircon	$ZrO_2.SiO_2$	65.6% ZrO_2 31.2% SiO_2 TiO_2, Fe_2O_3, P_2O_5 and Al_2O_3	Fine Tetragonal Light brown	4.68	Non cond Non mag

Mineral	Chemical formula	Composition	Grain size Crystal structure Colour	Sp.gr	Conductivity Susceptibility
Garnet	$3FeO, Al_2O_3$ SiO_2	$FeO, Al_2O_3,$ $SiO_2, Fe_2O_3,$ MgO and CaO	Coarse to fine Isometric Cherry red	4.11	Non cond Mag
Monazite	Phosphate of Thorium and rare earths	60% oxides of monazite 8%–10% oxides of thorium 27%–29% P_2O_5	Fine Monoclinic Yellowish green	5.24	Non cond Feebly mag
Sillimanite	$Al_2O_3SiO_2$	60% Al_2O_3 36.5% SiO_2 TiO_2, Fe_2O_3 and ZrO_2	Coarse Orthorhombic Colourless to white	3.25	Non cond Non mag
Leucoxene	Alteration product of Ilmenite	---	Medium to fine ---- White	3.50	Non cond Non mag

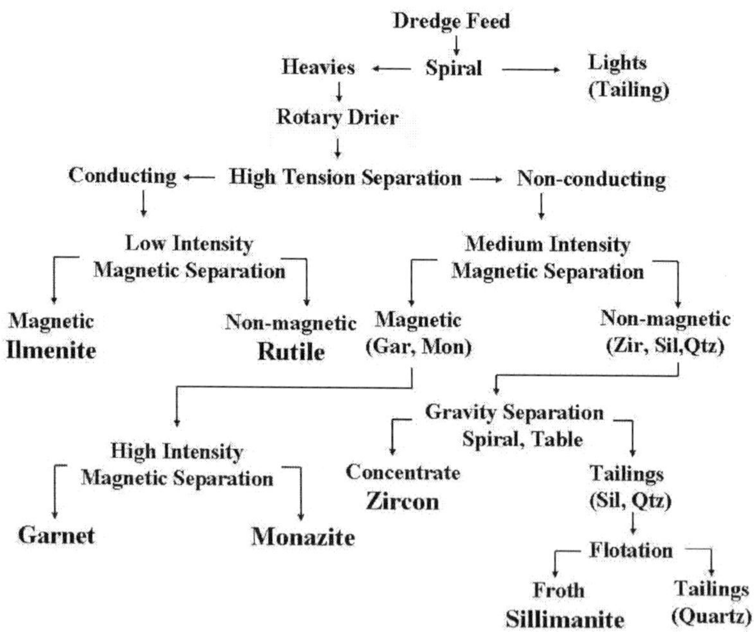

Fig. 194 Process flowsheet for Beach Sands Beneficiation Plant at Matikhalo (Source: Mineral resources & beneficiation plant practices in India, IIME Bhubaneswar chapter)

Different minerals present in beach sands are separated utilising the physical properties of those minerals as indicated in the above table 57. The dredged feed is pre-concentrated in spiral concentrator to remove quartz tailings. The concentrate containing heavy minerals is dried in a rotary drier and subjected to high-tension separation. The conducting minerals ilmenite and rutile are further separated by low-intensity magnetic separation. The non-conducting fraction is treated by medium-intensity magnetic separation. Garnet and monazite are obtained as magnetic fraction which is further separated by high-intensity magnetic separation. Non-magnetic fraction containing zircon, sillimanite and quartz is subjected to air tabling to obtain zircon as concentrate. Tailing of the air tabling is treated by the flotation operation where sillimanite is recovered as froth fraction. Flotation tailings are the quartz particles. Fig. 194 shows the process flowsheet.

20.2 PROCESSING OF ORES OF INDUSTRIAL MINERALS

1. Limestone

The most important and common use of limestone is the manufacture of cement. Limestone, either directly or on calcination to lime, also finds extensive use in metallurgical, chemical, sugar, glass, textile and other industries. The common gangue materials in limestone are clay, quartz, flint, chert and sometimes mica and talc. The beneficiation techniques employed are to tailor the composition to meet the specific requirements for any particular industry. Beneficiation is done to a limited extent and that too only in some cement plants. Other consumers use the limestone of select grades to suit their specifications. Beneficiation of limestone in cement industry is directed to achieve the following

1. Removal of free and coarse silica to decrease the grinding costs
2. Enough $CaCO_3$ to provide necessary and sufficient CaO on burning in rotary kiln
3. Removal of certain minerals to reduce the ratio of (Al_2O_3+Fe_2O_3+SiO_2) to $CaCO_3$

Apart from crushing and screening, no beneficiation is performed for limestone used in steel industry. At the Nandini mine which supplies limestone to Bhilai steel plant, run-of-quarry limestone is crushed to −80 mm in two stages by jaw and cone crushers followed by screening to separate −80+25 mm and −25 mm. The coarser fraction is used in a blast furnace and the finer fraction is utilised by other consumers.

2. Rock Phosphate

Rock phosphate is mainly used for the production of fertilizer. Rock phosphate beneficiation plant of Rajasthan State Mines and Minerals Limited is located at Jhamarkotra. The major minerals are sedimentary crypto crystalline apatite, dolomite and quartz along with some minor quantities of oligoclase, sillimanite etc.

ROM ore of −1000 mm is crushed in a jaw crusher followed by a short-head cone crusher in closed circuit with a double-deck screen of 200- and 60-mm openings. Undersize of the screen is further crushed in tertiary short head cone crusher in closed circuit with double deck screen of 60 mm and 13 mm opening. The final crushed product of –13 mm is grounded in ball mill in closed circuit with primary hydrocyclone. Overflow of primary hydrocyclone is again classified in secondary hydro-cyclone. The overflow of secondary cyclone is subjected to bulk flotation to float apatite and dolomite. The tailings are discarded as silicate tailings and concentrate is thickened. After the addition of phosphoric acid to the thickened concentrate, the pH is adjusted to 5–5.5 by adding sulphuric acid and subjected to flotation. Soap solution is added at three banks to float dolomite. The tailing of this flotation is the phosphate concentrate. It is thickened and filtered in disc filter. The dolomite rougher concentrate is cleaned two times to get carbonate concentrate. The process flowsheet is shown in Fig. 195.

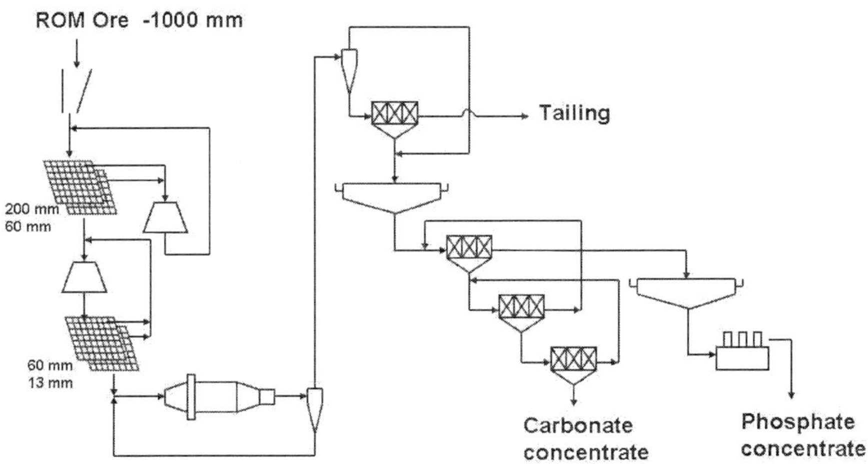

Fig. 195 Process flowsheet for Rock phosphate Beneficiation Plant at Jhamarkotra (Source: Mineral resources & beneficiation plant practices in India, IIME Bhubaneswar chapter)

3. Feldspar

The principal mineral raw materials for the manufacture of glass and ceramic products are silica in its different forms and alumino-silicates. While silicate minerals include quartz and other forms such as glass sand, the alumino-silicates comprise of the feldspars and clays. Feldspars for use in glass and ceramic industry should contain a minimum of 65%–72% SiO_2. The common impurities to be eliminated are the iron-bearing minerals such as garnet and mica. The quarried material is broken and feldspar is easily removed by handpicking. The picked feldspar is reduced in size by crushing in conventional crushers and subsequent grinding in pulverisers. For glazing purposes and for the manufacture of ceramic ware, iron contamination during size reduction is avoided by grinding in pebble mills using flint pebbles as the grinding media. Flotation also can be used for feldspar concentration.

4. Clay

Clay is a certain type of earthy material formed out of the decomposition and disintegration of hard rocks with a specific property of plasticity when mixed with water. Clays are classified as china clay, ball clay, fire clay and building-brick clay. China clay belongs to the kaolin group and is utilised in rubber, textile, paint, pharmaceutical and paper industries besides ceramic manufacturing. Major impurities in china clay are quartz, mica, feldspar and iron-oxide minerals.

The china clay beneficiation plant of Kerala Ceramics Ltd. at Kundara is shown in Fig. 196. The plant draws its ore from two captive mines at Kundara. Raw clay is mixed with water to form slurry in horizontal and vertical blungers that are basically knife washers. The resulting slurry is passed through rotary screens; thereafter, some coarser particles are removed by hydrocyclones which constitute tailings. The overflow slurry from hydrocyclone is collected in concrete settling tanks where Sodium silicate and sodium carbonate are added, and a pH of 8.5 is maintained. The slurry is then allowed to settle. After aobut 12 hours, slurry from top 160 centimetres is transferred to the leach section. Coarser particles which settle and remain in the tank are processed for filler-grade kaolin.

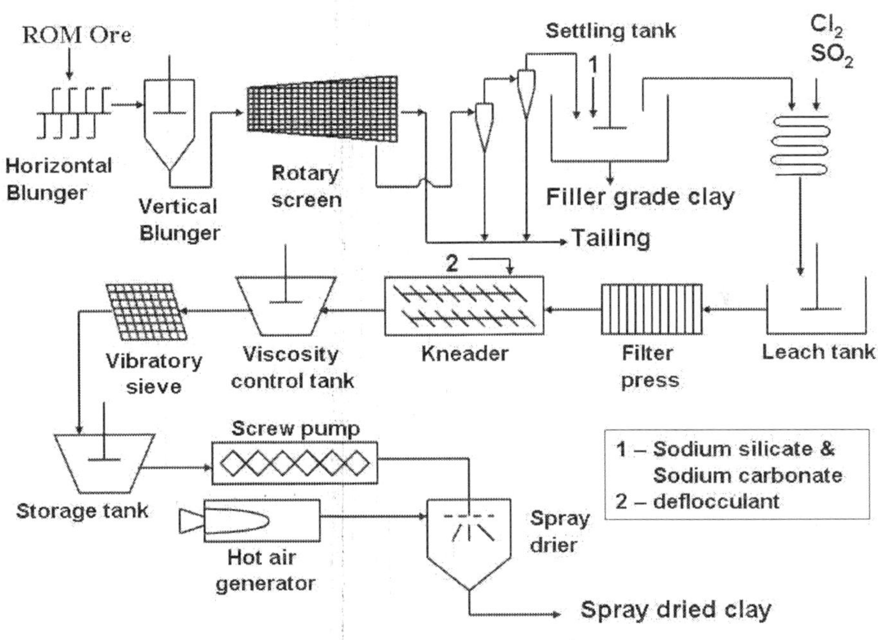

Fig. 196 Process flowsheet for China Clay Beneficiation at Kundara (Source: Mineral resources & beneficiation plant practices in India, IIME Bhubaneswar chapter)

In the leach tank, the kaolin is treated with Chlorine and sulphur dioxide to improve brightness and whiteness. Excess iron is also removed by treatment. Thus, treated slurry is sent to filter press for pressure filtration where water content of processed china-clay slurry is reduced to about 30%. Cake formed in filter press is made to slurry form in a kneader machine by the addition of a deflocculating agent. The slurry is sent to storage tank through viscosity control tank where slurry is made to 35% moisture. This slurry is pumped by a screw pump and sprayed to the drying chamber through which hot air is also passed. The fine droplets slowly descend through a stream of hot air. A fine powder is formed that is both free flowing and virtually dust free. This final product is stored in conical bins and conveniently packed.

5. Graphite

Graphite is carbon in its chemical composition. The graphite is associated with feldspar, mica, quartz, clay, goethite, limonite, carbonates, hematite, garnet and zircon. In certain deposits, wolframite and scheelite are also present.

Graphite Beneficiation Plant of Tamil Nadu Minerals Ltd. is located 10 km from Sivagangai. The raw material predominantly consists of quartz with subordinate amounts of graphite, clay, goethite, limonite, carbonates, feldspar and hematite. Pyrite is noticed in minor to trace amounts. Graphite is mostly medium to coarse (120–400 microns) grained and embedded in the quartz matrix. At places, graphite flakes are also present as clusters in the quartz matrix. Occasionally, graphite is associated with goethite. Graphite is also present as fine grained (18–60 microns) irregularly distributed grains in the quartz matrix. Quartz is fine to medium grained (90–360 microns) and mostly occurs as anhedral, elongated grains and forms the matrix to graphite and other minerals. The raw material consists of 10%–15% graphite.

The process flowsheet of graphite beneficiation plant at Sivagangai is shown in Fig. 197. ROM Ore is passed through a grizzly and screened on a double-deck vibrating screen with 50- and 10-mm openings. The +50-mm fraction is crushed in a jaw crusher and –50+10 mm fraction is crushed in an impactor. After size reduction, both the fractions are recycled to vibrating screen. The –10 mm material is wet ground in a rod mill with the addition of sodium carbonate and sodium silicate solution (10%). The pulp is then conditioned with reagents and subjected to rougher flotation in a bank of flotation cells at a pH of 8.5–9.0 which is maintained by addition of sodium carbonate solution. The tailings from the rougher bank are sent to another bank of cells for scavenging after addition of MIBC and LDO mixture.

The froth from the rougher cells is fed to a hydrocyclone which is in closed circuit with regrinding ball mill. Overflow of cyclone along with mill discharge is subjected to the first stage of cleaning in a flotation column. The froth product from the column is subjected to grinding in a ball mill and the mill product is further cleaned in a bank of flotation cells. The froth product from this bank is screened, washed, filtered using a horizontal belt filter and dried in a double-drum rotary drier to reduce the moisture from 30% to approximately 0.2% for flaky graphite, while the tailings are fed to the scavenger bank.

The undersize of the screen is fed to a fines regrinding ball mill through a cyclone. The overflow of the cyclone along with the mill discharge is fed to a flotation column for cleaning of the fines. The froth product from this column is re-cleaned in another column, the froth from recleaner column is thickened, filtered and dried for graphite fines.

The tailings from fine product cleaner and re-cleaner columns are fed to the scavenger column along with the tailings of the coarse cleaner and the froth product of the coarse scavenger cells. The froth from this scavenger column is fed to a secondary regrinding ball mill through a hydrocyclone. The discharge from this mill and the overflow from the cyclone are fed to the first cleaner column along with the discharge from the primary regrind mill for cleaning. The tailing from the scavenger column are recirculated to the rougher cells.

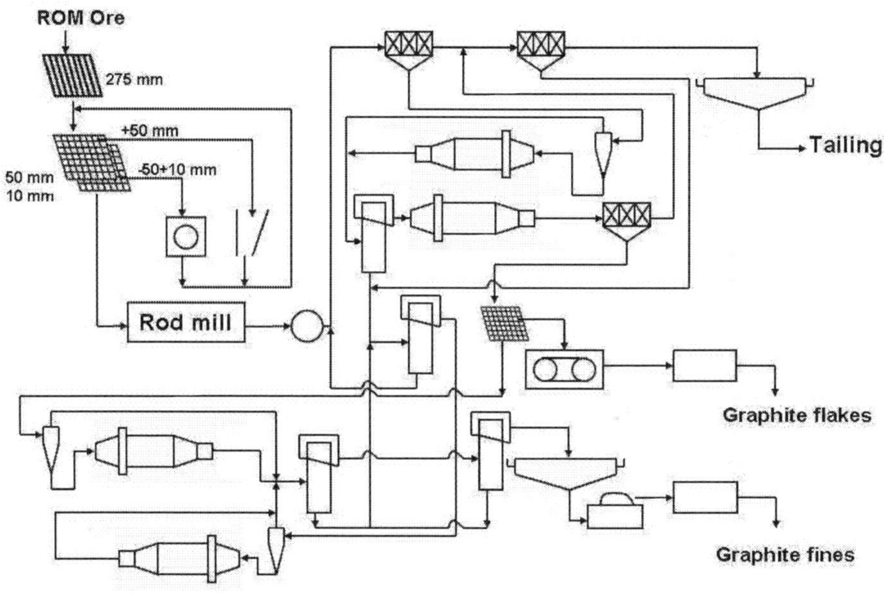

Fig. 197 Process flowsheet for graphite beneficiation at Sivagangai
(Source: Mineral resources & beneficiation plant practices in India, IIME Bhubaneswar chapter)

6. Diamond

Diamonds are economically essential to India. Cutting and polishing of rough diamonds is a large industry in the country. The National Mineral Development Corporation operates the only diamond mine in the country at Panna, producing diamonds consisting of mostly precious stones and partly for industrial use. The processing plant at Majhgaon (Panna) was essentially designed to handle soft yellow/white tuff excavated from the mine along with a hard-black material.

The mined ore is crushed to 18 mm in three stages by employing a primary jaw crusher and secondary and tertiary roll crushers. The crushed material (−18 mm) is subjected to washing to remove slimes of −1.2 mm fine tailings. The washed material is subjected to a 50 tph capacity HMS plant using Ferro Silicon as the separating media. About 99% of the feed report as float (coarse tailings), while 1% reports sink (concentrate). The sinks are dried in a rotary dryer and dried material is sized to +10 mm, −10 + 5 mm, −5+3.5 mm and −3.5 mm fractions. Each fraction is subjected to magnetic separation individually. The total non-magnetics yielding approximately 20% is subjected to an x-ray sorter, the rejects of which (about 90%) are processed on a grease table. Approximately 10% of the feed to the grease tables is recovered as concentrate. The products of the x-ray sorter and grease table are subjected to hand picking of diamonds. The process flowsheet is shown in Fig. 198.

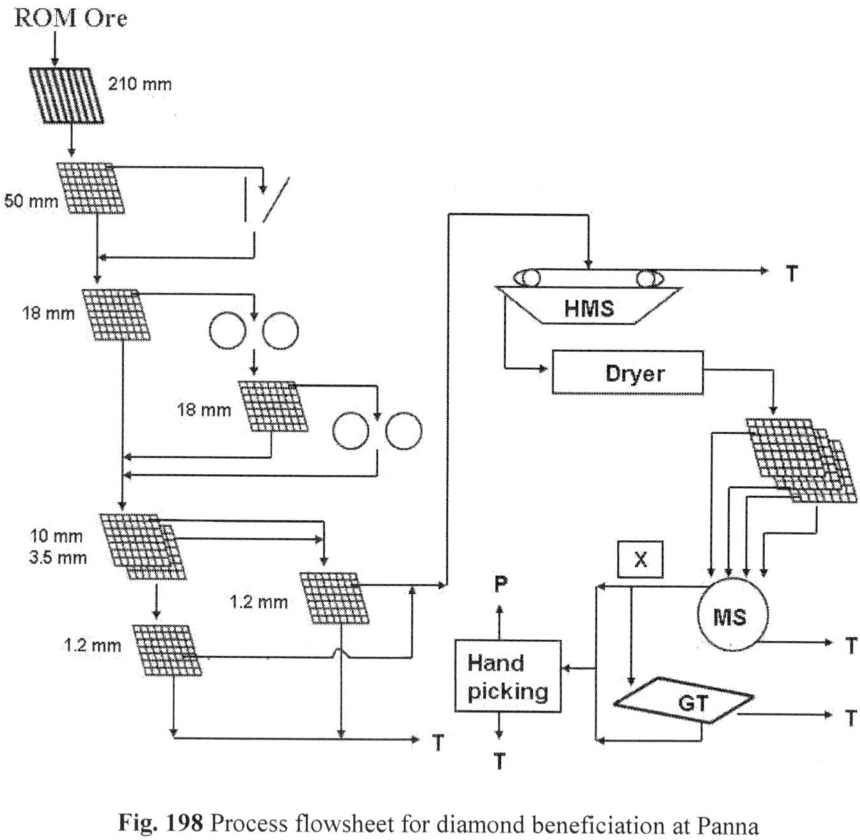

Fig. 198 Process flowsheet for diamond beneficiation at Panna (Source: Mineral resources & beneficiation plant practices in India, IIME Bhubaneswar chapter)

20.3 PROCESSING OF COAL

Coal is broadly classified into coking and non-coking coals. Coking coals are used for making coke for steel industry, and the chief use of non-coking coal is for the production of power in thermal power plants.

1. Coking coal

Different washing schemes have been adopted for different washeries to tackle various problems and to upgrade a wide variety of coking coal feed. The washing schemes which have so far been introduced in the existing coking coal washeries can be classified into three broad groups, as outlined below:

Group I: The washing scheme relating to this group covers mostly coals having comparatively easy washability characteristics. These ROM coals are initially crushed to 75 mm and then classified into –75 mm +25mm or –75 mm +13 mm and small-size –25 mm +0.5 mm or −13 mm +0.5 mm fractions for treatment in H.M. baths and that of small coals in Jigs or an HM cyclone. In some installations, the coarse-sized sinks separated from H.M. baths are crushed and rewashed in H.M. cyclone to recover additional cleans. The washeries at Jamadoba and West Bokaro more or less conform to this group.

Group II: Treatment under this group is similar to that of the first group, except that a pre-washer is incorporated to eliminate the free dirt, e.g. Dugda I and Patherdih washeries.

Group III: This scheme is meant for upgrading difficult-to-wash coals which are required to be crushed to below 20 mm or 13 mm prior to washing. Whenever so warranted, the entire coal or the fraction above 20 mm or 13 mm is first de-shaled and the floats are crushed and retreated for the recovery of cleans and middlings. In this case, the fraction above 0.5 mm is treated in HM cyclones and the slurry below 0.5mm in water only cyclones or flotation cells. Dugda II, Kathara, Swang and all recently commissioned washeries fall under this group.

Presently, there are 16 coking coal washeries operational in India. In majority of the washeries, the coal is washed by Heavy Medium Separation. This is because of the accuracy of separation at any predetermined specific gravity within the range of 1.2–2.50 and continuous maintenance of pre-selected value of specific gravity within ±0.005 of its value, which are the salient features of heavy medium separation. This accuracy of separation is all the more essential in view of the high percentage (30%) of near-gravity materials in Indian coals.

Out of six BCCL washeries, the **Bhojudih** washery, which is the oldest central washery where coal from different mines are washed, is the most successful coking coal washery owing to easy washability characteristics of coal and the use of an HM Bath for 75–25 mm coarse coal and Jigs for smalls worked well. Fig. 199 shows the simplified flowsheet of the Bhojudih Coal Washery.

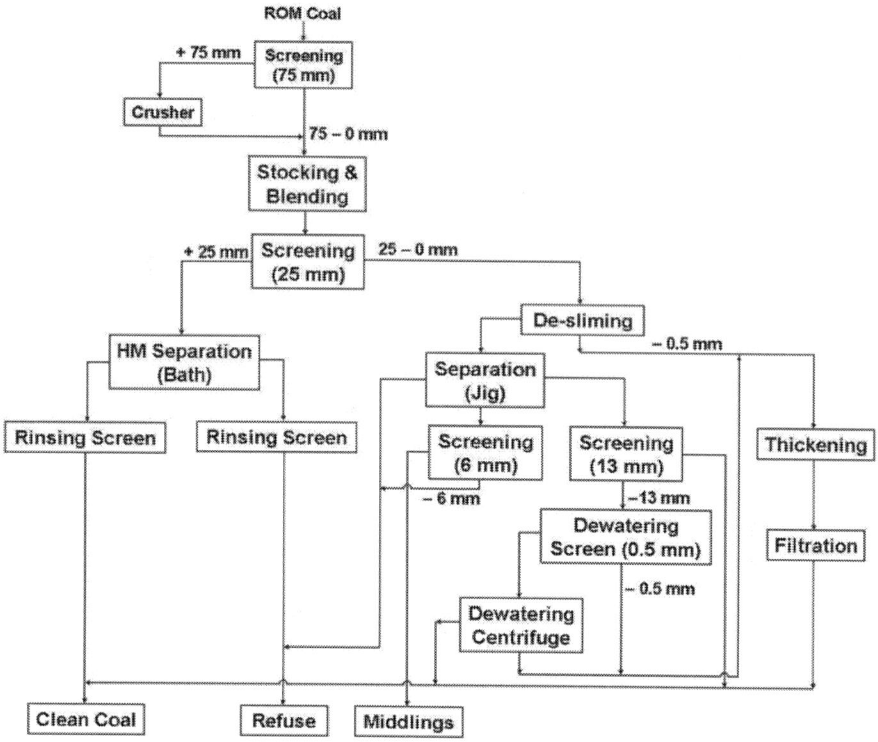

Fig. 199 Simplified process flowsheet of the Bhojudih coal washery

2. Non-coking coal

Non-coking coals are available in many locations in India. They cannot be directly used owing to their high ash content of about 45% and more. The non-coking coal used in power plants in India is mined by mechanised mining technique which produces small and fine sizes of coal and receives the coal from multiple sources. These are the major constraints in India for achieving efficiency besides high ash and moisture contents of coal. The average quality of non-coking coal has been progressively deteriorating due to inferior grades of coal reserves available from open cast mining and the high degree of mechanisation introduced in the large open cast mines.

Presently, about 60 non-coking coal washeries are in operation in the country. Coal India Limited operates five non-coking coal washeries. Among them, the Kargali washery was originally a coking coal washery and has undergone several changes since its inception. Coal feed size of –80 mm is screened at 12 mm. A two-stage Wemco Drum for coarse coal and a Jig for the smalls washing are used. Sinks of the first Jig is rewashed in a smaller capacity second Jig for reducing the overall costs. The flowsheet was subsequently modified to utilise two Baum jigs in parallel and increase the plant's throughput. An HM cyclone is incorporated to clean the sinks of Wemco Drum after crushing to −16 mm. In the latest flowsheet, a Batac Jig has replaced the HM cyclones and small coal Baum jigs. Wemco Drums perform the separation for coarse coal. A dry de-duster was installed in the early 1980s along with the Batac jig. Fig. 200 shows the latest flowsheet of the Kargali washery.

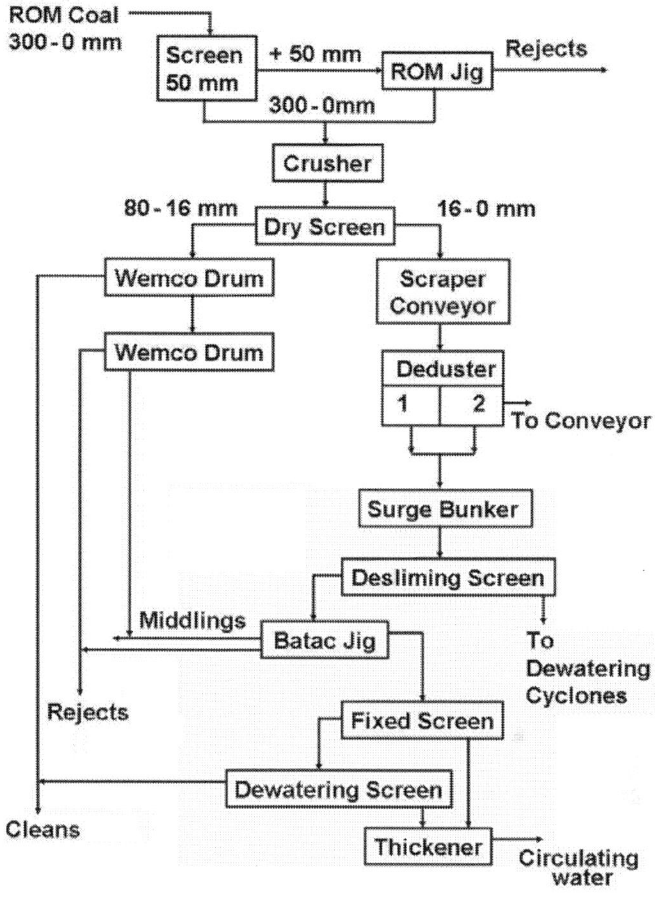

Fig. 200 Flowsheets of Kargali non-coking coal washery

Most non-coking coal washeries in India concentrated mainly on partial beneficiation, i.e. coarse coal beneficiation. The finer fraction (generally, of lower ash content but prone to higher moisture take up) is allowed to bypass the washing circuit. Since most of the free dirt and banded materials are included in the coarser fraction, beneficiation of this fraction and mixing of the washed product with the untreated smaller fraction gained acceptance for maintaining desired ash at 34%.

■■■■

Annexure
Experimental Procedure for determination of Bonds Work Index

Equipment and materials required

1. Standard 12 × 12″ ball mill rotating at 70 rpm with a mechanical counter to register the number of revolutions.
2. Steel balls of following sizes

Diameter (inches)	Number of balls	Approximate weight (g)
1.50	43	8730
1.25	67	7197
1.00	10	705
0.75	71	2058
0.50	94	1441
Total	285	20131

Calculated surface area = 842 in^2

(or)

Diameter (inch)	Number of balls	Approximate weight (g)
1.50	36	9900
1.00	130	7175
0.75	119	3020
Total	285	20095

3. Standard nest of sieves ranging from 6 to 200 mesh
4. Ro-tap sieve shaker
5. Approximately 30 kg of the given ore

Preparation of feed material

1. Crush approximately 5 kg of given ore to approximately one-fourth of an inch in stages to avoid excess fines.

Crush this in stages to all −6 mesh.

Mix the sample well.

2. Perform sieve analysis on a representative sample of the crushed feed and determine the feed size F, i.e. the size in microns through which 80% of the feed passes.

   Let this percent −200 mesh in the feed be 'x'

3. Place the feed sample in a 1,000-ml measuring cylinder.

   Compact by shaking to yield a volume of 700 ml.

   This unit volume of 700 ml is selected as the volume of the ore always present in the mill during closed-circuit grinding tests.

   Let its weight be W in grams.

Experiment

1. Load the mill with the balls and 700 ml of feed.
2. During first cycle, run the mill for N_1 (say 100) revolutions.
3. Unload the mill and sieve the ground material through 200 mesh sieve taking small portions at a time.

   Separate the +200 and −200 mesh materials.

   Weigh −200 mesh material. Let this weight be W_1 in grams.

4. Discord −200 mesh material.

   Add new feed of W_1 g to +200 mesh material and load this into the ball mill. (+200 mesh material is the circulating load)

   Weight of −200 mesh material in the feed $= \frac{Wx}{100}$

   Net weight of −200 mesh material formed during first cycle $= W_1 - \frac{Wx}{100}$

   Net weight of −200 mesh material produced per revolution

   (i.e. grindability in the first cycle) $G_1 = \dfrac{\left(W_1 - \dfrac{Wx}{100}\right)}{N_1}$

Closed circuit is achieved by screening through 200 mesh and replacing −200 mesh fraction by fresh addition of feed at the end of each cycle. For these tests, a circulating load of 250% is assumed. For this circulating load, the mill throughput will be 350% of the mill product and at equilibrium, mill product is equal to mill feed. Hence, the weight of the product to be produced in each cycle will be $\frac{100}{3.5}$ or 28.6% of the total mill feed.

This is called **Intended Product Passing or IPP (0.286W)**

For the second cycle, calculate the number of revolutions N_2 for which the mill is to run.

Net weight of −200 mesh material to be produced in the second cycle

$$= \text{IPP} - \text{weight of } -200 \text{ mesh in new feed}$$

$$= 0.286\text{ W} - \frac{W_1 x}{100}$$

Number of revolutions in second cycle N_2

$$= \frac{\text{Net weight of - 200 mesh material to be produced}}{\text{Grindability of the first cycle}}$$

$$= \frac{\left(0.286 - \dfrac{W_1 x}{100}\right)}{G_1}$$

5. Now run the mill for N_2 revolutions.

   Unload the mill and sieve the ground material through −200 mesh.

   Separate the +200 and −200 mesh materials.

   Weigh −200 mesh material. Let this weight be W_2 g.

6. Discord −200 mesh material

   Add new feed of W_2 g to +200 mesh material and load this into the ball mill. (+200 mesh material is the circulating load)

   Net weight of −200 mesh material produced during second cycle = $W_2 - \dfrac{W_1\, x}{100}$

   Grindability during second cycle $G_2 = \dfrac{\left(W_2 - \dfrac{W_1 x}{100}\right)}{N_2}$

   For the third cycle

   New feed to be added = W_2 g

   −200 mesh material in new feed added = $\dfrac{W_2 x}{100}$

   Net weight of −200 to be produced in third cycle = $0.286W - \dfrac{W_2 x}{100}$

   Number of revolutions in third cycle, $N_3 = \dfrac{\left(0.286W - \dfrac{W_2 x}{100}\right)}{G_2}$

7. Now run the mill for N_3 revolutions.

Unload the mill and sieve the ground material through −200 mesh.

Separate the +200 and −200 mesh materials.

Weigh −200 mesh material. Let this weight be W_3 in grams.

Proceed as before.

Repeat the above procedure at each successive cycle. The net production of material finer than 200 mesh usually shows a definite trend either increasing or decreasing with each succeeding cycle. Continue the grinding cycles until a reversal or levelling off has occurred. From the last three cycles take the average value and find out the equilibrium grams/mill revolution or the ball mill grindability (G). At equilibrium, 28.6% of the mill discharge should be finer than the 200 mesh and make up feed must be equal to the product material formed during the cycle. The combined product from the final three cycles forms the equilibrium test product.

Carryout a sieve analysis on this product and plot the results in a manner similar to that used for the feed. From this plot, determine the 80% passing size of the product (P).

Observations

1. Record the data as follows:

   Weight of 700 ml of feed = W =

   Percent −200 mesh material in feed = x =

   Weight of −200 mesh material in feed = $\frac{Wx}{100}$ =

   Weight of the product to be produced for 250% circulating load = 0.286 W =

   Record the experiment data in a tabular form in the following column-wise order:

   Cycle

   Number of revolutions

   −200 mesh material in the product (g)

   −200 mesh material in the feed (g)

   Net −200 mesh material produced (g)

   Grindability (G) (g/revolution)

   New feed to be added (g)

   −200 mesh material in new feed (g)

   Number of revolutions for next cycle

   Determine equilibrium grindability = G =

   (average value of last three cycles)

2. Calculate the work index of the ore by using the following formula:

$$W_i = \frac{4.45}{P_i^{0.22} G^{0.8} (P^{-0.5} - F^{-0.5})}$$

where W_i = work index

P_i = mesh of grind in microns (74 microns in this case)

G = ball mill grindability (g/revolution)

P = product size in microns

F = feed size in microns

■■■■

References

1. Read HH 1984 *Rutley's Elements of Mineralogy*, Delhi: CBS Publishers & Distributors.
2. Gaudin AM 2005 *Principles of Mineral Dressing*, New Delhi: Tata McGraw-Hill Publishing Company Ltd.
3. Taggart AF 1945 *Handbook of Mineral Dressing*, New York: John Wiley & Sons.
4. Atiq S et al. 2005 Beneficiation Studies on the Low-Grade Chromite of Muslim Bagh, Balochistan, Pakistan, *Pakistan Journal of Scientific and Industrial Research* **48**.
5. Chinnaiah 2004 Beneficiation studies of Manganese ore of Kumsi, Shimoga, Southern India, *Indian Journal of Applied Research* **4**.
6. Rittinger PR 1867 *Lehrbuch der Aufbereitungskunde*, Berlin.
7. Kick F 1885 *Das Gesetz der Proportionalem Widerstand und Seine Anwendung*, Leipzig.
8. Bond FC 1952 Third theory of comminution, *AIME Trans.*
9. Gaudin AM, Hukki RT 1944 Principles of comminution–Size and Surface Distribution, *A.I.M.E.*
10. Reynolds 1883 An Experimental Investigation of the Circumstances Which Determine Whether the Motion of Water Shall be Direct or Sinuous, and of the Law Resistance in Parallel Channels, Phil. Trans. Roy. Soc. London.
11. Stokes GG 1891 *Mathematical and Physical Paper III*, Cambridge University Press.
12. Rouse 1937 *Nomogram for the Settling Velocity of Sphere, Report of the Committee on Sedimentation, 1936-1937*, National Research Council.
13. Newton 1729 *The Mathematical Principles of Natural Philosophy, Book II, Translated into English by Andrew Motte.*
14. Brown GG et al 1950 *Unit Operations*, Wiley: New York.
15. Coe GD 1938 An explanation of washability curves for the interpretation of float-and-sink data on coal, *Information circular No.7045, US Bureau of Mines.*
16. Bird BM 1931 Interpretation of Float-and-sink Data, *Proceedings of the Third International Conference on Bituminous Coal, Pittsburgh. Vol.2.*
17. Sarkar GG, Bose RN, Mitra SK, Lahiri A 1962 *An Index for the comparison and correlation of washability characteristics of coal,* Presented to Fourth International Coal Preparation Congress, Harrogate (UK).

Further Readings

1. Wills BA, Napier-Munn T 2006 *Mineral Processing Technology*, Elsevier Science & Technology Books.
2. Gaudin AM 2010 *Principles of Mineral Dressing*, New Delhi: Tata Mc Graw Hill Publishing Company Ltd.
3. Gokhale KVGK, Rao TC 1983 *Ore Deposits of India (Their distribution and processing),* New Delhi: Affiliated East West Press Pvt. Ltd.
4. Jain SK 1986 *Ore Processing*, New Delhi: Oxford & IBH Publishing Co.
5. Kelly EG, Spottiswood EJ 1982 *Introduction to Mineral Processing*, New York John Wiley & Sons.
6. Pryor EJ 1974 *Mineral Processing, Third Edition,* London: Applied Science Publishers Limited.
7. Burt RO 1984 *Gravity Concentration Technology*, Amsterdam: Elsevier Scientific Publishing company.
8. Subba Rao DV 2011 *Mineral Beneficiation – A Concise Basic Course*, The Netherlands: CRC Press, Taylor & Francis.
9. Subba Rao DV, Gouricharan T 2016 *Coal Processing and Utilization*, The Netherlands: CRC Press, Taylor & Francis.
10. Subba Rao DV 2016 *Minerals and Coal Process Calculations*, The Netherlands: CRC Press, Taylor & Francis.
11. Venkatachalam S, Degaleesan SN 1982 *Laboratory experiments in Mineral Engineering*, New Delhi: Oxford & IBH Publishing Co.
12. Vijayendra HG 1995 *Handbook of Mineral Dressing*, New Delhi: Vikas Publishing House Pvt Ltd.

Subject Index